STATISTICAL AND COMPUTATIONAL METHODS IN DATA ANALYSIS

Determination of mean foot length
[Woodcut reproduced from KÖBEL (1570)]

Statistical and Computational Methods in Data Analysis

Second, revised edition

SIEGMUND BRANDT

Physics Department, Siegen University
Siegen, Germany

1976

NORTH-HOLLAND PUBLISHING COMPANY
AMSTERDAM · NEW YORK · OXFORD

Library of Congress Catalog Card Number: 77-113749

ISBN North-Holland 0 7204 0334 0

1st edition 1970
2nd printing 1973
2nd, revised edition 1976

Publishers:

NORTH-HOLLAND PUBLISHING COMPANY
AMSTERDAM · NEW YORK · OXFORD

Sole distributors for the U.S.A. and Canada:

ELSEVIER/NORTH-HOLLAND INC.
52 VANDERBILT AVENUE
NEW YORK, N.Y. 10017

Library of Congress Cataloging in Publication Data

Brandt, Siegmund.
Statistical and computational methods in data analysis.

Translation of Statistische Methoden der Datenanalyse.
Bibliography: p.
Includes index.
1. Probabilities. 2. Mathematical statistics.
I. Title.
[QA273.B86213 1976] 519.5 77-113749
ISBN 0-444-10893-9 (American Elsevier)

PRINTED IN THE NETHERLANDS

FROM THE PREFACE TO THE FIRST EDITION

The present book is based on lectures given in 1967/68 to physics students and particle physicists at Heidelberg University. It discusses those parts of mathematical statistics most relevant to data analysis. The book is intended for students and research workers in science, medicine, engineering and economics, who are faced with the problem of evaluating experimental data.

It is written from the standpoint of a user of mathematical statistics. Mathematical rigour is not overstressed. On the other hand, it does not merely list a number of recipes for different practical applications but attempts to explain the concepts and principles of the statistical methods discussed. Part of the presentation is influenced by lecture notes and review articles written by, and for, physicists [Böck, 1960; Orear, 1958; Solmitz, 1964].

A good basic knowledge of calculus is assumed. Other necessary mathematical tools, especially probability theory, are briefly reviewed. An essential factor of the presentation is the use of matrix notation. It provides a very compact presentation of many problems such as the least squares method. An introduction to matrix calculus is given in the appendix.

Since most complex data analysis problems are now tackled with the help of computers, FORTRAN programs are presented for several such cases. The essential features of the FORTRAN language are discussed in the appendix. It also contains a short library of matrix handling subprograms, based on a similar set of programs originally written at CERN, Geneva, by R. Böck.

It is hoped that the book will not only serve as an introduction to statistical data analysis but will also be used in everyday work. It therefore contains a few statistical tables and a short collection of the more important formulae for quick reference.

I am indebted to several of my Heidelberg colleagues for discussions, in particular to Dr. T. P. Shah who read the manuscript and made many valuable suggestions for improvements. My thanks are also due to Dr. A. G. C. Tenner (Amsterdam) for a very fruitful discussion on the organization of the book. Dr. H. Immich (Heidelberg) has kindly provided the examples 8-2, 11-1 and 11-2. Dr. H. Frenk (Wetzlar) made available a copy of the woodcut reproduced in front.

S. BRANDT

Heidelberg, January 1970

PREFACE TO THE SECOND EDITION

In the present edition the general concept of the book – short but sufficiently rigorous mathematical treatment, main emphasis on applications – has been left unchanged. However substantial additions have been made to the sector of statistical methods for direct application, in particular with computer programs. The main new sections are

- Elements of the Monte Carlo method (ch. 5, § 5.3)
- Rough numerical and graphical analysis of sampled data (ch. 6, § 8)
- FORTRAN program for linear regression (ch. 12, § 5)
- Time series analysis (ch. 13).

All of them contain FORTRAN programs. (The number of programs for direct application to statistical problems has been tripled.)

Furthermore the question of convolution of several distributions which can be rather cumbersome in practice, has been dealt with in greater detail and examples of convolution with the normal distribution are given. The chapter on sampling now contains a short section on very small samples.

Exercises are now given at the end of the chapters. Their solutions are outlined in a special section.

I should like to thank several of my collegues in Siegen for valuable discussions and suggestions, in particular Dr. W. Heinrich who read the manuscript of the new sections. On this occasion it is a pleasure for me also to acknowledge the excellent work of Dr. W. Wojcik and Prof. H. Yoshiki who made the translations for the Polish and Japanese editions of the book.

S. BRANDT

Siegen, June 1976

CONTENTS

APPENDICES

LIST OF EXAMPLES

LIST OF FORTRAN PROGRAMS

LIST OF FREQUENTLY USED SYMBOLS

$x, y, \xi, \eta, \ldots$	(ordinary) variables
$\boldsymbol{x}, \boldsymbol{y}, \boldsymbol{\xi}, \boldsymbol{\eta}, \ldots$	vectors of variables
$\mathsf{x}, \mathsf{y}, \xi, \eta, \ldots$	random variables
$\boldsymbol{\mathsf{x}}, \boldsymbol{\mathsf{y}}, \boldsymbol{\xi}, \boldsymbol{\eta}, \ldots$	vectors of random variables
$A, B, C, \ldots$	matrices
B	bias
$\mathrm{cov}(\mathsf{x}, \mathsf{y})$	covariance
F	variance quotient
$f(x)$	probability density
$F(x)$	distribution function
$E(\mathsf{x}) = \hat{x}$	mean, expectation value
H	hypothesis
H_0	null hypothesis
L, l	likelihood functions
$L(S_c, \lambda)$	operation characteristic
$M(S_c, \lambda)$	power function
M	function to be minimized
$P(A)$	probability of event A
$\mathsf{s}^2, \mathsf{s}_x^2$	sample variance
S	statistic, estimator
S_c	critical region
t	variable of Student's distribution
x_m	mode
$x_{0.5}$	median
x_q	fractile
$\bar{\mathsf{x}}$	sample mean
$\tilde{x}$	maximum likelihood or least squares estimator
λ	parameter of a distribution
$\varphi(t)$	characteristic function
$\phi(x), \psi(x)$	probability density and distribution function of normal distribution

$\phi_0(x)$, $\psi_0(x)$	probability density and distribution function of a normalized Gaussian distribution
$\sigma(x) = \Delta(x)$	standard deviation
$\sigma^2(x)$	variance
$\Omega(P)$	inverse of normal distribution

CHAPTER 1

INTRODUCTION

Every branch of experimental science, after passing through an early stage of qualitative description, concerns itself with quantitative studies of the phenomena of interest, i.e. measurements. Next to the design and the performance of the experiment, an important task is the accurate evaluation and the complete exploitation of the data obtained. Let us list a few typical problems.

1. The increase in the weight of test animals under the influence of various drugs is studied. After the application of drug A to 25 animals an average increase of 5% is observed. Drug B, used on 10 animals, yields 3%. Is drug A more effective? The averages 5% and 3% give practically no answer to this question, since the lower value may have been caused by a single animal that, for some reason, lost weight. One has therefore to study the *distribution* of individual weights and their dispersion around the average value. Moreover one has to decide whether the number of test animals used will enable one to differentiate between the effects of the two drugs with a certain accuracy.

2. In experiments on crystal growth the exact maintenance of the ratio of different components is essential. From a total of 500 crystals 20 are selected and analyzed. What conclusions can be drawn as to the composition of the remaining 480? This problem of *sampling* occurs for example in production control, reliability tests of automatic measuring devices and opinion polls.

3. A certain experimental result has been obtained. It has to be decided whether it contradicts some predicted theoretical value or previous experiments. The experiment is used for *hypothesis testing*.

4. A general law is known to describe the dependence of measured variables, but parameters of this law have to be obtained from experiment. In radioactive decay, for example, the number N of atoms that decay per second decreases exponentially with time: $N(t) = \text{const.} \times \exp(-\lambda t)$. The decay con-

stant λ and its measurement error are to be determined using a number of observations $N_1(t_1)$, $N_2(t_2)$ .... This problem of *parameter estimation* is perhaps the most interesting for many experimentalists.

From these examples some of the features of data analysis become apparent. We see in particular that the outcome of an experiment is not uniquely determined by the experimental procedure but is also subject to chance: it is a *random variable*. This stochastic tendency is either rooted in the nature of the experiment (test animals are necessarily different, radioactivity is a stochastic phenomenon), or it is a consequence of the inevitable uncertainties of the experimental equipment, i.e. the measurement errors. The next chapter is therefore devoted to reviewing the most important concepts of the theory of probability.

In chapters 3 and 4 random variables are introduced. The distribution of random variables is discussed and parameters, such as mean and variance are found to characterize these distributions. Special attention is given to the interdependence of several random variables. In chapter 5 a number of distributions is studied which are of special interest in applications, in particular the properties of the normal or Gaussian distribution are discussed in detail.

In practice a distribution has to be determined from a finite number of observations, i.e. a *sample*. Different cases of sampling are considered in chapter 6. FORTRAN programs are presented for a first rough numerical treatment and graphical display of empirical data. Functions of the sample, i.e. functions containing the individual observations, can be used to estimate the characteristic parameters of the distribution. The requirements that a good estimate should satisfy are derived. At this stage the quantity χ^2 is introduced. It is the sum of the squares of the deviation between observed and expected values and is therefore a suitable indicator of the quality of observation.

The *maximum likelihood method*, discussed in chapter 7, is the core of modern statistical analysis. It allows one to construct estimators with optimum properties. The method is discussed for the single- and multi-parameter cases and illustrated in a number of examples.

Chapter 8 is devoted to hypothesis testing. It contains the most commonly used *F*-, *t*- and χ^2-tests and outlines the general theory.

The *method of least squares*, which is perhaps the most widely used statistical procedure, is the subject of chapter 9. The special cases of direct, indirect and constrained measurements, often encountered in applications, are developed in detail before the general case is discussed. A FORTRAN program for general least squares problems is presented and its use is demonstrated in different examples. Every least squares problem can be expressed as the task of determining the minimum of a function of several variables. This is true of all parameter estimation based on the idea of maximum likelihood. In chapter 10 several computational methods are sketched to obtain such minima.

The *analysis of variance* (chapter 11) can be considered as an extension of the *F*-test. It is widely used in biological and medical research to study the dependence, or rather to test the independence, of a measured variable of experimental conditions expressed by other variables. For several variables rather complex situations can arise. Some simple numerical examples are calculated using a FORTRAN program.

Linear *regression*, the subject of chapter 12, is a special case of the least squares method and therefore already dealt with in chapter 9. Before the advent of computers usually only linear least squares problems were tractable. A special terminology, still used, has developed for this case. It seems therefore justified to devote a special chapter to this subject. At the same time it extends the treatment of chapter 9. For example the determination of confidence intervals for a solution and the relation between regression and analysis of variance are studied. A general FORTRAN program for linear regression is given and its use is shown in examples.

In the last chapter the elements of *time series analysis* are introduced. This method is used if data are given as a function of a controlled variable (usually time) and no theoretical prediction for the behaviour of the data as a function of the controlled variable is known. It is used to try to reduce the statistical fluctuation of the data without destroying the genuine dependence on the controlled variable. Since the computational work in time series analysis is rather awkward a FORTRAN program is also given.

CHAPTER 2

PROBABILITIES

2-1. Experiments, events, sample space

Since in this book we are concerned with the analysis of data originating from experiments, we will have to state first what we mean by an experiment and its result. As for laboratory work, we define an experiment to be the strict following of a prescribed procedure, as a consequence of which a quantity or a set of quantities is obtained which constitute the result. These quantities are continuous (temperature, length, current) or discrete (number of particles, birthday of a person, one of three possible colours) in nature. Now, no matter how accurately all conditions of the prescription are maintained, the results of repetitions of an experiment will in general differ. This is caused either by the intrinsic statistical nature of the phenomenon under investigation or by the finite accuracy of measurement. The possible results will therefore always span a finite region for each quantity. The totality of these regions for all quantities that constitute the result of an experiment build the *sample space* of that experiment. Since it is difficult and often impossible, to determine exactly the accessible regions for the quantities measured in a particular experiment, the sample space actually used is larger and contains the sample space proper as a subspace. We shall use this looser concept of a sample space.

Example 2-1: *Sample space spanned by continuous variables*
In the production of resistors it is important to maintain the values R (electric resistance measured in Ω) and N (maximum heat dissipation measured in W) at given values. The sample space for R and N could be a plane spanned by axes labelled R and N. Since both quantities are always positive the first quadrant of this plane would again be a sample space.

Example 2-2: *Sample space spanned by discrete variables*
In practice the exact values of R and N are unimportant as long as they are contained within a certain interval about the nominal value (e.g. $99\ \mathrm{k\Omega} < R < 101\ \mathrm{k\Omega}$, $0.49\ \mathrm{W} < N < 0.60\ \mathrm{W}$). If this is the

case we shall say that the resistor has the properties R_n, N_n. If the value falls below (above) the lower (upper) limit then we shall substitute the suffix n by $-$ $(+)$. The possible values of resistance and heat dissipation are therefore R_-, R_n, R_+ and N_-, N_n, N_+. The sample space now consists of 9 points

$$
\begin{array}{lll}
R_- N_-, & R_- N_n, & R_- N_+, \\
R_n N_-, & R_n N_n, & R_n N_+, \\
R_+ N_-, & R_+ N_n, & R_+ N_+.
\end{array}
$$

Often one or more particular subspaces of the sample space are of special interest. In example 2-2 for instance the point $R_n N_n$ represents the case where the resistors meet the production specifications. We can give such subspaces names, e.g. A, B, ..., and say that if the result of an experiment falls into one such subspace then the *event* A (or B, C, ...) has occurred. If A has not occurred we speak of the *complementary event* $\bar{A}$ (not A). The whole sample space corresponds to an event that will occur in every experiment. We call it E. In the rest of this chapter we shall define what we mean by the probability of the occurrence of an event.

2-2. The concept of probability

Let us consider the simplest experiment namely the tossing of a coin. Like the throwing of dice or certain problems with playing cards it is of no practical interest but is valuable for didactic purposes. What is the probability that a "regular" coin shows "heads" when tossed once? Our intuition suggests that this probability equals $\frac{1}{2}$. It is based on the assumption that all points in sample space (there are only 2 points: "heads" and "tails") are equally probable and on the convention that we give the event E (here: "heads" *or* "tails") the probability one. This way of determining probabilities can be applied only to perfectly symmetric experiments and is therefore of very little practical use. (It is, however, of great importance in statistical physics and quantum statistics where the equal probabilities of all allowed states is an essential postulate of very successful theories.)

If no such perfect symmetry exists – which will even be the case with normal "physical" coins – the following procedure seems reasonable. In a large number N of experiments the event A is observed to occur n times.

We define

$$P(A) = \lim_{N\to\infty} \frac{n}{N} \tag{2.1}$$

as the probability of the occurrence of the event A. This somewhat loose *frequency definition* of probability is sufficient for practical purposes, although it is mathematically unsatisfactory. One of the difficulties with this definition is the need for an infinity of experiments which are of course impossible to perform and even difficult to conceive. Although we shall in fact use the frequency definition in this book, we should like to indicate the basic concepts of an axiomatic theory of probability due to KOLMOGOROV [1933]. The minimal set of axioms generally used is the following.

(a) To each event A there corresponds a non-negative number, its probability

$$P(A) \geqslant 0. \tag{2.2}$$

(b) The event E has unit probability

$$P(E) = 1. \tag{2.3}$$

(c) If A and B are *mutually exclusive* events then the probability of A *or* B $(A + B)$ is

$$P(A + B) = P(A) + P(B). \tag{2.4}$$

From these axioms* one obtains immediately the following useful results. From (b) and (c)

$$P(A + \bar{A}) = P(A) + P(\bar{A}) = 1, \tag{2.5}$$

and furthermore with (a)

$$0 \leqslant P(A) \leqslant 1. \tag{2.6}$$

From (c) one can easily obtain the more general theorem for mutually exclusive events $A, B, C, \ldots$

$$P(A + B + C + \ldots) = P(A) + P(B) + P(C) + \ldots. \tag{2.7}$$

It should be noted that the only way of combining events which has so far

* Sometimes the definition (3.1) is introduced as a fourth axiom.

been mentioned is the *exclusive or*. Any other combination has to be built from it. In throwing a die A may signify even, B odd, C less than 4 dots, D 4 or more dots. Of interest is the probability for the event *A or C*, which are obviously not exclusive. One forms *A and C* (AC) as well as AD, BC and BD, which are mutually exclusive, and finds for A or C (sometimes written $A \dotplus C$) the expression $AC + AD + BC$.

It is interesting to note that the axioms do not prescribe a method for obtaining a particular probability $P(A)$.

Finally it should be pointed out that the word probability is often used in common language in a sense that is different or even opposed to that considered by us. We mean *subjective probability*. Here the probability of an event is given by the measure of our belief in its occurrence. An example of this is: "The probability that a tunnel will be constructed under the English Channel before 1984 is $\frac{1}{3}$". As another example consider the case of a certain track in nuclear emulsion which could have been left by a proton or pion. One often says "The track was caused by a pion with probability $\frac{1}{2}$". But since the event had already taken place and only one of the two kinds of particle could have caused that particular track, the probability in question is either 0 or 1, although unknown.

2-3. Rules of probability calculus; conditional probability

The result of an experiment may have the property A. We now ask: What is the probability that it also has the property B, i.e. the probability of B under the condition A. We define this *conditional probability* as

$$P(B \mid A) = \frac{P(A\,B)}{P(A)}. \tag{3.1}$$

It follows that

$$P(A\,B) = P(A)\,P(B \mid A). \tag{3.2}$$

From the example shown in fig. 2-1 it can be seen that this definition is reasonable. Here the event *A or B* occurs if a point lies inside the region *A or B* respectively. If the point happens to lie in the overlapping region, the event AB (*A and B*) occurs. Let the area of the different regions be proportional to the probabilities of the corresponding events. Then the probability of B under the condition A is the quotient of the areas AB and A, i.e. in particular 1 if A is contained in B and zero if the overlapping area vanishes.

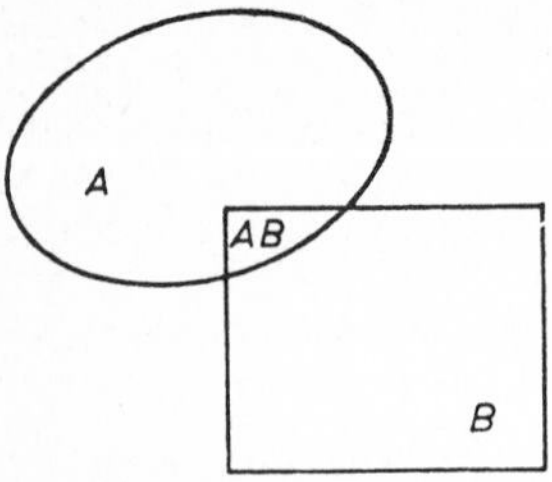

Fig. 2-1. Illustration of conditional probability.

Using conditional probability we can now formulate the *rule of total probability*. An experiment may lead to one of n possible mutually exclusive events

$$E = A_1 + A_2 + \dots + A_n. \tag{3.3}$$

The probability for the occurrence of any event with the property B is

$$P(B) = \sum_{i=1}^{n} P(A_i)\, P(B \mid A_i), \tag{3.4}$$

as can be seen easily from eqs. (3.2) and (2.7).

We can now also define the *independence of events*. Two events A and B are said to be independent if the knowledge that A has occurred does not change the probability for B and vice versa, i.e. if

$$P(B \mid A) = P(B), \tag{3.5}$$

or, with eq. (3.2),

$$P(A\,B) = P(A)\, P(B). \tag{3.6}$$

In general several decompositions of the type (3.3)

$$\begin{aligned} E &= A_1 + A_2 + \dots + A_n, \\ E &= B_1 + B_2 + \dots + B_m, \\ &\vdots \\ E &= Z_1 + Z_2 + \dots + Z_l, \end{aligned} \tag{3.7}$$

are independent if for all possible combinations $\alpha, \beta, \dots, \omega$ the condition

$$P(A_\alpha B_\beta \cdots Z_\omega) = P(A_\alpha)\, P(B_\beta) \cdots P(Z_\omega) \tag{3.8}$$

is fulfilled.

Exercises

2-1: *Determination of probability from symmetry*

There are n students in a class. What is the probability that at least two have the same birthday? Solve the problem by answering the following questions.

(a) What is the number N of possibilities of distributing the n birthdays in the year (365 days)?

(b) What is the number N' so that all n birthdays are different?

(c) What, then, is the probability P_{diff} of them being different?

(d) Finally, what is the probability P that at least two are not different?

2-2: *Probability of non-exclusive events*

The probabilities $P(A)$, $P(B)$ and $P(AB) \neq 0$ are given for the non-exclusive events A and B. What is the probability $P(A + B)$ of observing A or B? As an example compute the probability that a card taken at random from a pack of 52 is an ace or a diamond.

2-3: *Dependent or independent events*

Are the events A and B that a card drawn from a pack is an ace or a diamond independent

(a) if an ordinary pack of 52 is used?

(b) if a joker is added to the pack?

[Use eq. (3.6).]

2-4: *Complementary events*

Prove that $\bar{A}$ and $\bar{B}$ are independent if A and B are independent. Use the result of exercise 2-2 to compute $P(\bar{A}\bar{B})$ in terms of $P(A)$, $P(B)$ and $P(AB)$.

2-5: *Probabilities drawn from large and small populations*

A box contains a large number (>1000) of coins. They are divided into 3 species A, B and C which make up 20%, 30% and 50% of the total.

(a) What are the probabilities $P(A)$, $P(B)$, $P(C)$ of picking a coin of type A, B or C if one coin is taken at random? What are the probabilities $P(AB)$, $P(AC)$, $P(BC)$, $P(AA)$, $P(BB)$, $P(CC)$, P(2 equal coins), P(2 different coins) for picking two coins?

(b) What are the probabilities if 10 coins (2 of type A, 3 of type B and 5 of type C) are in the box?

CHAPTER 3

RANDOM VARIABLES; DISTRIBUTIONS OF A RANDOM VARIABLE

3-1. Random variables

We will not consider now the probability of observing particular events, but the events themselves and try to find a particularly simple way of classifying them. We can for instance associate the event "heads" with the number 0 and the event "tails" with the number 1. Generally we can classify the events of the decomposition (2-3.3) by associating each event A_i with the real number i. In this way each event can be characterized by one of the possible values of a *random variable*. Random variables can be discrete or continuous. We denote them by symbols like x, y, ...

Example 3-1: *Discrete random variable*
It may be of interest to study the number of coins still in circulation as a function of their age. It is obviously most convenient to use the year of issue stamped on each coin directly as the (discrete) random variable, e.g. x = ..., 1873, 1874, 1875, ...

Example 3-2: *Continuous random variable*
All processes of measurement or production are subject to smaller or larger imperfections or fluctuations that lead to variations in the result, which is therefore described by one or several random variables. Thus the values of electrical resistance and maximum heat dissipation characterizing a resistor in example 2-2 are (continuous) random variables.

3-2. Distributions of one random variable

From the classification of events we return to probability considerations. We consider the random variable x and a real number x, which can assume any value between $-\infty$ and $+\infty$, and study the probability for the event $(\mathsf{x} < x)$. This probability is a function of x and is called the (cumulative) *distribution function* of x

$$F(x) = P(\mathsf{x} < x). \tag{2.1}$$

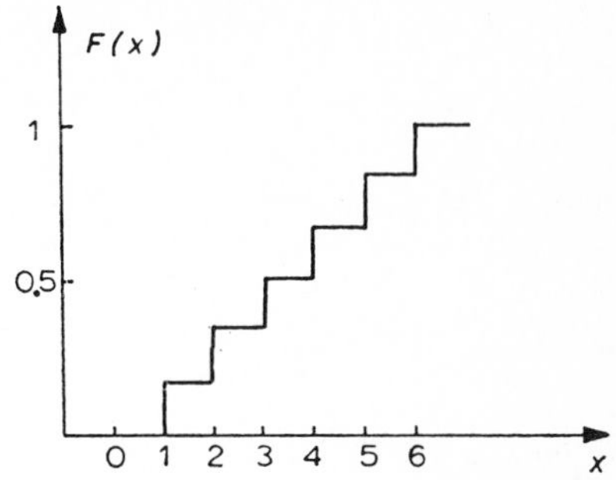

Fig. 3-1. Distribution function for the tossing of an ideally symmetric die.

If x can assume only a finite number of discrete values, e.g. the number of dots on the faces of a die, then the distribution function is a step function. It is shown in fig. 3-1 for the example mentioned above. Obviously any distribution function is *monotonic* and *non-decreasing.*

Because of eq. (2-2.3) in the limit $x \to \infty$ it becomes

$$\lim_{x\to\infty} F(x) = \lim_{x\to\infty} P(\mathsf{x} < x) = P(E) = 1. \tag{2.2}$$

Applying eq. (2-2.5) to eq. (2.1) we get

$$P(\mathsf{x} \geqslant x) = 1 - F(x) = 1 - P(\mathsf{x} < x) \tag{2.3}$$

and therefore

$$\lim_{x\to-\infty} F(x) = \lim_{x\to-\infty} P(\mathsf{x} < x) = 1 - \lim_{x\to-\infty} P(\mathsf{x} \geqslant x),$$
$$\lim_{x\to-\infty} F(x) = 0. \tag{2.4}$$

Of special interest are distribution functions $F(x)$ which are continuous and possess a derivative. The first derivative

$$f(x) = \frac{\mathrm{d}F(x)}{\mathrm{d}x} = F'(x) \tag{2.5}$$

is called the *probability density (function)* of x. It is a measure of the probability of the event $(x \leqslant \mathsf{x} < x + \mathrm{d}x)$. From eqs. (2.1) and (2.5) it is obvious that

$$P(\mathsf{x} < a) = F(a) = \int_{-\infty}^{a} f(x)\,\mathrm{d}x, \tag{2.6}$$

$$P(a \leqslant x < b) = \int_a^b f(x)\,\mathrm{d}x = F(b) - F(a), \tag{2.7}$$

and especially that

$$\int_{-\infty}^{\infty} f(x)\,\mathrm{d}x = 1. \tag{2.8}$$

A trivial example of a continuous distribution is given by the angular position of the hand of a watch read at random intervals. We obtain a constant probability density (fig. 3-2).

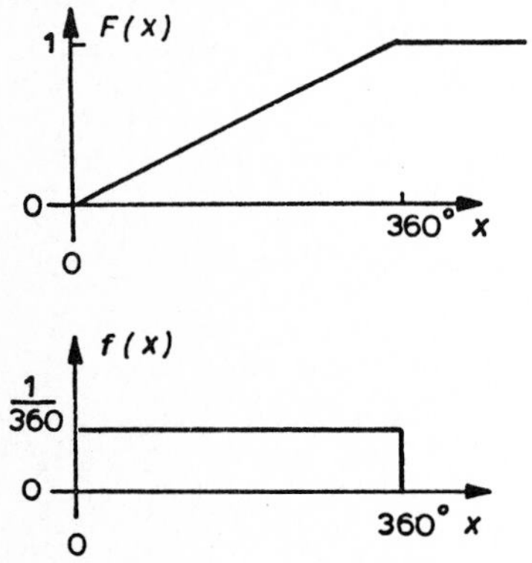

Fig. 3-2. Distribution function and probability density for the angular position of a watch hand.

3-3. Functions of one random variable, expectation value, variance, moments

In addition to the distribution of a random variable x, we are often interested in the distribution of a function of x. Such a function of a random variable is also a random variable

$$y = H(x). \tag{3.1}$$

The variable y then possesses a distribution function and probability density in the same way as x.

In the two simple examples of the last section we were able to state the distribution function immediately because of the symmetric nature of the problems. Usually this is not possible. Instead, we have to obtain it from experiment. Often we are limited to determining a few characteristic parameters instead of the complete distribution.

The *mean* or *expectation value* of a random variable is the sum of all possible values x_i of x multiplied by their respective probabilities

$$E(x) = \hat{x} = \sum_{i=1}^{n} x_i \, P(x = x_i). \tag{3.2}$$

Note that $\hat{x}$ is not a random variable but it is exactly defined. Correspondingly the expectation value of a function (3.1) is defined to be

$$E\{H(x)\} = \sum_{i=1}^{n} H(x_i) \, P(x = x_i). \tag{3.3}$$

In the case of a continuous random variable (with a differentiable distribution function), then by analogy we define

$$E(x) = \hat{x} = \int_{-\infty}^{\infty} x f(x) \, dx, \tag{3.4}$$

and

$$E\{H(x)\} = \int_{-\infty}^{\infty} H(x) f(x) \, dx. \tag{3.5}$$

If we choose especially

$$H(x) = (x - c)^l, \tag{3.6}$$

we obtain the expectation values

$$\alpha_l = E\{(x - c)^l\}, \tag{3.7}$$

which are called the lth *moments* of the variable about the point c. Of special interest are the *moments about the mean*

$$\mu_l = E\{(x - \hat{x})^l\}. \tag{3.8}$$

The lowest moments are obviously

$$\mu_0 = 1, \qquad \mu_1 = 0. \tag{3.9}$$

The quantity

$$\mu_2 = \sigma^2(x) = \text{var}(x) = E\{(x - \hat{x})^2\} \tag{3.10}$$

is the lowest moment that contains information about the average deviation of the variable x from its mean. It is called the *variance* of x.

We now try to visualize the practical meaning of expectation value and variance of a random variable x. Let us consider the measurement of some quantity, for example, the length x_0 of a small crystal using a microscope.

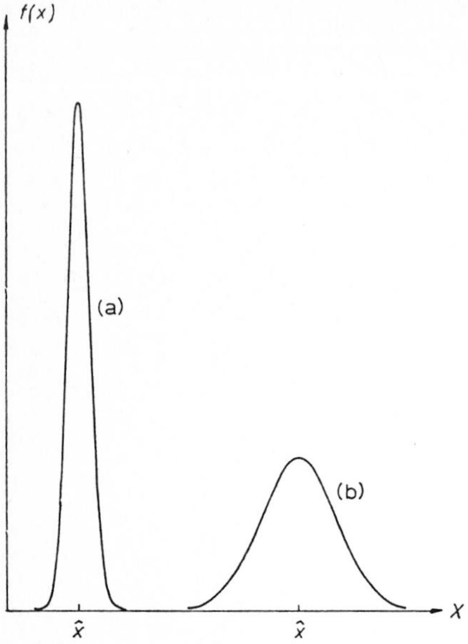

Fig. 3-3. Distribution with small variance (a) and large variance (b).

Due to the influence of different factors, such as the imperfections of the different components of the microscope and observational errors, repetitions of the measurement will yield slightly different results x. The individual measurements will, however, tend to group themselves in the neighbourhood of the true value of the length to be measured, i.e. it will be more probable to find a value of x near to x_0 than far from it, providing no systematic biases exist. The probability density of x will therefore have a bell-shaped form as sketched in fig. 3-3, although it need not be symmetric. It seems reasonable – especially in the case of a symmetric probability density – to interpret the expectation value (3.4) as the true value. It is interesting to note that (3.4) has the mathematical form of a centre of gravity, i.e. that $\hat{x}$ can be visualized as the abscissa of the centre of gravity of the surface under the curve describing the probability density $f(x)$.

The variance (3.10)

$$\sigma^2(x) = \int_{-\infty}^{\infty} (x - \hat{x})^2 f(x)\, dx,$$

which has the form of a moment of inertia, is a measure for the width or dispersion of the probability density about the mean $\hat{x}$. If it is small, the individual measurements lie near to $\hat{x}$ (fig. 3-3a); if it is large they will in general be more distant from the mean (fig. 3-3b). The positive square root of the variance

$$\sigma = +\sqrt{\sigma^2(x)} \tag{3.11}$$

is called the *standard deviation* (sometimes the *dispersion*) of x. Like the variance itself it is a measure of the average deviation of the measurements x from the expectation value.

Since the standard deviation has the same dimension as x (in our example both have the dimension length) it is identified with the *error* of measurement

$$\sigma(x) = \Delta x.$$

This definition of measurement error is discussed in more detail in ch. 5, §§7–12. It should be noted that the definitions (3.4) and (3.10) do not directly provide a way of calculating the mean of the measurement error since the probability density describing a measurement is in general unknown.

The third moment about the mean, μ_3, is sometimes called skewness. We prefer to define the dimensionless parameter

$$\gamma = \mu_3/\sigma^3 \tag{3.12}$$

to be the *skewness* of x. It contains information about a possible difference between positive and negative deviation from the mean. For symmetric distributions γ vanishes. It is positive (negative) if the distribution is skew to the right (left) of the mean.

We will now obtain a few important rules about means and variances. In the case in which

$$H(x) = cx, \qquad c = \text{const.}, \tag{3.13}$$

it follows immediately that

$$\begin{aligned} E(cx) &= c\,E(x), \\ \sigma^2(cx) &= c^2\,\sigma^2(x), \end{aligned} \tag{3.14}$$

and therefore

$$\sigma^2(x) = E\{(x-\hat{x})^2\} = E\{x^2 - 2x\hat{x} + \hat{x}^2\} = E(x^2) - \hat{x}^2. \tag{3.15}$$

We now consider the special function

$$u = \frac{x-\hat{x}}{\sigma(x)}. \tag{3.16}$$

It has the expectation value

$$E(u) = \frac{1}{\sigma(x)} E(x - \hat{x}) = \frac{1}{\sigma(x)} (\hat{x} - \hat{x}) = 0 \tag{3.17}$$

and the variance

$$\sigma^2(u) = \frac{1}{\sigma^2(x)} E\{(x - \hat{x})^2\} = \frac{\sigma^2(x)}{\sigma^2(x)} = 1. \tag{3.18}$$

The function u – which is also a random variable – has especially simple properties which make its use in more involved calculations preferable. We will call such a variable (having zero mean and unit variance) a *reduced variable*. It is also called a *standardized, normalized* or *dimensionless* variable.

Although a distribution is mathematically most easily described by its expectation value, variance and higher moments (in fact any distribution can be completely specified by these quantities, cf. ch. 5, §6), it is often convenient to introduce further definitions so as to visualize better the form of a distribution. The *mode* x_m of a distribution is defined as that value of the random variable that corresponds to the highest probability

$$P(x = x_m) = \max. \tag{3.19}$$

If the distribution has a differentiable probability density the mode, which then corresponds to its maximum, is easily determined by the conditions

$$\frac{d}{dx} f(x) = 0, \qquad \frac{d^2}{dx^2} f(x) < 0. \tag{3.20}$$

In many cases only one maximum exists; the distribution is said to be *unimodal*. Otherwise it is called *multimodal*.

The *median* $x_{0.5}$ of a distribution is defined as that value of the random variable for which the distribution function equals one half:

$$F(x_{0.5}) = P(x < x_{0.5}) = 0.5. \tag{3.21}$$

In the case of a continuous probability density eq. (3.21) takes the form

$$\int_{-\infty}^{x_{0.5}} f(x)\,dx = 0.5, \tag{3.22}$$

i.e. the median divides the total range of the random variable into two regions with equal probability.

It is clear from these definitions that in the case of a unimodal distribution with continuous probability density, which is symmetric about its maximum,

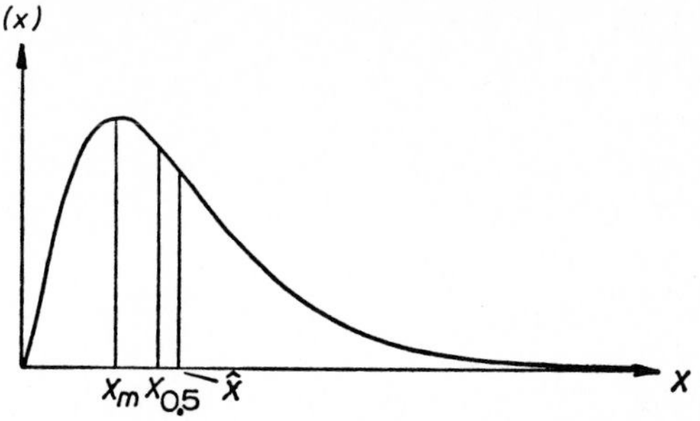

Fig. 3-4. Mode (x_m), mean ($\hat{x}$) and median ($x_{0.5}$) of an asymmetric distribution.

the values of mean, mode, and median coincide. This is, however, not the case for asymmetric distributions (fig. 3-4).

The definition (3.21) can easily be generalized. The quantities $x_{0.25}$ and $x_{0.75}$ defined by

$$F(x_{0.25}) = 0.25, \qquad F(x_{0.75}) = 0.75 \tag{3.23}$$

are called lower and upper *quartiles*. Similarly we can define *deciles* $x_{0.1}$, $x_{0.2}$, ..., $x_{0.9}$ or in general *fractiles* (also called *quantiles*) x_q by

$$F(x_q) = \int_{-\infty}^{x_q} f(x)\, \mathrm{d}x = q \tag{3.24}$$

with $0 < q < 1$.

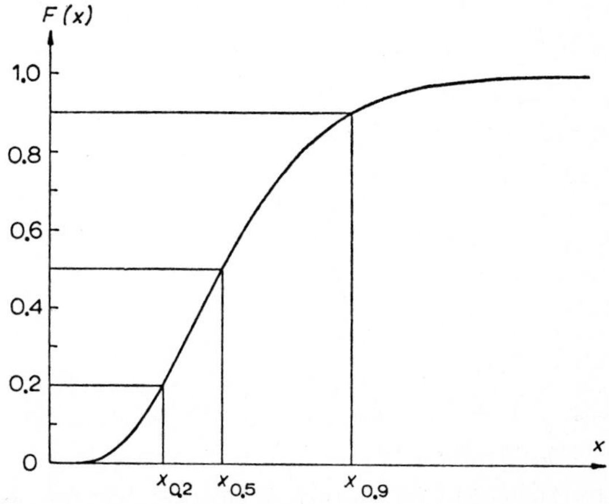

Fig. 3-5. Median and fractiles of a continuous distribution.

The definition of fractiles is most easily visualized from fig. 3-5. In a diagram of the distribution function $F(x)$ the fractile x_q can be read off as the abcissa corresponding to the value q on the ordinate. The fractile x_q regarded as a function of the probability q is simply the inverse of the distribution function

3-4. Chebychev's inequality

Clearly the values of a random variable x are somewhere in the neighbourhood of the mean $\hat{x}$. Deviations from $\hat{x}$ are less probable the larger they are compared with the standard deviation. This fact is expressed quantitatively by Chebychev's inequality.

If k is any positive real number then the probability of an absolute deviation from the mean being larger than k times the standard deviation is less than k^{-2}

$$P(|x - \hat{x}| > k\sigma) < k^{-2}. \tag{4.1}$$

We now prove the inequality for the case of a continuous variable. Let us define $g(t)$ with $t = (x - \hat{x})^2$ as the probability density of the variable $t = (x - \hat{x})^2$. Then

$$P = P(|x - \hat{x}| > k\sigma) = P((x - \hat{x})^2 > k^2\sigma^2),$$

$$P = \int_{k^2\sigma^2}^{\infty} g(t)\, \mathrm{d}t,$$

and

$$\sigma^2 = E\{(x - \hat{x})^2\} = \int_{-\infty}^{\infty} t\, g(t)\, \mathrm{d}t,$$

$$\sigma^2 = \int_{0}^{k^2\sigma^2} t\, g(t)\, \mathrm{d}t + \int_{k^2\sigma^2}^{\infty} t\, g(t)\, \mathrm{d}t. \tag{4.2}$$

Since the integration extends over positive values of t only and since $g(t)$, being a probability density, is positive definite the value of each integral will be at least as large as the value that would be obtained by setting the factor t equal to the lower limit of the integration. From eq. (4.2) we therefore obtain immediately

$$\sigma^2 > k^2\sigma^2 \int_{k^2\sigma^2}^{\infty} g(t)\,\mathrm{d}t = k^2\sigma^2 P,$$

which is an equivalent expression for Chebychev's inequality (4.1).

In general this inequality is rather weak. In fact in most cases of interest

$$P(|x < \hat{x}| > k\sigma) \ll k^{-2}$$

(for an example see ch. 5, §9).

Exercises

3-1: *Mean, variance and skewness of a discrete distribution*

The throwing of a die yields the possible results $x_i = 1, 2, \ldots, 6$. For an ideally symmetric die one has $p_i = P(x_i) = \frac{1}{6}$, $i = 1, 2, \ldots, 6$. Determine the expectation value $\hat{x}$, the variance $\sigma^2(x) = \mu_2$ and the skewness γ of the distribution

(a) for an ideally symmetric die,

(b) for a die with $p_1 = \frac{1}{6}$, $p_2 = \frac{1}{12}$, $p_3 = \frac{1}{12}$, $p_4 = \frac{1}{6}$, $p_5 = \frac{3}{12}$, $p_6 = \frac{3}{12}$.

3-2: *Mean, mode, median and variance of a continuous distribution*

Consider a probability density function $f(x)$ of triangular shape as in fig. 3-6 given by

$$f(x) = 0, \quad x < a, \quad x \geqslant b$$

$$f(x) = \frac{2}{(b-a)(c-a)}(x-a), \quad a < x \leqslant c$$

$$f(x) = \frac{2}{(b-a)(b-c)}(b-x), \quad c < x \leqslant b.$$

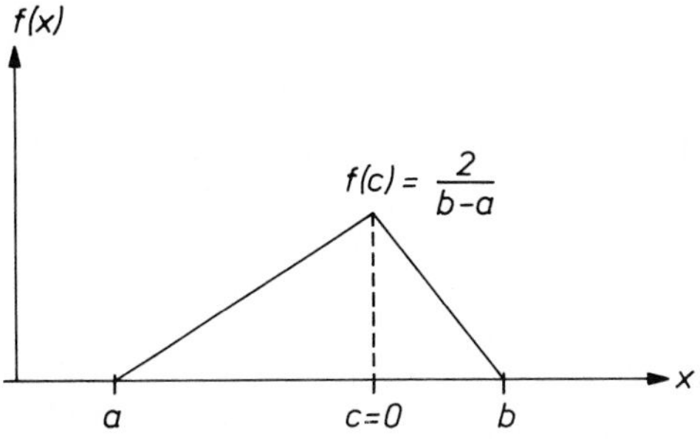

Fig. 3-6. Triangular probability density.

Determine mean $\hat{x}$, mode x_m, median $x_{0.5}$, variance $\sigma^2(x)$ of the distribution. For simplicity of calculation choose $c = 0$ (this corresponds to a substitution of x by $x' = x - c$). For determination of the median assume $b \leqslant -a$. Specialize the results for the symmetrical case $a = -b$ and for $a = -2b$ (the latter case is drawn in fig. 3-5).

3-3: *Distribution with infinite variance*

Show that the "Cauchy-distribution" $f(x) = 1/\pi(1 + x^2)$ has infinite variance. Use the symmetry of the distribution to determine the expectation value $\hat{x}$. (For a discussion of the Cauchy distribution see exercise 4-1.)

CHAPTER 4

DISTRIBUTIONS OF SEVERAL RANDOM VARIABLES

4-1. Distribution function and probability density of two variables; conditional probability

We now consider 2 random variables x and y and ask for the probability that both $\mathsf{x} < x$ and $\mathsf{y} < y$. As in the case of one variable we expect the existence of a distribution function (see fig. 4-1)

$$F(x, y) = P(\mathsf{x} < x, \mathsf{y} < y). \tag{1.1}$$

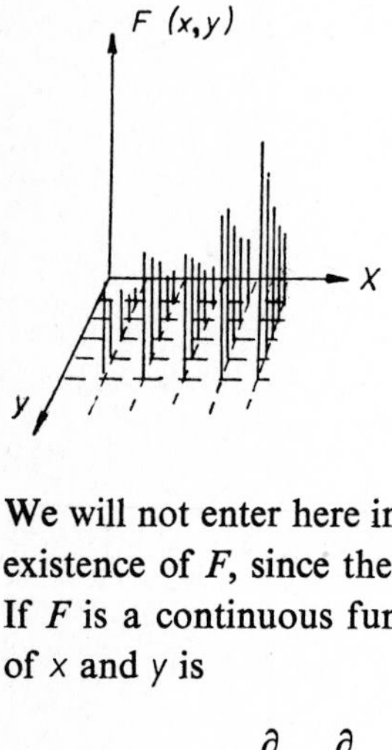

Fig. 4-1. Distribution function of two variables.

We will not enter here into axiomatic details and into the conditions for the existence of F, since these are always fulfilled in cases of practical interest. If F is a continuous function of x and y then the *joint probability density* of x and y is

$$f(x, y) = \frac{\partial}{\partial x}\frac{\partial}{\partial y} F(x, y). \tag{1.2}$$

Then

$$P(a \leqslant \mathsf{x} < b, c \leqslant \mathsf{y} < d) = \int_a^b \int_c^d f(x, y)\, \mathrm{d}x\, \mathrm{d}y. \tag{1.3}$$

Often we face the following experimental problem. As the result of a measurement we obtain a set of 2 random variables x, y, i.e., with many measurements we determine approximately the joint distribution function $F(x, y)$, but only the probability behaviour of y (irrespective of x) is of interest. (For example, the probability density for the appearance of a certain infective disease might be given as a function of date and geographic location. For some investigations the dependence on the time of year might be without interest.) We integrate eq. (4.3) over the whole range of y and obtain

$$P(a \leqslant \mathsf{x} < b, -\infty < \mathsf{y} < \infty) = \int_a^b \left[\int_{-\infty}^{\infty} f(x, y)\,\mathrm{d}y\right]\mathrm{d}x = \int_a^b g(x)\,\mathrm{d}x.$$

Here

$$g(x) = \int_{-\infty}^{\infty} f(x, y)\,\mathrm{d}y \tag{1.4}$$

is a probability density of x. It is called *marginal distribution* of x. The corresponding distribution of y is

$$h(y) = \int_{-\infty}^{\infty} f(x, y)\,\mathrm{d}x. \tag{1.5}$$

Analogous to the independence of events (eq. 2-3.6)) we can now define the *independence of random variables*. The variables x and y are independent if

$$f(x, y) = g(x)\,h(y). \tag{1.6}$$

Using the marginal distribution we can also define conditional probability for y under the condition that x is known

$$P(y \leqslant \mathsf{y} < y + \mathrm{d}y \mid x \leqslant \mathsf{x} \leqslant x + \mathrm{d}x). \tag{1.7}$$

We define the conditional probability density

$$f(y \mid x) = \frac{f(x, y)}{g(x)}. \tag{1.8}$$

Then the probability (1.7) is given by

$$f(y \mid x)\,\mathrm{d}y.$$

Also the rule of total probability can now be expressed for distributions:

$$h(y) = \int_{-\infty}^{\infty} f(x, y)\,\mathrm{d}x = \int_{-\infty}^{\infty} f(y \mid x)\, g(x)\,\mathrm{d}x. \tag{1.9}$$

For independent variables, i.e. if eq. (1.6) holds, it is obvious that

$$f(y \mid x) = \frac{f(x, y)}{g(x)} = \frac{g(x)\, h(y)}{g(x)} = h(y). \tag{1.10}$$

This was expected since, in the case of independent variables, any constraint on one variable cannot contribute information about the probability distribution of the other.

4-2. Expectation values, variances, covariances and correlation coefficient

Analogous to eq. (3-3.5) we define the expectation value of a function $H(x, y)$ to be

$$E\{H(x, y)\} = \int_{-\infty}^{\infty} \int_{-\infty}^{\infty} H(x, y) f(x, y)\,\mathrm{d}x\,\mathrm{d}y. \tag{2.1}$$

The variance of $H(x, y)$ is

$$\sigma^2\{H(x, y)\} = E\{[H(x, y) - E(H(x, y))]^2\}. \tag{2.2}$$

For the simple case $H(x, y) = ax + by$, eq. (2.1) results in

$$E(ax + by) = aE(x) + bE(y). \tag{2.3}$$

We now choose

$$H(x, y) = x^l y^m, \qquad (l, m \text{ non-negative integer}). \tag{2.4}$$

The expectation values of such functions are the lmth *moments* of x, y about the origin:

$$\lambda_{lm} = E(x^l y^m). \tag{2.5}$$

If we choose more generally

$$H(x, y) = (x - a)^l (y - b)^m, \tag{2.6}$$

the expectation values

$$\alpha_{lm} = E\{(x - a)^l (y - b)^m\} \tag{2.7}$$

are the lmth moments about the point a, b. Of special interest are the moments about the point λ_{10}, λ_{01}

$$\mu_{lm} = E\{(x - \lambda_{10})^l (y - \lambda_{01})^m\}. \tag{2.8}$$

As in the case of one variable the lower moments have a special significance

$$\begin{aligned} \mu_{00} &= \lambda_{00} = 1, \\ \mu_{10} &= \mu_{01} = 0; \\ \lambda_{10} &= E(x) = \hat{x}, \\ \lambda_{01} &= E(y) = \hat{y}; \\ \mu_{11} &= E\{(x - \hat{x})(y - \hat{y})\} = \operatorname{cov}(x, y), \\ \mu_{20} &= E\{(x - \hat{x})^2\} = \sigma^2(x), \\ \mu_{02} &= E\{(y - \hat{y})^2\} = \sigma^2(y). \end{aligned} \tag{2.9}$$

We can now express the variance of $(ax + by)$ in terms of these quantities

$$\begin{aligned} \sigma^2(ax + by) &= E\{[(ax + by) - E(ax + by)]^2\} \\ &= E\{[a(x - \hat{x}) + b(y - \hat{y})]^2\} \\ &= E\{a^2(x - \hat{x})^2 + b^2(y - \hat{y})^2 + 2ab(x - \hat{x})(y - \hat{y})\}, \end{aligned}$$

$$\sigma^2(ax + by) = a^2\sigma^2(x) + b^2\sigma^2(y) + 2ab \operatorname{cov}(x, y). \tag{2.10}$$

In deriving (2.10) we have made use of relation (3-3.14). As another example we consider

$$H(x, y) = xy. \tag{2.11}$$

In this case we have to assume the independence of x and y in order to obtain the expectation value. Then

$$\begin{aligned} E(xy) &= \int_{-\infty}^{\infty} \int_{-\infty}^{\infty} x\, y\, g(x)\, h(y)\, \mathrm{d}x\, \mathrm{d}y \\ &= \left(\int_{-\infty}^{\infty} x\, g(x)\, \mathrm{d}x \right) \left(\int_{-\infty}^{\infty} y\, h(y)\, \mathrm{d}y \right), \end{aligned} \tag{2.12}$$

i.e.

$$E(xy) = E(x)\, E(y). \tag{2.13}$$

While the quantities $E(x)$, $E(y)$, $\sigma^2(x)$, $\sigma^2(y)$ are very similar to those obtained in the case of one variable we have still to elucidate the meaning of $\mu_{11} = \operatorname{cov}(x, y)$. The concept of *covariance* is of considerable importance for the understanding of many of our subsequent problems. From its definition we see that cov (x, y) is positive if values $x > \hat{x}$ $(x < \hat{x})$ appear preferentially together with values $y > \hat{y}$ $(y < \hat{y})$. On the other hand cov (x, y) is negative if in general $x > \hat{x}$ $(x < \hat{x})$ implies $y < \hat{y}$ $(y > \hat{y})$. If finally the knowledge of the value of x does not give us additional information about the probable position of y, the covariance vanishes. Fig. 4-2 illustrates these cases in graphical form.

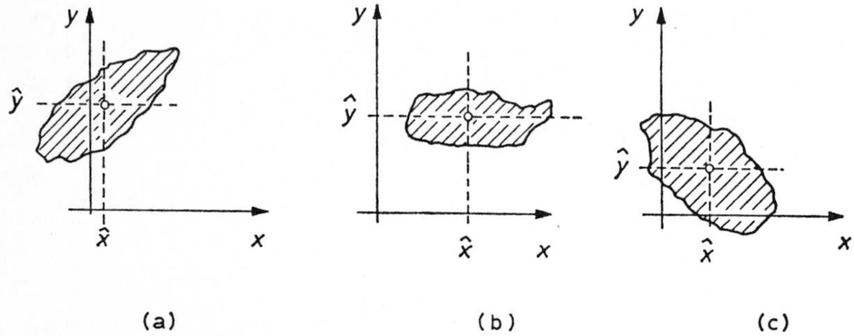

Fig. 4-2. Covariance between variables. (a) cov $(x, y) > 0$; (b) cov $(x, y) \approx 0$; (c) cov $(x, y) < 0$.

It is often convenient to use the *correlation coefficient*

$$\rho(x, y) = \frac{\operatorname{cov}(x, y)}{\sigma(x)\,\sigma(y)}, \tag{2.14}$$

rather than the covariance. Both covariance and correlation coefficient offer a – necessarily crude – measure of the mutual dependence of x and y. To investigate this further we now consider two reduced variables u and v in the sense of eq. (3-3.16) and determine the variance of their sum according to eq. (2.10)

$$\sigma^2(u + v) = \sigma^2(u) + \sigma^2(v) + 2\rho(u, v)\,\sigma(u)\,\sigma(v).$$

From eq. (3-3.18) we know that

$$\sigma^2(u) = \sigma^2(v) = 1.$$

Therefore

$$\sigma^2(u + v) = 2(1 + \rho(u, v)), \tag{2.15}$$

and correspondingly

$$\sigma^2(u - v) = 2(1 - \rho(u, v)). \tag{2.16}$$

Since for each variance $\sigma^2 \geqslant 0$ it follows that

$$-1 \leqslant \rho(u, v) \leqslant 1.$$

It is easy to show that

$$\rho(u, v) = \rho(x, y). \tag{2.17}$$

Therefore finally

$$-1 \leqslant \rho(x, y) \leqslant 1. \tag{2.18}$$

We now investigate the limiting cases $\rho = \pm 1$. For $\rho(u, v) = 1$ the variance $\sigma^2(u - v)$ vanishes according to eq. (2.16), i.e. the random variable $(u - v)$ is a constant

$$u - v = \frac{x - \hat{x}}{\sigma(x)} - \frac{y - \hat{y}}{\sigma(y)} = \text{const.} \tag{2.19}$$

This equation is fulfilled if

$$y = a + bx, \tag{2.20}$$

where a is any constant and b is a positive constant. Therefore in the case of a linear dependence (b positive) between x and y the correlation coefficient takes the value $+1$. Correspondingly one finds $\rho(x, y) = -1$ for a negative linear dependence (b negative).

We would expect that the covariance of two independent variables vanishes. With (1.6) we find indeed

$$\begin{aligned}\operatorname{cov}(x, y) &= \int_{-\infty}^{\infty}\int_{-\infty}^{\infty} (x - \hat{x})(y - \hat{y})\, g(x)\, h(y)\, \mathrm{d}x\, \mathrm{d}y \\ &= \left(\int_{-\infty}^{\infty} (x - \hat{x})\, g(x)\, \mathrm{d}x\right)\left(\int_{-\infty}^{\infty} (y - \hat{y})\, h(y)\, \mathrm{d}y\right) \\ &= 0.\end{aligned}$$

4-3. More than two variables; vector and matrix notation

By analogy with (1.1) we define the *distribution function of n variables* $x_1, x_2, \ldots, x_n$

$$F(x_1, x_2, \ldots, x_n) = P(x_1 < x_1, x_2 < x_2, \ldots, x_n < x_n). \tag{3.1}$$

If the function F fulfils certain requirements concerning continuity and partial differentiability with respect to the x_i then the *joint probability density* is

$$f(x_1, x_2, \ldots, x_n) = \frac{\partial^n}{\partial x_1 \, \partial x_2 \cdots \partial x_n} F(x_1, x_2, \ldots, x_n). \tag{3.2}$$

The *marginal distribution*

$$g_r(x_r) = \int_{-\infty}^{\infty} \int_{-\infty}^{\infty} \cdots \int_{-\infty}^{\infty} f(x_1, x_2, \ldots, x_n) \, dx_1 \, dx_2 \cdots dx_{r-1} \, dx_{r+1} \cdots dx_n \tag{3.3}$$

can be interpreted as the probability density of a single variable x_r. If $H(x_1, x_2, \ldots, x_n)$ is a function of the variables then the *expectation value* of H *is*

$$E\{H(x_1, x_2, \ldots, x_n)\} =$$

$$= \int_{-\infty}^{\infty} \int_{-\infty}^{\infty} \cdots \int_{-\infty}^{\infty} H(x_1, x_2, \ldots, x_n) f(x_1, x_2, \ldots, x_n) \, dx_1 \, dx_2 \cdots dx_n. \tag{3.4}$$

With $H = x_r$ we obtain

$$E(x_r) = \int_{-\infty}^{\infty} \int_{-\infty}^{\infty} \cdots \int_{-\infty}^{\infty} x_r f(x_1, x_2, \ldots, x_n) \, dx_1 \, dx_2 \cdots dx_n,$$

$$E(x_r) = \int_{-\infty}^{\infty} x_r \, g_r(x_r) \, dx_r. \tag{3.5}$$

The variables are *independent* if

$$f(x_1, x_2, \ldots, x_n) = g_1(x_1) \, g_2(x_2) \cdots g_n(x_n). \tag{3.6}$$

We can now also construct a *joint marginal distribution* for a number l out of the n variables* by extending the integration in eq. (3.3) only over the remaining $n - l$ variables

$$g(x_1, x_2, \ldots, x_l) = \int_{-\infty}^{\infty} \int_{-\infty}^{\infty} \cdots \int_{-\infty}^{\infty} f(x_1, x_2, \ldots, x_n) \, dx_{l+1} \cdots dx_n. \tag{3.7}$$

* We can assume that these are the variables $x_1, x_2, \ldots, x_l$ without any loss of generality.

These l variables are independent if

$$g(x_1, x_2, \ldots, x_l) = g_1(x_1)\,g_2(x_2)\cdots g_l(x_l). \tag{3.8}$$

The *moments* of order $l_1, l_2, \ldots, l_n$ about the origin are the expectation values of functions

$$H = x_1^{l_1}\, x_2^{l_2} \cdots x_n^{l_n}, \tag{3.9}$$

i.e.

$$\lambda_{l_1 l_2 \ldots l_n} = E(x_1^{l_1}\, x_2^{l_2} \cdots x_n^{l_n}), \tag{3.10}$$

in particular

$$\begin{aligned} \lambda_{100\ldots 0} &= E(x_1) = \hat{x}_1, \\ \lambda_{010\ldots 0} &= E(x_2) = \hat{x}_2, \\ &\vdots \\ \lambda_{000\ldots 1} &= E(x_n) = \hat{x}_n. \end{aligned} \tag{3.11}$$

The moments about $(\hat{x}_1, \hat{x}_2, \ldots, \hat{x}_n)$ are

$$\mu_{l_1 l_2 \ldots l_n} = E\{(x_1 - \hat{x}_1)^{l_1}(x_2 - \hat{x}_2)^{l_2} \cdots (x_n - \hat{x}_n)^{l_n}\}. \tag{3.12}$$

Then the variances of the x_i are

$$\begin{aligned} \mu_{200\ldots 0} &= E\{(x_1 - \hat{x}_1)^2\} = \sigma^2(x_1), \\ \mu_{020\ldots 0} &= E\{(x_2 - \hat{x}_2)^2\} = \sigma^2(x_2), \\ &\vdots \\ \mu_{000\ldots 2} &= E\{(x_n - \hat{x}_n)^2\} = \sigma^2(x_n). \end{aligned} \tag{3.13}$$

The moment about $(\hat{x}_1, \hat{x}_2, \ldots, \hat{x}_n)$ with $l_i = l_j = 1$ and $l_k = 0\ (i \neq k \neq j)$ is the *covariance* between the variables x_i and x_j

$$c_{ij} = \operatorname{cov}(x_i, x_j) = E\{(x_i - \hat{x}_i)(x_j - \hat{x}_j)\}. \tag{3.14}$$

It proves useful to represent the n variables $x_1, x_2, \ldots, x_n$ as components of a vector $\boldsymbol{x}$ in an n-dimensional space. We can then write the distribution function (3.1)

$$F = F(\boldsymbol{x}), \tag{3.15}$$

and the corresponding probability density (3.2)

$$f(\boldsymbol{x}) = \frac{\partial}{\partial \boldsymbol{x}} F(\boldsymbol{x}). \tag{3.16}$$

The expectation value of a function $H(x)$ simplifies to

$$E\{H(x)\} = \int H(\boldsymbol{x}) f(\boldsymbol{x}) \, \mathrm{d}\boldsymbol{x}. \tag{3.17}$$

Variances and covariances can now be collected into a matrix. It is called the *covariance matrix*

$$C = \begin{pmatrix} c_{11} & c_{12} & \cdots & c_{1n} \\ c_{21} & c_{22} & \cdots & c_{2n} \\ \vdots & & & \\ c_{n1} & c_{n2} & \cdots & c_{nn} \end{pmatrix}. \tag{3.18}$$

Here the c_{ij} are defined by (3.14). The diagonal elements are the variances $c_{ii} = \sigma^2(x_i)$. Since

$$c_{ij} = c_{ji}, \tag{3.19}$$

the covariance matrix is symmetric. If we now also write the expectation values as a vector

$$E(x) = \hat{x}, \tag{3.20}$$

we see that each element of the covariance matrix

$$c_{ij} = E\{(x_i - \hat{x}_i)(x_j - \hat{x}_j)\}$$

can be interpreted as the element ij of the (dyadic) product of the row vector $(x - \hat{x})^{\mathrm{T}}$ with the column vector $(x - \hat{x})$. The notation* is

$$\boldsymbol{x}^{\mathrm{T}} = (x_1, x_2, \ldots, x_n), \qquad \boldsymbol{x} = \begin{pmatrix} x_1 \\ x_2 \\ \vdots \\ x_n \end{pmatrix}.$$

The covariance matrix is therefore simply

$$C = E\{(x - \hat{x})(x - \hat{x})^{\mathrm{T}}\}. \tag{3.21}$$

4-4. Transformation of variables

As already mentioned in ch. 3, §3 a function of a random variable is itself a random variable

$$y = y(x).$$

* For details of the matrix notation used see appendix B.

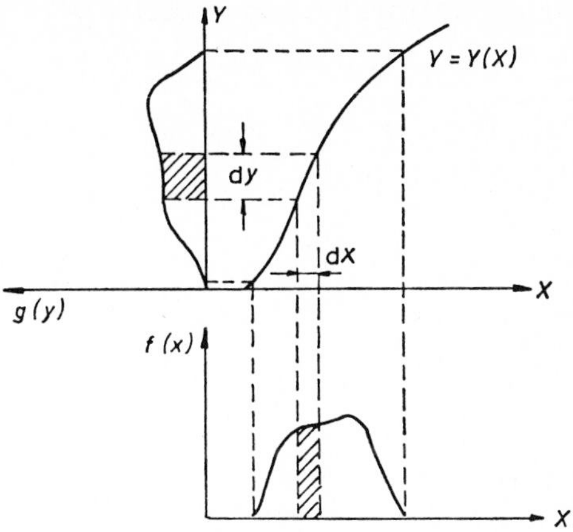

Fig. 4-3. Transformation of variable from x to y.

We now ask: what is the probability density $g(y)$ if the density $f(x)$ is known? The problem is illustrated in fig. 4-3. Evidently the probability $g(y)\,\mathrm{d}y$ that y is in the small interval $\mathrm{d}y$ is equal to the probability $g(x)\,\mathrm{d}x$ that x is contained in the "corresponding" interval $\mathrm{d}x$. This correspondence is indicated in fig. 4-3. We find that

$$\mathrm{d}y = \left|\frac{\mathrm{d}y}{\mathrm{d}x}\right| \mathrm{d}x \quad \text{or} \quad \mathrm{d}x = \left|\frac{\mathrm{d}x}{\mathrm{d}y}\right| \mathrm{d}y,$$

respectively. We have to take absolute values since the quantities $\mathrm{d}x$, $\mathrm{d}y$ are intervals with no directional properties. Only in this way can the probabilities $f(x)\,\mathrm{d}x$, $g(y)\,\mathrm{d}y$ always be kept positive. The relation between probability densities is therefore

$$g(y) = \left|\frac{\mathrm{d}x}{\mathrm{d}y}\right| f(x). \tag{4.1}$$

We see immediately that $g(y)$ is defined only in the case of a single-valued function $y = y(x)$ since only then is the differential quotient in (4.1) uniquely defined. Multivalued functions, for example $y = \sqrt{(x)}$, have to be treated specially. One would, in this case, consider only $y = +\sqrt{(x)}$. Eq. (4.1) also guarantees the normalization of the probability distribution of y to unity

$$\int_{-\infty}^{\infty} g(y)\,\mathrm{d}y = \int_{-\infty}^{\infty} f(x)\,\mathrm{d}x = 1.$$

Let us now consider the transition from 2 independent variables x and y to the new variables

$$u = u(x, y), \qquad v = v(x, y). \tag{4.2}$$

The problem is to find the function J that connects the probability densities $f(x, y)$ and $g(u, v)$

$$g(u, v) = f(x, y) \left| J\left(\frac{x, y}{u, v}\right) \right| . \tag{4.3}$$

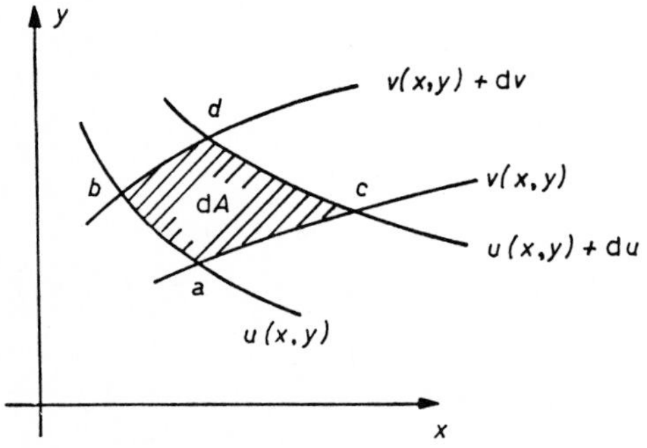

Fig. 4-4. Transformation of variables from x, y to u, v.

In fig. 4-4, the x–y plane is sketched with two lines each for u = const. and v = const. respectively. They define an element with respect to the variables u, v which corresponds to the area element $\mathrm{d}x\mathrm{d}y$ for the original variables. Since we are dealing with an "infinitesimal" element we can calculate its area by considering it as a small parallelogram with the vertices a, b, c, d (fig. 4-4). The coordinates of the first 3 vertices are

$$\begin{aligned} x_a &= x(u, v), & y_a &= y(u, v), \\ x_b &= x(u, v + \mathrm{d}v), & y_b &= y(u, v + \mathrm{d}v), \\ x_c &= x(u + \mathrm{d}u, v), & y_c &= y(u + \mathrm{d}u, v). \end{aligned}$$

Making use of a Taylor series expansion, we can write

$$x_b = x(u, v) + \frac{\partial x}{\partial v}\,\mathrm{d}v, \qquad y_b = y(u, v) + \frac{\partial y}{\partial v}\,\mathrm{d}v,$$

$$x_c = x(u, v) + \frac{\partial x}{\partial u}\,\mathrm{d}u, \qquad y_c = y(u, v) + \frac{\partial y}{\partial u}\,\mathrm{d}u.$$

Apart from the sign, which is not of interest because of the modulus used in eq. (4.3), the area of a parallelogram is equal to the determinant

$$dA = \begin{vmatrix} 1 & x_a & y_a \\ 1 & x_b & y_b \\ 1 & x_c & y_c \end{vmatrix},$$

i.e.,

$$dA = \frac{\partial x}{\partial u}\,du\,\frac{\partial y}{\partial v}\,dv - \frac{\partial y}{\partial u}\,du\,\frac{\partial x}{\partial v}\,dv$$

or, written as a determinant of second order,

$$dA = \begin{vmatrix} \dfrac{\partial x}{\partial u} & \dfrac{\partial y}{\partial u} \\ \dfrac{\partial x}{\partial v} & \dfrac{\partial y}{\partial v} \end{vmatrix} du\,dv = J\left(\frac{x, y}{u, v}\right) du\,dv. \tag{4.4}$$

The determinant

$$J\left(\frac{x, y}{u, v}\right) = \begin{vmatrix} \dfrac{\partial x}{\partial u} & \dfrac{\partial y}{\partial u} \\ \dfrac{\partial x}{\partial v} & \dfrac{\partial y}{\partial v} \end{vmatrix} \tag{4.5}$$

is called the *Jacobian* of the transformation (4.2).

In the general case of n variables $\mathsf{x} = (\mathsf{x}_1, \mathsf{x}_2, \ldots, \mathsf{x}_n)$ and the transformation

$$\begin{aligned} \mathsf{y}_1 &= \mathsf{y}_1(\mathsf{x}), \\ \mathsf{y}_2 &= \mathsf{y}_2(\mathsf{x}), \\ &\vdots \\ \mathsf{y}_n &= \mathsf{y}_n(\mathsf{x}), \end{aligned} \tag{4.6}$$

the probability density of the transformed variables is

$$g(\mathbf{y}) = J\left(\frac{\mathbf{x}}{\mathbf{y}}\right) f(\mathbf{x}). \tag{4.7}$$

Here the Jacobian is

$$J\left(\frac{\boldsymbol{x}}{\boldsymbol{y}}\right) = J\left(\frac{x_1, x_2, \ldots, x_n}{y_1, y_2, \ldots, y_n}\right) = \begin{vmatrix} \dfrac{\partial x_1}{\partial y_1} & \dfrac{\partial x_2}{\partial y_1} & \cdots & \dfrac{\partial x_n}{\partial y_1} \\ \dfrac{\partial x_1}{\partial y_2} & \dfrac{\partial x_2}{\partial y_2} & \cdots & \dfrac{\partial x_n}{\partial y_2} \\ \vdots & & & \\ \dfrac{\partial x_1}{\partial y_n} & \dfrac{\partial x_2}{\partial y_n} & \cdots & \dfrac{\partial x_n}{\partial y_n} \end{vmatrix} . \tag{4.8}$$

Condition for the existence of $g(\boldsymbol{y})$ is of course again the uniqueness of all partial derivatives that compose J.

4-5. Linear and orthogonal transformations; propagation of errors

In practice we deal frequently with linear transformations of variables. The reason is mainly that they are particularly easy to handle and we try therefore to approximate other transformations by linear ones using Taylor series expansion techniques.

The functions $y = (y_1, y_2, \ldots, y_r)$ may be linear in the n variables $x = (x_1, x_2, \ldots, x_n)$, i.e.

$$\begin{aligned} y_1 &= a_1 + t_{11} x_1 + t_{12} x_2 + \ldots + t_{1n} x_n, \\ y_2 &= a_2 + t_{21} x_1 + t_{22} x_2 + \ldots + t_{2n} x_n, \\ &\vdots \\ y_r &= a_r + t_{r1} x_1 + t_{r2} x_2 + \ldots + t_{rn} x_n, \end{aligned} \tag{5.1}$$

or, in matrix notation,

$$y = Tx + \boldsymbol{a}. \tag{5.2}$$

From a generalization of (2.3) we obtain for the expectation values of the variables y

$$E(y) = \hat{y} = T\hat{x} + \boldsymbol{a}. \tag{5.3}$$

Together with (3.21) this leads us to the covariance matrix of the transformed variables y

$$\begin{aligned} C_y &= E\{(y - \hat{y})(y - \hat{y})^{\mathrm{T}}\} \\ &= E\{(Tx + \boldsymbol{a} - T\hat{x} - \boldsymbol{a})(Tx + \boldsymbol{a} - T\hat{x} - \boldsymbol{a})^{\mathrm{T}}\} \\ &= E\{T(x - \hat{x})(x - \hat{x})^{\mathrm{T}}T^{\mathrm{T}}\} \\ &= TE\{(x - \hat{x})(x - \hat{x})^{\mathrm{T}}\}T^{\mathrm{T}}, \end{aligned}$$

$$C_y = TC_xT^{\mathrm{T}}. \tag{5.4}$$

With the help of (5.4) we can now formulate the well known *law of the propagation of errors*. Let the expectation values $\hat{x}_i$ be measured, and let us also assume that their errors, i.e. the standard deviations of the x_i, are known. We now wish to know what the error in given functions $y(x)$ is. If the errors of x are comparatively small, then the probability density $f(\mathbf{x})$ will be significantly different from zero only in a small neighbourhood around $\hat{\mathbf{x}}$ of the order of the standard deviation. We can therefore expand

$$y_i = y_i(\hat{x}) + \left(\frac{\partial y_i}{\partial x_1}\right)_{\mathbf{x}=\hat{\mathbf{x}}} (x_1 - \hat{x}_1) + \ldots + \left(\frac{\partial y_i}{\partial x_n}\right)_{\mathbf{x}=\hat{\mathbf{x}}} (x_n - \hat{x}_n)$$

$$+ \text{ terms of higher order,}$$

or in matrix notation

$$y = y(\hat{\mathbf{x}}) + T(x - \hat{\mathbf{x}}) + \text{terms of higher order,} \tag{5.5}$$

with

$$T = \begin{pmatrix} \frac{\partial y_1}{\partial x_1} & \frac{\partial y_1}{\partial x_2} & \cdots & \frac{\partial y_1}{\partial x_n} \\ \frac{\partial y_2}{\partial x_1} & \frac{\partial y_2}{\partial x_2} & \cdots & \frac{\partial y_2}{\partial x_n} \\ \vdots & & & \\ \frac{\partial y_r}{\partial x_1} & \frac{\partial y_r}{\partial x_2} & \cdots & \frac{\partial y_r}{\partial x_n} \end{pmatrix}_{\mathbf{x}=\hat{\mathbf{x}}}. \tag{5.6}$$

If we neglect terms of higher than the first order, and introduce the matrix T of partial derivatives into eq. (5.4), we obtain the law of error propagation. We note in particular that the errors of y, i.e. the diagonal elements of C_y, are determined not only by the errors (or the variances) of x, but that the covariances between different x_i also contribute in an essential way to the errors of y. If covariances are not taken into account in an error propagation, the results cannot be relied upon.

The covariances can only be neglected when they vanish anyway, i.e. in the case of independent original variables x. In this case C_x reduces to a diagonal matrix. The diagonal elements of C_y then have the simple form

$$\sigma^2(y_i) = \sum_{j=1}^{n} \left(\frac{\partial y_i}{\partial x_j}\right)^2_{\mathbf{x}=\hat{\mathbf{x}}} \sigma^2(x_j). \tag{5.7}$$

If we now call the standard deviation, i.e. the positive square root of the variance, the error of the corresponding quantity and denote it by the symbol Δ, eq. (5.7) leads immediately to the formula

$$\Delta y_i = \sqrt{\left[\sum_{j=1}^{n} \left(\frac{\partial y_i}{\partial x_j}\right)^2 (\Delta x_j)^2\right]}, \tag{5.8}$$

known commonly as the law of the propagation of errors. It cannot be stressed too strongly that this expression is incorrect in cases of non-vanishing covariances. This is illustrated in the following example.

Example 4-1: *Propagation of errors and covariance*

In a cartesian coordinate system a point x, y is measured. The measurement is performed with a coordinatograph whose error in y is 3 times larger than the one in x. The measurements of x and y are independent. We therefore can set the covariance matrix (up to a factor common to all elements)

$$C_{xy} = \begin{pmatrix} 1 & 0 \\ 0 & 9 \end{pmatrix}.$$

We now evaluate the errors or the covariance matrix in polar coordinates

$$r = \sqrt{(x^2 + y^2)}, \qquad \varphi = \arctan \frac{y}{x}.$$

The transformation matrix (5.6) in this case is

$$T = \begin{pmatrix} \dfrac{x}{r} & \dfrac{y}{r} \\ -\dfrac{y}{r^2} & \dfrac{x}{r^2} \end{pmatrix}.$$

To simplify the numerical calculations we consider only the point $(x, y) = (1,1)$. Then

$$T = \begin{pmatrix} \dfrac{1}{\sqrt{2}} & \dfrac{1}{\sqrt{2}} \\ -\dfrac{1}{2} & \dfrac{1}{2} \end{pmatrix},$$

and therefore

$$C_{r\varphi} = \begin{pmatrix} \dfrac{1}{\sqrt{2}} & \dfrac{1}{\sqrt{2}} \\ -\dfrac{1}{2} & \dfrac{1}{2} \end{pmatrix} \begin{pmatrix} 1 & 0 \\ 0 & 9 \end{pmatrix} \begin{pmatrix} \dfrac{1}{\sqrt{2}} & -\dfrac{1}{2} \\ \dfrac{1}{\sqrt{2}} & \dfrac{1}{2} \end{pmatrix} = \begin{pmatrix} 5 & \dfrac{4}{\sqrt{2}} \\ \dfrac{4}{\sqrt{2}} & \dfrac{5}{2} \end{pmatrix}.$$

We can now return to the original cartesian coordinate system

$$x = r\cos\varphi, \qquad y = r\sin\varphi,$$

by the transformation

$$T' = \begin{pmatrix} \cos\varphi & -r\sin\varphi \\ \sin\varphi & r\cos\varphi \end{pmatrix} = \begin{pmatrix} \frac{1}{\sqrt{2}} & -1 \\ \frac{1}{\sqrt{2}} & 1 \end{pmatrix}.$$

We obtain

$$C_{xy} = \begin{pmatrix} \frac{1}{\sqrt{2}} & -1 \\ \frac{1}{\sqrt{2}} & 1 \end{pmatrix} \begin{pmatrix} 5 & \frac{4}{\sqrt{2}} \\ \frac{4}{\sqrt{2}} & \frac{5}{2} \end{pmatrix} \begin{pmatrix} \frac{1}{\sqrt{2}} & \frac{1}{\sqrt{2}} \\ -1 & 1 \end{pmatrix} = \begin{pmatrix} 1 & 0 \\ 0 & 9 \end{pmatrix},$$

as expected. If we had used formula (5.8) instead, i.e. if we had neglected the covariances in the transformation from r, φ to x, y, the result would have been

$$C'_{xy} = \begin{pmatrix} \frac{1}{\sqrt{2}} & -1 \\ \frac{1}{\sqrt{2}} & 1 \end{pmatrix} \begin{pmatrix} 5 & 0 \\ 0 & \frac{5}{2} \end{pmatrix} \begin{pmatrix} \frac{1}{\sqrt{2}} & \frac{1}{\sqrt{2}} \\ -1 & 1 \end{pmatrix} = \begin{pmatrix} 5 & 0 \\ 0 & 5 \end{pmatrix},$$

which is different from the original covariance matrix. This example stresses the importance of covariances since it is obviously not possible to change errors of measurements by simply transforming back and forth between coordinate systems.

Finally we discuss a special form of linear transformation. We consider the case of exactly n functions y of the n variables x. In addition we require $a = 0$ in eq. (5.2). We then have

$$\boldsymbol{y} = R\,\boldsymbol{x}, \tag{5.9}$$

where R is a square matrix. We now require that the transformation (5.9) leaves the modulus of a vector invariant

$$\boldsymbol{y}^2 = \sum_{i=1}^{n} y_i^2 = \boldsymbol{x}^2 = \sum_{i=1}^{n} x_i^2. \tag{5.10}$$

Using eq. (B-2.20) this can be written

$$y^T y = (Rx)^T (Rx) = x^T R^T Rx = x^T x.$$

This requires

$$R^T R = I,$$

or, in components,

$$\sum_{i=1}^{n} r_{ik} r_{il} = \delta_{kl} = \begin{cases} 0 & \text{for} \quad l \neq k, \\ 1 & \text{for} \quad l = k. \end{cases} \tag{5.11}$$

A transformation of the type (5.9) which fulfils condition (5.11) is called *orthogonal*. We now consider the determinant of the transformation matrix

$$D = \begin{vmatrix} r_{11} & r_{12} & \dots & r_{1n} \\ r_{21} & r_{22} & \dots & r_{2n} \\ \vdots & & & \\ r_{n1} & r_{n2} & \dots & r_{nn} \end{vmatrix}$$

and form its square. Using the rules of determinant algebra we obtain from eq. (5.11)

$$D^2 = \begin{vmatrix} 1 & 0 & \dots & 0 \\ 0 & 1 & \dots & 0 \\ \vdots & & & \\ 0 & 0 & \dots & 1 \end{vmatrix},$$

i.e. $D = \pm 1$. The determinant D is also the Jacobian of the transformation (5.9)

$$J\left(\frac{y}{x}\right) = \pm 1. \tag{5.12}$$

If we multiply the system (5.9) from the left with R^T we obtain

$$R^T y = R^T R \, x,$$

which because of (5.11) reduces to

$$x = R^T y. \tag{5.13}$$

The inverse of an orthogonal transformation is described simply by the transposed transformation matrix.

An important property of any linear transformation of the type

$$y_1 = r_{11} x_1 + r_{12} x_2 + \dots + r_{1n} x_n$$

is the following. By constructing additional functions $y_2, y_3, \ldots, y_n$ of equivalent structure it can be extended to yield an orthogonal transformation if only the condition

$$\sum_{i=1}^{n} r_{1i}^2 = 1$$

is fulfilled.

Exercises

4-1: *Transformation of a single variable*
(a) Suppose a gun is mounted so that it can rotate around a vertical axis in front of a long wall at unit distance (fig. 4-5). It fires at randomly chosen angles θ in the range $-\frac{1}{2}\pi \leqslant \theta < \frac{1}{2}\pi$, i.e. the probability density is constant in $\theta: f(\theta) = c$. From (3-2.8) it then follows that $c = 1/\pi$.

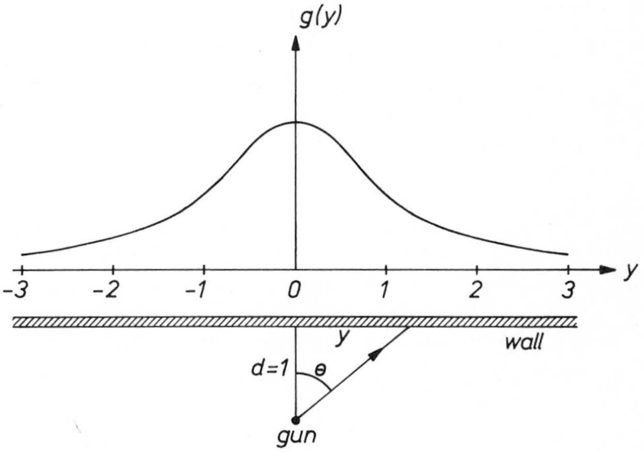

Fig. 4-5. A model for the Cauchy distribution.

Determine the probability density $g(y)$ which describes the distribution of the bullet holes in the wall and which is called the "Cauchy distribution" (cf. exercise 3-3).
(b) In appendix D it is shown that

$$\int_{-\infty}^{\infty} \exp(-x^2/2)\mathrm{d}x = \sqrt{(2\pi)}.$$

Use the transformation $y = x/\sigma$ to show that

$$\int_{-\infty}^{\infty} \exp(-x^2/2\sigma^2)\mathrm{d}x = \sigma\sqrt{(2\pi)}.$$

4-2: *Transformation of several variables*
A "bivariate normal distribution" (cf. ch. 5, § 10) can have the form

$$f(x, y) = \frac{1}{2\pi\sigma_x\sigma_y} \exp\left(-\frac{1}{2}\frac{x^2}{\sigma_x^2} - \frac{1}{2}\frac{y^2}{\sigma_y^2}\right).$$

(a) Determine the marginal distributions $f(x)$, $f(y)$ using the result of exercise 4-1(b).
(b) Are x and y independent variables?
(c) Transform the distribution $f(x, y)$ to variables $u = x\cos\phi + y\sin\phi$, $v = y\cos\phi - x\sin\phi$ (the coordinate system u, v has the same origin as the system x, y but is rotated with respect to it by the angle ϕ). Hint: Show that the transformation is orthogonal and make use of eq. (5.12).
(d) Show that u and v are independent variables only if $\phi = 90°$, $180°$, $270°$ or for $\sigma_x = \sigma_y = \sigma$.
(e) Consider the case $\sigma_x = \sigma_y = \sigma$, i.e.

$$f(x) = \frac{1}{2\pi\sigma^2} \exp\left[-\frac{1}{2\sigma^2}(x^2 + y^2)\right].$$

Transform the distribution to polar coordinates as described in example 4-1, determine the marginal distributions $g(r)$ and $g(\phi)$ and show that r and ϕ are independent.

4-3: *Propagation of errors*
The time T for one period of the motion of a pendulum is given by $T = 2\pi\sqrt{(l/g)}$, l being the length of the pendulum and g the acceleration of gravity. Compute g and Δg using measured values of $l = 99.8$ cm, $\Delta l = 0.3$ cm, $T = 2.03$ sec, $\Delta T = 0.05$ sec and assuming no correlation between the measurements of l and T.

4-4: *Covariance and correlation*
Mass and velocity of an object are denoted by m and v, their measurement errors by $\Delta m = \sqrt{[\sigma^2(m)]}$ and $\Delta v = \sqrt{[\sigma^2(v)]}$. The measurements are assumed to be independent, i.e. $\text{cov}(m, v) = 0$. It is also assumed that the relative errors are known to be constant, i.e. $\Delta m/m = a$, $\Delta v/v = b$.
(a) Consider the momentum $p = mv$ and the kinetic energy $E = \frac{1}{2}mv^2$ of the object and compute $\sigma^2(p)$, $\sigma^2(E)$, $\text{cov}(p, E)$ and the correlation $\rho(p, E)$. Discuss $\rho(p, E)$ for the special cases $a = 0$ and $b = 0$. Hint: Form the vectors $\mathbf{x} = (m, v)$ and $\mathbf{y} = (p, E)$. Then approximate $\mathbf{y} = \mathbf{y}(\mathbf{x})$ as a linear transformation by using eq. (5.5) and finally compute the covariance matrix with the help of (5.4).
(b) Experimental values for E, p and the covariance matrix of both are know. Compute the mass m and its error by propagation of errors. Use the result obtained in (a) to verify $\Delta m = am$. Note that the correct result is obtained only if $\text{cov}(p, E)$ is taken into account.

CHAPTER 5

SOME IMPORTANT DISTRIBUTIONS AND THEOREMS

We shall now discuss in detail some special distributions. This chapter could therefore be regarded as a collection of examples. However, these distributions are of great practical importance. Moreover their study will lead us to a few important theorems.

5-1. Binomial and multinomial distributions

An experiment may have only two possible mutually exclusive outcomes and therefore be characterized by the simple decomposition

$$E = A + \bar{A}. \tag{1.1}$$

The outcomes may have the probabilities

$$P(A) = p, \qquad P(\bar{A}) = 1 - p = q. \tag{1.2}$$

The result of the experiment may be represented by a random variable x_i which takes the value 1 (or 0) if the event A (or $\bar{A}$) occurs. The index i numbers the individual experiments within a series. We now repeat the experiment n times and consider the probability distribution of the variable

$$x = \sum_{i=1}^{n} x_i.$$

The probability that the first k experiments yield A and all others $\bar{A}$ is according to eq. (2-3.8)

$$p^k q^{n-k}.$$

The event "k times A in n experiments" can occur in

$$\binom{n}{k} = n!/[k!(n-k)!]$$

different ways* according to the sequence of appearance of A and $\bar{A}$. The probability of this event is therefore

* Cf. appendix C.

$$W_k^n = \binom{n}{k} p^k q^{n-k}. \tag{1.3}$$

Let us now consider the mean and variance of x. We first determine these quantities for the variable x_i of a single experiment. According to (3-3.2) we find that

$$E(x_i) = 1 \cdot p + 0 \cdot q, \tag{1.4}$$

and that

$$\sigma^2(x_i) = E\{(x_i - p)^2\} = (1 - p)^2 p + (0 - p)^2 q,$$

$$\sigma^2(x_i) = pq. \tag{1.5}$$

Now from a generalization of (4-2.3) we obtain for $x = \Sigma\, x_i$

$$E(x) = \sum_{i=1}^{n} p = np, \tag{1.6}$$

and from (4-2.10), since because of the independence of the x_i all covariances vanish,

$$\sigma^2(x) = npq. \tag{1.7}$$

The distribution W_k^n is shown in fig. 5-1a for fixed p and different n and in fig. 5-1b for fixed n and different p. Finally in fig. 5-1c, n and b are varied but the product np is kept fixed. The figures will help us to discover similarities between the *binomial distribution* W_n^k (so called because of the two possible outcomes of a singular experiment) and other distributions.

A logical extension of the binomial distribution deals with experiments where more than 2 different outcomes are possible. Instead of (1.1) we then have

$$E = A_1 + A_2 + \ldots + A_l. \tag{1.8}$$

Let the probabilities for the mutually exclusive events A_j be given by

$$P(A_j) = p_j, \qquad \sum_{j=1}^{l} p_j = 1, \tag{1.9}$$

and again let n experiments be performed. The probability of always finding k_j events of the kind A_j is

$$W^n_{(k_1, k_2, \ldots k_l)} = \frac{n!}{\prod_{j=1}^{l} k_j!} \prod_{j=1}^{l} p_j^{k_j}. \tag{1.10}$$

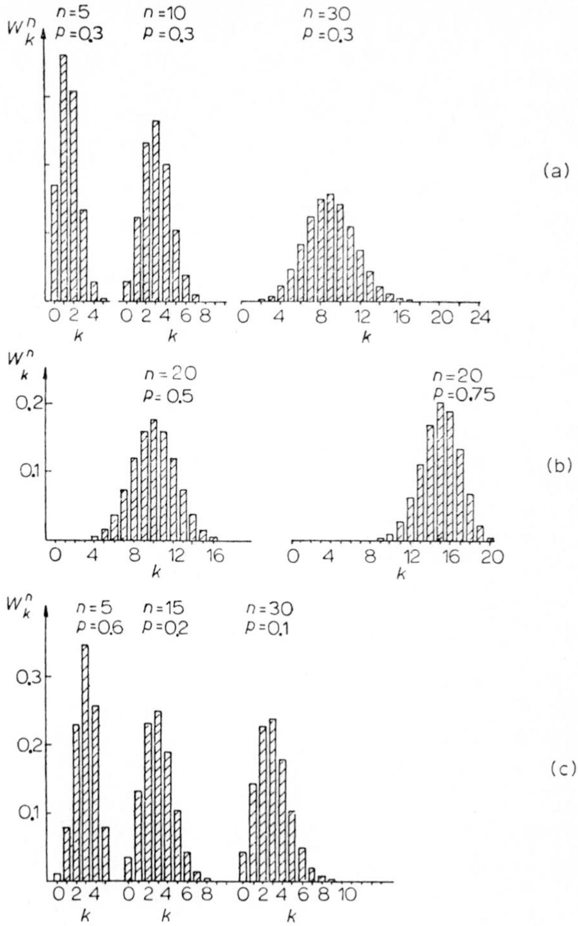

Fig. 5-1. Binomial distributions. (a) for fixed p and different values of n, (b) for fixed n and different values of p, (c) for different values of n and p but with np = const.

We leave the proof of (1.10) to the reader. The distribution (1.10) is called the *multinomial distribution.* If we define x_{ij} to be 1 if the ith experiment yields an event A_j and 0 otherwise then, with

$$x_j = \sum_{i=1}^{n} x_{ij},$$

we obtain the expectation value

$$E(x_j) = \hat{x}_j = np_j, \tag{1.11}$$

The elements of the covariance matrix of the x_j are

$$c_{ij} = np_i(\delta_{ij} - p_j). \tag{1.12}$$

We find that the matrix has nonvanishing off-diagonal elements. This was to be expected since the x_j are not independent because of relation (1.9).

5-2. Frequency; the law of large numbers

Usually the probabilities for the different types of events, e.g. p_j in the case of the multinomial distribution, are not known but have to be obtained from experiment. We determine the *frequency* of events A_j in n experiments

$$h = \frac{1}{n} \sum_{i=1}^{n} x_{ij} = \frac{1}{n} x_j. \tag{2.1}$$

Unlike the probability, the frequency is a random variable since it depends on the result of the particular n experiments under investigation. Using (1.11), (1.12), and (3-3.14) we obtain

$$E(h) = \hat{h} = E\left(\frac{x_j}{n}\right) = p_j, \tag{2.2}$$

and

$$\sigma^2(h) = \sigma^2\left(\frac{x_j}{n}\right) = \frac{1}{n^2}\sigma^2(x_j) = \frac{1}{n} p_j (1 - p_j). \tag{2.3}$$

The product $p_j(1 - p_j)$ in eq. (2.3) takes at most the value $\frac{1}{4}$. We see therefore that the expectation value of the frequency of an event is equal to the probability of its occurrence, and that the standard deviation of this expectation value shrinks beyond any given limit only if the number of experiments becomes sufficiently large. For finite n it is at most of order $1/\sqrt{n}$. This property of the frequency is usually known as the *law of large numbers*. Clearly it is the reason for the frequency definition of probability in eq. (2-2.1).

Frequently the purpose of experimental investigation is just to determine the probability of the occurrence of a certain type of event. According to (2.2) we can use the frequency as an approximation to the probability. The square of the error of this approximation is then inversely proportional to the number of individual experiments. This kind of error, which originates from the fact that only a finite number of experiments can be performed, is called the *statistical error*. It is of prime importance for applications that are concerned with the counting of individual events, e.g. nuclear particles passing through a counter, animals with certain traits in heredity experiments, defective items in production control, etc.

Example 5-1: *Statistical error*
It may be roughly known from previous experiments that a fraction R of about 1/200 of a sample of dew flies (drosophila) develops a certain property A if exposed to a given dose of X-ray irradiation. An experiment to determine the ratio R with an accuracy of 1 % is planned. How large has the original sample to be chosen to achieve this accuracy? We use eq. (2.3) and have $p_j = 0.005$, $(1 - p_j) \approx 1$. We want to adjust n such that $\sigma(h)/h = 200\ \sigma(h) = 0.01$. Therefore $\sigma(h) = 0.00005$ and $\sigma^2(h) = 0.25 \times 10^{-8}$. Eq. (2.3) becomes

$$0.25 \times 10^{-8} = \frac{1}{n} \times 0.005,$$

which yields

$$n = 2 \times 10^6.$$

A total of two million dew flies would have to be used. This is practically impossible. To determine the ratio R with an accuracy of 10% would require 20000 flies.

5-3. Hypergeometric distribution

Although we shall later introduce rigorously the concept of random sampling we will now discuss a typical problem of sampling. We consider a container – we shall not break with the habit of mathematicians of calling such a container an *urn* – with K white and $L = N - K$ black beads. We want to determine the probability that by drawing n beads (without replacing them) we will find exactly k white and $l = n - k$ black ones. The problem is rendered difficult by the fact that the drawing of a bead of a particular colour changes the ratio of white and black beads and therefore influences the outcome of the next draw. We therefore cannot follow the reasoning used in the case of the binomial distribution. We know however that there are $\binom{N}{n}$ equivalent possibilities of drawing n beads out of N. The probability of hitting upon one of these possibilities is therefore $1 \Big/ \binom{N}{n}$.

Correspondingly there are $\binom{K}{k}$ and $\binom{L}{l}$ possibilities of drawing k out of K white and l out of L black beads, respectively. The required probability is therefore

$$W_k = \frac{\binom{K}{k}\binom{L}{l}}{\binom{N}{n}}, \tag{3.1}$$

As in §1 we define the random variable

$$x = \sum_{i=1}^{n} x_i$$

with $x_i = 1$ or 0 if the ith move yields the result white or black, respectively (in other words we define k to be the random variable x).

To calculate the expectation value of x we cannot simply add up the expectation values of the x_i since these are not independent. We must therefore return to definition (3-3.2)

$$E(x) = \frac{1}{\binom{N}{n}} \sum_{i=1}^{n} i \binom{K}{i} \binom{N-K}{n-i}$$

$$= \frac{(N-n)!\,n!}{N!} \sum_{i=1}^{n} \frac{i\,K!\,(N-K)!}{i!\,(K-i)!\,(n-i)!\,(N-K-n+i)!}$$

$$= \frac{n(n-1)!\,(N-n)!}{N(N-1)!} \sum_{i=1}^{n}$$

$$\frac{K!\,(N-K)!}{(i-1)!\,(K-1-(i-1))!\,(n-1-(i-1))!\,(N-K-n+1+(i-1))!}$$

We now set $i - 1 = j$ and obtain

$$E(x) = n\,\frac{K}{N}\,\frac{(n-1)!\,(N-n)!}{(N-1)!} \sum_{j=0}^{n-1}$$

$$\frac{(K-1)!\,(N-K)!}{j!\,(K-1-j)!\,(n+1-j)!\,(N-K-n+1+j)!}$$

$$= n\,\frac{K}{N}\,\frac{1}{\binom{N-1}{n-1}} \sum_{j=0}^{n-1} \binom{K-1}{j} \binom{N-K}{n-1-j}.$$

Using eq. (C-5) we finally get

$$E(x) = n \frac{K}{N}. \tag{3.2}$$

The calculation of the variance is along the same lines but rather lengthy. The result is

$$\sigma^2(x) = \frac{nK(N-K)(N-n)}{N^2(N-1)}. \tag{3.3}$$

A few examples of the hypergeometric distribution (3.1) are given in fig. 5-2.

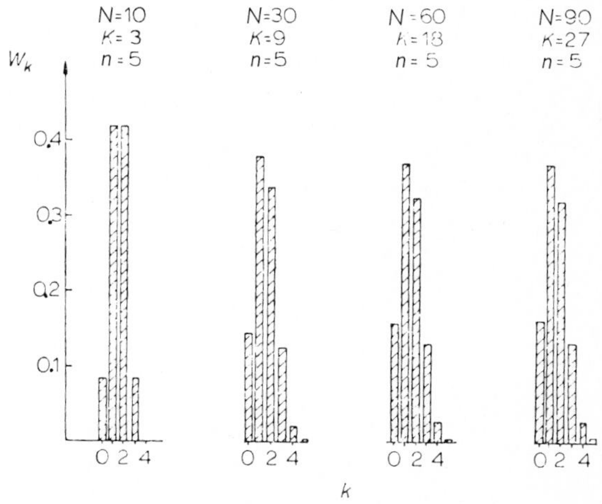

Fig. 5-2. Hypergeometric distributions.

In the case of $n \ll N$ the result of one draw will influence the outcome of the next one only slightly. We therefore expect that in this case W_k behaves similarly to a binomial distribution with $p = K/N$ and $q = (N - K)/N$. The expectation value is in fact

$$E(x) = n \frac{K}{N} = np,$$

as in the case of the binomial distribution. The variance (3.3) can be written

$$\sigma^2(x) = \frac{npq(N-n)}{N-1},$$

which for $n \ll N$ becomes

$$\sigma^2(x) = npq.$$

This relation between binomial and hypergeometric distributions is also apparent from a comparison of fig. 5-2 with the left part of fig. 5-1a.

There are manifold applications of the hypergeometric distribution. Polls, quality control etc. are based on the drawing of an object without replacing it in the original sample. The distribution can be extended in two directions. First we can of course consider more properties instead of just 2 (white and black beads). This leads us to a similar transition as the one from the binomial to the multinomial distribution. The original sample (population) contains N elements each of which possesses one of l properties

$$N = N_1 + N_2 + \ldots + N_l.$$

The probability of finding n_j ($j = 1, 2, \ldots, l$) elements of each type in a total of n draws is

$$W_{n_1, n_2, \ldots, n_l} = \frac{\binom{N_1}{n_1}\binom{N_2}{n_2}\cdots\binom{N_l}{n_l}}{\binom{N}{n}}, \tag{3.4}$$

by analogy with (3.1).

Another extension of the hypergeometric distribution is obtained in the following way. We saw earlier that consecutive drawings ceased to be independent because the beads were not replaced. If now each time we draw a bead of one type we place in addition a fixed number of beads of that type in the urn, this dependence of one drawing on the previous one is enhanced. In this way we obtain *Polya's distribution*. It is of importance in the study of epidemic diseases, where the appearance of a case of the disease enhances the probability of future cases.

Example 5-2: *Application of the hypergeometric distribution for the determination of zoological populations*

From a pond K fish are taken and marked. They are then returned to the pond. After a short while n fish are caught, k of which are marked. This process is described by the hypergeometric distribution if the pond contains N fish of which K were marked in the first step. We shall return to the problem of how to estimate quantitatively the population N of the pond in example 7-3.

5-4. Poisson distribution

The study of fig. 5-1c suggests that if n tends to infinity, but at the same time $np = \lambda$ is kept constant the binomial distribution approaches a fixed distribution. We rewrite eq. (1.3)

$$W_k^n = \binom{n}{k} p^k q^{n-k}$$

$$= \frac{n!}{k!(n-k)!} \left(\frac{\lambda}{n}\right)^k \frac{\left(1 - \frac{\lambda}{n}\right)^n}{\left(1 - \frac{\lambda}{n}\right)^k}$$

$$= \frac{\lambda^k}{k!} \frac{n(n-1)(n-2) \ldots (n-k+1)}{n^k} \frac{\left(1 - \frac{\lambda}{n}\right)^n}{\left(1 - \frac{\lambda}{n}\right)^k}$$

$$= \frac{\lambda^k}{k!} \left(1 - \frac{\lambda}{n}\right)^n \frac{\left(1 - \frac{1}{n}\right)\left(1 - \frac{2}{n}\right) \ldots \left(1 - \frac{k-1}{n}\right)}{\left(1 - \frac{\lambda}{n}\right)^k}.$$

In the limit of very large n each of the finite number of factors in the term on the far right approaches 1. Since

$$\lim_{n \to \infty} \left(1 - \frac{\lambda}{n}\right)^n = e^{-\lambda},$$

we finally obtain

$$\lim_{n \to \infty} W_k^n = f(k) = \frac{\lambda^k}{k!} e^{-\lambda}. \tag{4.1}$$

The quantity $f(k)$ is the probability density of the Poisson distribution. In fig. 5-3 this distribution is plotted for different values of λ. This should be compared with fig. 5-1c. Like the other distributions we have discussed so far it is defined only for integer values of k. The total probability, i.e. the probability of observing any value k has of course to be unity:

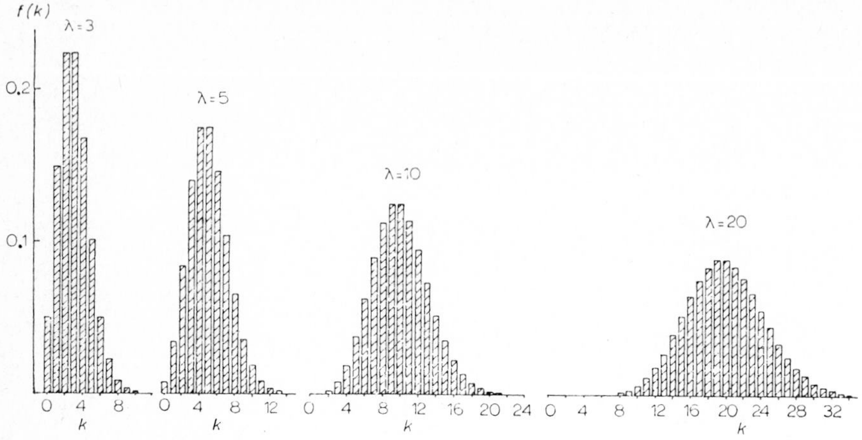

Fig. 5-3. Poisson distributions.

$$\sum_{k=0}^{\infty} f(k) = \sum_{k=0}^{\infty} \frac{e^{-\lambda}\lambda^k}{k!}$$

$$= e^{-\lambda}\left(\lambda + \frac{\lambda^2}{2!} + \frac{\lambda^3}{3!} + \ldots\right)$$

$$= e^{-\lambda} e^{\lambda},$$

$$\sum_{k=0}^{\infty} f(k) = 1, \tag{4.2}$$

since the expression in brackets is nothing else than the exponential e^{λ}.

We now want to determine the expectation value, variance and skewness of the Poisson distribution. From the definition (3-3.2) we find (writing x for the random variable k)

$$E(x) = \sum_{k=0}^{\infty} k \frac{\lambda^k}{k!} e^{-\lambda} = \sum_{k=1}^{\infty} k \frac{\lambda^k}{k!} e^{-\lambda}$$

$$= \sum_{k=1}^{\infty} \frac{\lambda \lambda^{k-1}}{(k-1)!} e^{-\lambda} = \lambda \sum_{j=0}^{\infty} \frac{\lambda^j}{j!} e^{-\lambda},$$

and therefore with eq. (4.2)

$$E(x) = \lambda. \tag{4.3}$$

Let us now consider $E(x^2)$. Following the same procedure we get

$$E(x^2) = \sum_{k=1}^{\infty} k^2 \frac{\lambda^k}{k!} e^{-\lambda} = \lambda \sum_{k=1}^{\infty} k \frac{\lambda^{k-1}}{(k-1)!} e^{-\lambda}$$

$$= \lambda \sum_{j=0}^{\infty} (j+1) \frac{\lambda^j}{j!} e^{-\lambda} = \lambda \left(\sum_{j=0}^{\infty} j \frac{\lambda^j}{j!} e^{-\lambda} + 1 \right),$$

$$E(x^2) = \lambda(\lambda + 1). \tag{4.4}$$

Following (3-3.15) we get the variance of x from eqs. (4.3) and (4.4)

$$\sigma^2(x) = E(x^2) - \{E(x)\}^2 = \lambda(\lambda+1) - \lambda^2, \tag{4.5}$$

$$\sigma^2(x) = \lambda. \tag{4.6}$$

Let us now look at the skewness of the Poisson distribution. Following ch. 3, § 3 we find readily that

$$\mu_3 = E\{(x - \hat{x})^3\} = \lambda.$$

With this the skewness (3-3.12) becomes

$$\gamma = \frac{\mu_3}{\sigma^3} = \frac{\lambda}{\lambda^{\frac{3}{2}}} = \lambda^{-\frac{1}{2}}, \tag{4.7}$$

i.e. the Poisson distribution becomes increasingly symmetric as λ increases. This is also evident in fig. 5-3.

We have obtained the Poisson distribution from the binomial distribution with large n but constant $\lambda = np$, i.e. small p. We therefore expect it to apply to processes in which a large number of events occur but of which only very few have a certain property of interest to us (i.e. a large number of "trials" but few "successes").

Example 5-3: *Poisson distribution and independence of radioactive decays* We consider the very large number n of atomic nuclei in a radioactive source and ask for the probability $\lambda(n, \Delta t)$ that one of these decays in a time interval Δt. This probability will remain constant if n is multiplied by a constant c provided that now only decays in the time interval $\Delta t/c$ are considered. The reason for this is that the decays of the individual atoms occur independently and that therefore $\lambda(n, \Delta t) = np(\Delta t)$, where $p(\Delta t)$ is the probability that an individual atom will decay within Δt. (We consider time elements that are very small compared to the half life of the nuclei. Therefore $p(\Delta t)$ is directly proportional to

Δt.) We therefore expect that the frequency f_k ($k = 0, 1, 2, \ldots$) with which k decays are observed in a certain interval of time follows the Poisson distribution. Conversely we can reason that, if this frequency is in fact described by a Poisson distribution, the individual decays do indeed happen independently. In this way the statistical nature of radioactive decay was established in a famous experiment by Rutherford and Geiger. The actual numerical data of this experiment are presented in many textbooks on statistics, e.g. FREEMAN [1963], FISZ [1963].

Similarly the frequency of finding k stars per element of the celestial sphere or k raisins per volume element of a fruit cake is distributed according to the Poisson law, but not however the frequency of finding k animals of a given species per element of area, at least if these animals live in herds, since in this case the assumption of independence is not fulfilled.

As a quantitative example of the Poisson distribution most textbooks discuss the number of Prussian cavalrymen killed during a period of 20 years by horse kicks, an example originally due to BORTKIEWICZ [1898]. We prefer to turn our attention to a somewhat less macabre example taken from a lecture of DE SOLLA PRICE [1963].

Example 5-4: *Poisson distribution and independence of scientific discoveries*

The author first constructs the model of an apple tree with 1000 apples and of 1000 pickers with bandaged eyes who try at the same time to pick an apple each. Since we are dealing with a model they do not hinder each other but it can happen that two or several of them will

TABLE 5-1
Poisson distribution and simultaneous discovery

Number of simultaneous discoveries	Cases of simultaneous discovery	Prediction of Poisson distribution
0	Indeterminate	368
1	No data	368
2	179	184
3	51	61
4	17	15
5	6	3
$\geqslant 6$	8	1

grab the same apple. The number of apples picked by k persons ($k = 0, 1, 2, \ldots$) then follows the Poisson distribution. De Solla Price reports that also the number of scientific discoveries made twice, three times, etc. independently is distributed according to the Poisson law (table 5-1). He concludes that scientists, like the blind apple-pickers, care little about the activities of their colleagues since "scientists have a strong urge to write papers but only a relatively mild one to read them".

Let us now return to the relation between the binomial and the Poisson distribution from which the discussion in this section started. We can assume that even in cases in which the limit of an infinite number of experiments ($n \to \infty$) is not reached the two distributions are very close. For one example this is demonstrated in table 5-2. Therefore if tables of the binomial distribution are not available (which was generally the case before the advent of computers), and for sufficiently large n and small p, the Poisson distribution with $\lambda = np$ can be used also in cases where strictly speaking only the binomial distribution would be appropriate.

TABLE 5-2

Various binomial distributions with $np = 3$, compared to Poisson distribution with $\lambda = 3$

k	Binomial distribution W_k^n						Poisson distribution $f(k)$
	$p = 0.5$ $n = 6$	$p = 0.2$ $n = 15$	$p = 0.1$ $n = 30$	$p = 0.05$ $n = 60$	$p = 0.02$ $n = 150$	$p = 0.01$ $n = 300$	$\lambda = 3$
0	0.0156	0.0352	0.0424	0.0461	0.0483	0.0490	0.0498
1	0.0937	0.1319	0.1413	0.1455	0.1478	0.1486	0.1494
2	0.2344	0.2309	0.2276	0.2259	0.2248	0.2244	0.2240
3	0.3125	0.2501	0.2361	0.2298	0.2263	0.2252	0.2240
4	0.2344	0.1876	0.1771	0.1724	0.1697	0.1689	0.1680
5	0.0937	0.1032	0.1023	0.1016	0.1011	0.1010	0.1008
6	0.0156	0.0430	0.0474	0.0490	0.0499	0.0501	0.0504
7	0.0000	0.0138	0.0180	0.0199	0.0209	0.0213	0.0216
8	0.0000	0.0035	0.0058	0.0069	0.0076	0.0079	0.0081
9	0.0000	0.0007	0.0016	0.0021	0.0025	0.0026	0.0027
10	0.0000	0.0001	0.0004	0.0006	0.0007	0.0008	0.0008
11	0.0000	0.0000	0.0001	0.0001	0.0002	0.0002	0.0002
12	0.0000	0.0000	0.0000	0.0000	0.0000	0.0000	0.0001

5-5. Uniform distribution and an application: the Monte Carlo method

5-5.1. *Probability density expectation value variance*

So far we have only discussed distributions of one or (in the case of the multinomial distribution) several *discrete* variables. We now consider the simplest case of a *continuous* distribution function. It has in fact already been mentioned as an example in ch. 3, §2. The *uniform distribution* is defined as follows: The probability density is constant in the interval $a \leqslant x < b$ and vanishes outside this region:

$$\begin{aligned} f(x) &= c, & a \leqslant x < b, \\ f(x) &= 0, & x < a, x \geqslant b. \end{aligned} \tag{5.1}$$

Since the total probability is normalized to 1 (eq. (3-2.8)) we get

$$\int_{-\infty}^{\infty} f(x)\,\mathrm{d}x = c\int_a^b \mathrm{d}x = c(b-a) = 1,$$

and therefore

$$\begin{aligned} f(x) &= \frac{1}{b-a}, & a \leqslant x < b, \\ f(x) &= 0, & x < a, x \geqslant b. \end{aligned} \tag{5.2}$$

The distribution function then becomes

$$\begin{aligned} F(x) &= \int_a^x \frac{\mathrm{d}x}{b-a} = \frac{x-a}{b-a}, & a \leqslant x < b, \\ F(x) &= 0, & x < a, \\ F(x) &= 1, & x \geqslant b. \end{aligned} \tag{5.3}$$

On symmetry grounds we expect the expectation value to be the arithmetic mean of the limits. In fact eq. (3-3.4) yields

$$E(x) = \hat{x} = \frac{1}{b-a}\int_a^b x\,\mathrm{d}x = \frac{1}{2}\,\frac{1}{(b-a)}(b^2 - a^2) = \frac{b+a}{2}. \tag{5.4}$$

Correspondingly we obtain from (3-3.10)

$$\sigma^2(x) = \tfrac{1}{12}(b - a)^2. \tag{5.5}$$

The uniform distribution is by itself not of too great practical importance. Being the simplest continuous distribution, it is however very convenient in calculations. It is therefore often advantageous to transform a given distribution into the uniform distribution or, conversely, to obtain it by a transformation of variables from the uniform distribution. This technique is used extensively in the "Monte Carlo" method.

5-5.2. *Generation of uniformly distributed random numbers by computers*

So far in this book we have discussed the observation of random variables but not mechanisms for their generation. In many applications however it is useful to have a chain of values of a randomly distributed variable x (c.f. examples 5-5 and 5-6). Since operations will be performed on very many such *random numbers* it is most convenient to have them available directly in the computer. Although the correct way would be to measure some statistical process, e.g. the time between two decays in a radioactive source, and transfer the result to the computer, the numbers are usually computed directly by the computer. Since the computer works strictly deterministically such numbers are not subject to chance but can be predicted. Therefore they are called *pseudorandom.*

There is yet another problem. Since the numbers used in computers have a finite number of digits (usually they are binary numbers with k digits, where $16 \leqslant k \leqslant 60$), the variable x is not continuous but discrete and has the form

$$x = x^{(0)}2^0 + x^{(1)}2^1 + \ldots + x^{(k-1)}2^{k-1}. \tag{5.6}$$

If it is possible to select all $x^{(j)}$ at random from a distribution which yields the values 0 and 1 with the probability $\frac{1}{2}$ (dual distribution), the numbers x are described by a discrete distribution and assume any of the 2^k values

$$x = 0, 1, 2, \ldots, 2^{k-1} \tag{5.7}$$

with equal probability. By forming $x' = x/2^k$ an equally probable population would result at $2k$ discrete values in the interval $0 \leqslant x' < 1$. Such a distribution is called *quasiuniform.*

Since, as mentioned, computers work deterministically each random number will be a function of its predecessors

$$x_j = f(x_{j-1}, x_{j-2}, \ldots, x_1). \tag{5.8}$$

A simple function is*

$$x_j = (ax_{j-1} + c) \text{ modulo } 2^k. \tag{5.9}$$

Of course (5.9) leads to a repetition of each random number after at most 2^k steps. It can be proved that for certain conditions on a, c and k this maximum period is reached [HAMMERSLEY and HANDSCOMB (1964)].

The following short FORTRAN program generates quasiuniform random numbers with a period of 4096. The initial random number is defined by a data statement. The following ones are obtained according to eq. (5.9) with $a = 5$ and $c = 13$. In practice much larger periods are used. All computer installations have random number generators in their program libraries. These should always be used since they are optimized in period and speed to the installations.

Program 5-1: *Function* RAND *generating quasirandom numbers*

```
      FUNCTION RAND (DUMMY)
      DATA IX/1/
      IX=5*IX+13
   10 IF (IX—4096) 20, 20, 30
   30 IX=IX—4096
      GO TO 10
   20 RAND=IX/4096
      RETURN
      END
```

5-5.3. *Generation of any distribution by transformation of the uniform distribution*

If x is described by a uniform distribution of the type

$$f(x) = 1, \quad 0 \leqslant x < 1, \qquad f(x) = 0, \quad x < 0, \quad x \geqslant 1 \tag{5.10}$$

and y is a random variable described by the probability density $g(y)$ the transformation eq. (4-4.1) reduces to

$$g(y)\mathrm{d}y = \mathrm{d}x. \tag{5.11}$$

Using the distribution function $G(y)$ which is connected to $g(y)$ by $\mathrm{d}G(y)/\mathrm{d}y = g(y)$ we can rewrite (5.11) in the form

* The notation a modulo m signifies the remainder of the division a/m, e.g. 11 modulo 5 = 1.

$$dx = g(y)dy = dG(y) \tag{5.12}$$

or by integration

$$x = G(y) = \int_{-\infty}^{y} g(t)dt. \tag{5.13}$$

The significance of eq. (5.13) is the following. If a random number x is drawn from a uniform distribution in the interval between 0 and 1 and the function $x = G(y)$ is inverted, then a random number y is obtained which is distributed with probability density $g(y)$. This is illustrated in fig. 5-4(a). The probability of selecting a random number x between x and $x + dx$ is equal to the probability of obtaining a value $y(x)$ between y and $y + dy$.

The relation (5.13) can also be used to generate a discrete probability distribution. This is illustrated in fig. 5-4(b). The random variable y can assume the values $y_1, y_2, \ldots, y_n$ with the probabilities $P(y_1), P(y_2), \ldots, P(y_n)$. According to eq. (3-2.1) the distribution function is $G(y) = P(y < y)$. Constructing the step function $x = G(y)$ according to this equation yields the values

$$x_j = G(y_j) = \sum_{k=1}^{i} P(y_k) \tag{5.14}$$

which lie between 0 and 1. Therefore random numbers following the discrete distribution $G(y)$ can be produced by selection of random numbers x distributed uniformly between 0 and 1 and generation of the number y_j if x falls within the interval $x_{j-1} \leqslant x < x_j$.

Example 5-5: *Monte Carlo method for simulation*
Physical situations that are determined by statistical processes can be simulated using random numbers in the computer. Examples are the automobile traffic in a given network of streets or the behaviour of neutrons in a nuclear reactor. The so-called *Monte Carlo method* was originally developed for the latter problem by von Neumann and Ulam. Changing parameters of the distributions of random numbers is equivalent to a change of the physical situation. In this way the influence of new streets or alterations in the reactor can be studied without expensive and time consuming physical changes. As a simple example we want to generate a random variable t distributed as the time of decay of a radioactive nucleus. We have

$$g(t) = \frac{1}{\tau} e^{-t/\tau}, \quad t \geqslant 0 \quad \text{and} \quad g(t) = 0, \quad t < 0 \tag{5.15}$$

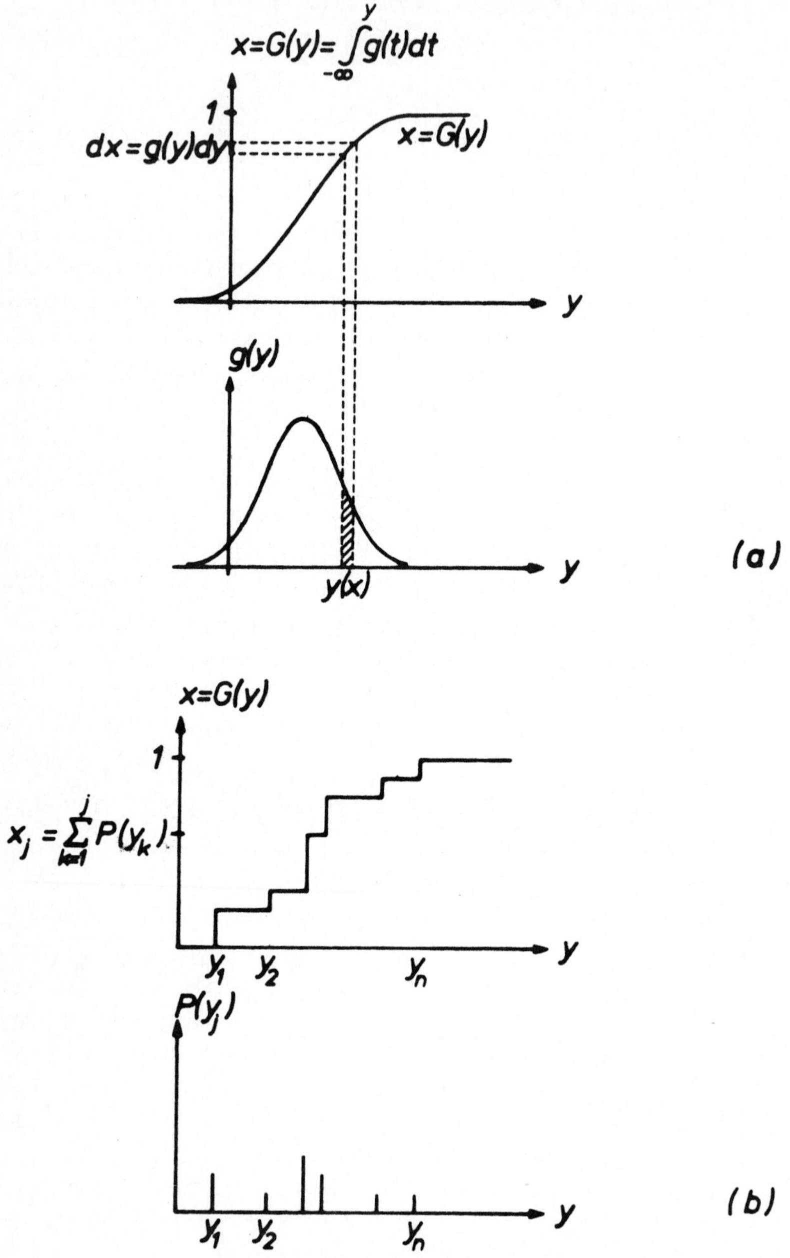

Fig. 5-4. Transformation from a uniformly distributed variable x to a variable y with the distribution function $G(y)$. The variable y can be continuous (a) or discrete (b).

$$x = G(t) = \frac{1}{\tau} \int_{t'=0}^{t} g(t')\mathrm{d}t' = -[\mathrm{e}^{-t'/\tau}]_0^1 = 1-\mathrm{e}^{-t/\tau}. \tag{5.16}$$

Now, if x is a uniformly distributed random number between 0 and 1, so is $x-1$. We can therefore write $x = -\exp(-t/\tau)$ or

$t = -\tau \ln x.$

Random decay times t can be generated by producing random numbers taking their natural logarithm and multiplying it by the negative mean life $-\tau$.

The method described above can be applied whenever the distribution function $G(y)$ is known. If however only the probability density $g(y)$ is given and the integration (5.13) cannot be performed one can still generate random numbers distributed as $g(y)$ in the interval $a \leqslant y \leqslant b$. One begins by determining the maximum $g_{\max}$ of $g(y)$ in this interval. One then picks a random number y_r from a uniform distribution of the range $a \leqslant y_r < b$ (cf. exercise 5-3(a)). Next a random number g_r is picked from a uniform distribution with a range between 0 and $g_{\max}$. Finally y_r is kept if $g_r < g(y_r)$ and discarded otherwise. Repeating this technique one obtains random numbers distributed according to $g(y)$, since the probability of keeping y_r is proportional to $g(y_r)$.

Example 5-6: *Generation of normally distributed random numbers*
Random numbers following the normalized Gaussian distribution (cf. § 8) with mean a and standard deviation σ can be generated in the range between $a - 5\sigma$ and $a + 5\sigma$ using the following simple FUNCTION subprogram.

Program 5-2: *Function* GAURND *generating normally distributed random numbers with mean* AM *and standard deviation* S

```
   FUNCTION GAURND (AM,S)
   S2=2.*S**2
   B=1./SQRT(6.283*S)
10 X=AM+5.*S*(2.*RAND(D)—1.)
   Y=RAND (D)*B
   GAUSS=B*EXP(—((X—AM)**2/S2))
   IF (GAUSS—Y)10,30,30
30 GAURND=X
   RETURN
   END
```

Example 5-7: Monte Carlo method for integration
As a second example of an application of random numbers we present a method of numerical integration. We begin with the special integral

$$I_0 = \int_0^1 g(x)\,dx, \tag{5.17}$$

where the integrand $g(x)$ varies in the region $0 \leqslant g(x) \leqslant 1$ as indicated in fig. 5-5. The integral I_0 is equal to the surface of the area ω, which is

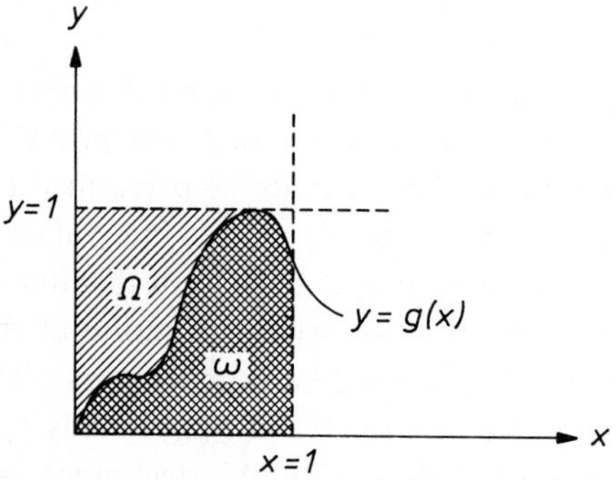

Fig. 5-5. Monte Carlo integration of $y = g(x)$.

bounded by the x-axis, the curve $y = g(x)$ and the two straight lines $x = 0$ and $x = 1$. The larger area Ω bounded by the x-axis, the y-axis and the straight lines $x = 1$ and $y = 1$ is equal to 1. Therefore if we pick N pairs of random numbers (x, y) both selected from a uniform distribution with the boundaries 0 and 1 and find n times the point (x, y) inside the area ω we have

$$I_0 = \lim_{N\to\infty} \frac{n}{N}. \tag{5.18}$$

Of course the method is easily generalized to

$$I = \int_{x=a}^{x=b} f(x)\,dx. \tag{5.19}$$

Using a new integration variable z related to x by

$$x = a + (b - a)z, \quad dx = (b - a)dz$$

we can rewrite (5.19) in the form

$$I = (b - a) \int_{z=0}^{z=1} f(a + (b - a)z)\,\mathrm{d}z. \tag{5.20}$$

Although the integration limits are now those of eq. (5.17) the integrand is not yet bounded between 0 and 1. This can be achieved by replacing $f(x)$ by

$$g(x) = \frac{f(x) - f_{\min}}{f_{\max} - f_{\min}},$$

where $f_{\min}$ and $f_{\max}$ are the smallest and largest values of the function $f(x)$ in the integration interval, respectively. Then

$$I = F_\Omega \int_0^1 g(z)\,\mathrm{d}z + F_0 = F_\Omega I_0 + F_0 \tag{5.21}$$

with

$$F_\Omega = (f_{\max} - f_{\min})(b - a), \quad F_0 = f_{\min}(b - a).$$

Now, since only a finite number of random numbers can be generated, I_0 in eq. (5.18) can be obtained only approximately. For finite N we obtain

$$\tilde{I}_0 = n/N.$$

It is a measure of the probability p of finding a pair of random numbers in the region ω. According to eq. (2.3) we have for the variance of $\tilde{I}_0$

$$\sigma^2(\tilde{I}_0) = \frac{p(1 - p)}{N} = \frac{1}{N^2}\left(n - \frac{n^2}{N}\right).$$

As the error in the determination of $\tilde{I}_0$ we use its square root

$$\Delta(\tilde{I}_0) = \frac{1}{N}\sqrt{\left(n - \frac{n^2}{N}\right)}. \tag{5.22}$$

Using (5.21) and (4-5.8) we finally obtain

$$\tilde{I} = F_\Omega \tilde{I}_0 + F_0,$$

$$\Delta(I) = F_\Omega \Delta(\tilde{I}_0). \tag{5.23}$$

Whereas for one-dimensional integrals other numerical integration techniques are often preferable the Monte Carlo method is in many cases the most convenient way to perform multiple integrations of the form

$$I = \int_{u_1}^{u_2} f(u)\,\mathrm{d}u \int_{v_1(u)}^{v_2(u)} g(u, v)\,\mathrm{d}v \cdots \int_{z_1(u,v,\ldots,y)}^{z_2(u,v,\ldots,y)} k(u, v, \cdots, z)\,\mathrm{d}z. \tag{5.24}$$

Before performing Monte Carlo integration on a large scale the specialized literature should be consulted since as in the case of other numerical methods ineffective use can result in unnecessary consumption of computer time. Finally we present a short program.

Program 5-3: *Subroutine* MTCINT *performing a one-dimensional Monte Carlo integration*

```
      SUBROUTINE MTCINT (XINT, DELINT, DELREL, A, B, FMIN,
     *FMAX)
      XN=0
      XNN=0
      I=0
      IX=0
      DELF=FMAX—FMIN
      DELX=B—A
    5 X=A+DELX*RAND(XX)
      G=(F(X)—FMIN)/DELF
      Y=RAND(XX)
      IF (Y—G) 10,20,20
   10 XN=XN+1.
   20 XNN=XNN+1.
      I=I+1
      IF (I—1000) 5,30,30
   30 I=0
      XINTO=XN/XNN
      DELIO=SQRT(XN—XN**2/XNN)/XNN
      DELINT=DELF*DELX*DELIO
      XINT=DELF*DELX*XINTO+FMIN*DELX
      IF (DELINT/XINT—DELREL) 40,40,5
   40 RETURN
      END
```

In the program the following variable names are used XINT $= I$, DELINT $= \Delta(I)$, XINTO $= I_0$, DELIO $= \Delta(I_0)$, XN $= n$, XNN $= N$. The program is written as a subroutine where the limits a, b of integration, the extreme values $f_{\min}, f_{\max}$ of the integrand and the predetermined relative error $\Delta I/I$ of the integration are input arguments. Since the function $f(x)$ itself is defined in a special FUNCTION subprogram the integration program can be used universally. The actual Monte Carlo sampling is done in the first part of the program. The second part serves to determine whether the required accuracy has been reached already. In order to save computer time this part is gone through only after every thousand samplings.

As a numerical example we want to compute with an accuracy of 1 % the integral

$$I = \frac{1}{\sqrt{(2\pi)}} \int_{-1}^{+1} e^{-\frac{1}{2}x^2} dx,$$

over the normalized Gaussian distribution. Obviously the function subroutine has the form

```
      FUNCTION F(X)
      F=EXP(—0.5*X**2)/2.5066283
      RETURN
      END
```

The main program

```
C     MAIN PROGRAM FOR INTEGRATION
C
      FMAX=F(0.)
      FMIN=F(1.)
      CALL MTCINT(XINT,DELINT,0.01,—1.,1.,FMIN,FMAX)
      WRITE (6,1000) XINT,DELINT
 1000 FORMAT (10H INTEGRAL=,F10.5, 7H ERROR=,F10.5)
      END
```

prints the following result

```
 INTEGRAL= 0.67576 ERROR= 0.00484
```

5-6. The characteristic function of a distribution

So far we have only considered real random variables. In fact in ch. 3, §1 we have introduced the concept of a random quantity as a real number associated with an event. Without changing this concept we can formally construct a *complex random variable* from two real ones by writing

$$z = x + \mathrm{i}\, y. \tag{6.1}$$

As its expectation value we define

$$E(z) = E(x) + \mathrm{i}\; E(y). \tag{6.2}$$

By analogy with real variables complex random variables are independent if real and imaginary parts are independent among themselves.

If x is a real random variable with distribution function $F(x)$ and probability density $f(x)$, we define its *characteristic function* to be the expectation value of the quantity $\exp(\mathrm{i}tx)$

$$\varphi(t) = E\{\exp(\mathrm{i}tx)\}. \tag{6.3}$$

Eq. (6.3) shows that in the case of a continuous variable the characteristic function is a Fourier integral with its known transformation properties

$$\varphi(t) = \int_{-\infty}^{\infty} \exp(\mathrm{i}tx) f(x)\, \mathrm{d}x. \tag{6.4}$$

For a discrete variable we get instead from (3-3.2)

$$\varphi(t) = \sum_i \exp(\mathrm{i}tx_i)\, P(x = x_i). \tag{6.5}$$

We now consider the moments of x about the origin

$$\lambda_n = E(x^n) = \int_{-\infty}^{\infty} x^n f(x)\, \mathrm{d}x, \tag{6.6}$$

and find that they can be obtained simply by differentiating the characteristic function n times at the point $t = 0$:

$$\varphi^{(n)}(t) = \frac{\mathrm{d}^n \varphi(t)}{\mathrm{d}t^n} = \mathrm{i}^n \int_{-\infty}^{\infty} x^n \exp(\mathrm{i}tx) f(x)\, \mathrm{d}x,$$

and therefore

$$\varphi^{(n)}(0) = \mathrm{i}^n \lambda_n. \tag{6.7}$$

If we now introduce the simple coordinate translation

$$y = x - \hat{x}, \tag{6.8}$$

and construct the characteristic function of y

$$\varphi_y(t) = \int_{-\infty}^{\infty} \exp\{\mathrm{i}\, t(x - \hat{x})\} f(x - \hat{x})\, \mathrm{d}x, \tag{6.9}$$

then of course its nth derivative is (up to a power of i) equal to the nth moment of x about the expectation value (cf. eq. (3-3.8))

$$\varphi_y^{(n)}(0) = \mathrm{i}^n \mu_n = \mathrm{i}^n E\{(x - \hat{x})^n\}, \tag{6.10}$$

i.e. especially

$$\sigma^2(x) = -\varphi_y''(0). \tag{6.11}$$

Inverting the Fourier transform (6.4) we see that it is possible to obtain the probability density from the characteristic function

$$f(x) = \frac{1}{2\pi} \int_{-\infty}^{\infty} \exp(-\mathrm{i}tx)\, \varphi(t)\, \mathrm{d}t. \tag{6.12}$$

It is possible to show that a distribution is determined *uniquely* by its characteristic function. This is the case even for discrete variables where eq. (6.12) is replaced by

$$F(b) - F(a) = \frac{\mathrm{i}}{2\pi} \int_{-\infty}^{\infty} \frac{\exp(\mathrm{i}tb) - \exp(-\mathrm{i}ta)}{t} \varphi(t)\, \mathrm{d}t, \tag{6.13}$$

since then no probability density is defined. Often it is more convenient to use the characteristic function rather than the original distribution. Because of the unique relation between the two it is possible to switch back and forth between them at any place in the course of a derivation.

We now consider the sum of two independent random variables

$$w = x + y.$$

Its characteristic function is

$$\varphi_w(t) = E[\exp\{it(x + y)\}] = E\{\exp(itx)\exp(ity)\}.$$

Generalizing relation (4-2.13) to complex variables we obtain

$$\varphi_w(t) = E\{\exp(itx)\}\, E\{\exp(ity)\} = \varphi_x(t)\,\varphi_y(t), \tag{6.14}$$

i.e. the characteristic function of a sum of independent random variables is equal to the product of their respective characteristic functions.

Example 5-8: *Addition of two Poisson-distributed variables using the characteristic function*

According to eqs. (6.5) and (4.1) the characteristic function of a Poisson distribution is

$$\varphi(t) = \sum_{k=0}^{\infty} \exp(itk)\frac{\lambda^k}{k!}\exp(-\lambda) = \exp(-\lambda)\sum_{k=0}^{\infty}\frac{(\lambda\exp(it))^k}{k!}$$

$$= \exp(-\lambda)\exp(\lambda\,\mathrm{e}^{it}) = \exp\{\lambda(\mathrm{e}^{it} - 1)\}. \tag{6.15}$$

We now form the characteristic function of the sum of two independent Poisson distributions with means λ_1 and λ_2

$$\varphi_{\text{sum}}(t) = \exp\{\lambda_1(\mathrm{e}^{it} - 1)\}\exp\{\lambda_2(\mathrm{e}^{it} - 1)\}$$

$$= \exp\{(\lambda_1 + \lambda_2)(\mathrm{e}^{it} - 1)\}. \tag{6.16}$$

This is again of the form (6.15). Therefore the distribution of the sum of independent Poisson distributions is again a Poisson distribution. Its mean is the sum of the means of the individual distributions.

5-7. The Laplace model of errors

In 1783 Laplace made the following remarks concerning the origin of errors of observation. Let the true value of a quantity to be measured be m_0. Now let the measurement be disturbed by a large number n of independent causes, each resulting in a disturbance of magnitude ε. For each disturbance there exists equal probability that a variation in the measured value, in either direction, i.e. $+\varepsilon$ or $-\varepsilon$, will occur. The measurement error is then composed of the sum of individual disturbances. It is clear that in this model the probability distribution of measurement errors will be given by the binomial distribution. It is interesting nevertheless to follow the model somewhat further, since it leads directly to the famous Pascal triangle.

In fig. 5-6 the probability distribution is developed following the model. The starting point is that with no disturbance the probability of measuring m_0 is equal to one. With one disturbance this probability is split equally between the neighbouring possibilities $m_0 + \varepsilon$ and $m_0 - \varepsilon$. The same happens with every further disturbance. Of course the individual probabilities leading to the same measured value have to be added. Each line of the resulting triangle contains the distribution W_k^n ($k = 0, 1, \ldots, n$) of eq. (1.3) for the case $p = q = \frac{1}{2}$. Multiplied by $(p^k q^{n-k})^{-1} = 2^n$ it turns into a line of binomial coefficients of Pascal's triangle (cf. Appendix C).

Number of disturbances	Observed value $(m_0 + \cdots)$						
n	-3ε	-2ε	$-\varepsilon$	0	$+\varepsilon$	$+2\varepsilon$	$+3\varepsilon$
0				1			
1			$\frac{1}{2}$		$\frac{1}{2}$		
2		$\frac{1}{4}$		$\frac{1}{4}$, $\frac{1}{4}$		$\frac{1}{4}$	
3	$\frac{1}{8}$		$\frac{1}{8}$, $\frac{1}{4}$; $\frac{3}{8}$	$\frac{1}{2}$	$\frac{1}{4}$, $\frac{1}{8}$; $\frac{3}{8}$		$\frac{1}{8}$

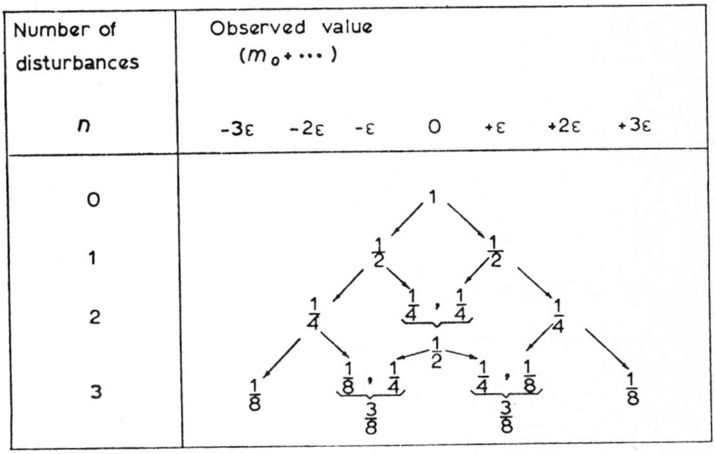

Fig. 5-6. Connection of Laplace model and binomial distribution.

We now go to the limit of very many disturbances $n \to \infty$. Unlike the similar procedure in the case of the Poisson distribution we keep p and q constant. Although in the Laplace model $p = q = \frac{1}{2}$ we deal here with the general case of any p. From eqs. (1.3) and (6.5) we get the characteristic function of the binomial distribution

$$\varphi(t) = \sum_{j=0}^{n} \exp(itj) \binom{n}{j} p^j q^{n-j}.$$

Using the binomial theorem (C-6) this can be reduced to

$$\varphi(t) = \{\exp(it)\,p + q\}^n. \tag{7.1}$$

Instead of x we now use the reduced variable

$$u = \frac{x - \hat{x}}{\sigma(x)} = \frac{x - np}{\sigma},$$

which we can consider as the sum of two independent random variables x/σ and $-\hat{x}/\sigma$, even though the second term is in fact constant. Its characteristic function is then according to eq. (6.14)

$$\varphi_u(t) = \exp\left(-\frac{itnp}{\sigma}\right)\left\{\exp\left(\frac{it}{\sigma}\right)p + q\right\}^n.$$

Taking the logarithm (and using $p = 1 - q$) we get

$$\ln \varphi_u(t) = -\frac{itnp}{\sigma} + n\ln\left[1 + p\left\{\exp\left(\frac{it}{\sigma}\right) - 1\right\}\right].$$

Since $t/\sigma = t/\surd(npq)$ approaches zero as $n \to \infty$, we can expand the expression in brackets and obtain

$$\ln \varphi_u(t) = -\frac{itnp}{\sigma} + n\ln\left[1 + p\left\{\frac{it}{\sigma} - \frac{1}{2}\left(\frac{t}{\sigma}\right)^2 + \ldots\right\}\right].$$

We now use the expansion

$$\ln(1 + \varepsilon) = \varepsilon - \frac{\varepsilon^2}{2} + \frac{\varepsilon^3}{3} - \ldots,$$

and find

$$\begin{aligned}\ln \varphi_u(t) &= -\frac{itnp}{\sigma} + n\left\{\frac{itp}{\sigma} - \frac{1}{2}\left(\frac{t}{\sigma}\right)^2 (p - p^2) + \mathcal{O}(\sigma^{-3})\right\} \\ &= -\tfrac{1}{2}t^2\,\frac{np(1 - p)}{\sigma^2} + \mathcal{O}(\sigma^{-3}) \\ &= -\tfrac{1}{2}t^2 + \mathcal{O}(\sigma^{-3}).\end{aligned}$$

In the limit of $n \to \infty$ the logarithm approaches $-\frac{1}{2}t^2$ and therefore the characteristic function becomes

$$\varphi_u(t) = \exp(-\tfrac{1}{2}t^2). \tag{7.2}$$

To obtain the corresponding probability density we use eq. (6.12)

$$f(u) = \frac{1}{2\pi} \int_{-\infty}^{\infty} \exp(-itu) \exp(-\tfrac{1}{2}t^2)\, dt$$

$$= \frac{1}{2\pi} \int_{-\infty}^{\infty} \exp\{-\tfrac{1}{2}(t + iu)^2 - \tfrac{1}{2}u^2\}\, dt. \tag{7.3}$$

We choose $r = (t + iu)$ as a new integration variable and get

$$f(u) = \frac{1}{2\pi} \exp(-\tfrac{1}{2}u^2) \int_{-\infty + iu}^{\infty + iu} \exp(-\tfrac{1}{2}r^2)\, dr.$$

The integration extends along a line parallel to the real axis, but because of the "good-natured" behaviour of the integrand at infinity, it can be performed along the real axis. Using eq. (D-4) we finally get

$$f(u) = \frac{1}{2\pi} \exp(-\tfrac{1}{2}u^2) \sqrt{2}\, \Gamma(\tfrac{1}{2}) = \frac{1}{\sqrt{(2\pi)}} \exp(-\tfrac{1}{2}u^2). \tag{7.4}$$

We obtained the distribution (7.4) from the binomial distribution using a reduced variable, i.e. zero mean and unit variance. We therefore expect these properties from the new distribution. This is just what we find:

$$E(u) = \hat{u} = \frac{1}{\sqrt{(2\pi)}} \int_{-\infty}^{\infty} u \exp(-\tfrac{1}{2}u^2)\, du = \Big[-\exp(-\tfrac{1}{2}u^2) \Big]_{-\infty}^{\infty} = 0, \tag{7.5}$$

and using relation (6.11)

$$\sigma^2(u) = -\varphi''(0) = 1. \tag{7.6}$$

5-8. Normal distribution

A more general form of eq. (7.4) is

$$f(x) = \phi(x) = \frac{1}{\sqrt{(2\pi)}\, b} \exp\left(-\frac{(x - a)^2}{2b^2} \right). \tag{8.1}$$

The expectation value and variance are

$$\hat{x} = a, \qquad \sigma^2(x) = b^2. \tag{8.2}$$

This distribution is the limit that is approached by a binomial with mean and variance (8.2) in the case $n \to \infty$. It is of central importance in mathematical statistics and especially in the theory of errors, and it is called *normal* or *Gaussian distribution*. Because of its importance we will have to study its properties in detail and we shall deal with its quantitative behaviour. As already indicated in (8.1) we use the special symbol $\phi(x)$ to indicate the probability density of the normal distribution. The *standardized normal distribution*, i.e. the normal distribution with zero mean and unit variance which we have already met in eq. (7.4) will be denoted by

$$\phi_0(x) = \frac{1}{\sqrt{(2\pi)}} \exp(-\tfrac{1}{2}x^2). \tag{8.3}$$

From eq. (6.4) it follows that the characteristic function of ϕ is

$$\varphi(t) = \frac{1}{\sqrt{(2\pi)}\, b} \int_{-\infty}^{\infty} \exp(itx) \exp\left(-\frac{(x-a)^2}{2b^2}\right) \mathrm{d}x.$$

By completing the quadratic as in eq. (7.3) we find that

$$\varphi(t) = \exp(ita) \exp(-\tfrac{1}{2}b^2t^2). \tag{8.4}$$

For $a = 0$ this leads to the interesting theorem:

> The characteristic function of a normal distribution with zero mean is itself a normal distribution with zero mean. The product of the variances of both distributions is one.

Considering the sum of independent normal distributions we obtain from (6.14) that its characteristic function is again of the form (8.4). A sum of independent normal distributions is again normally distributed. Similar properties have already been found in example 5-8 for the Poisson distribution.

To determine the higher moments of the Gaussian distribution we make use of (6.10). The characteristic function transformed to the variable $y = x - \hat{x}$ is

$$\varphi_y(t) = \exp(-\tfrac{1}{2}b^2t^2).$$

Then

$$\mu_n = \frac{1}{\mathrm{i}^n} \frac{\mathrm{d}^n}{\mathrm{d}t^n} \left(\exp(-\tfrac{1}{2}b^2t^2)\right)_{t=0}, \tag{8.5}$$

in particular

$$\mu_1 = 0, \qquad \mu_2 = b^2 = \sigma^2,$$

$$\mu_3 = 0, \qquad \mu_4 = 3\sigma^4,$$

and in general (for integer k)

$$\mu_{2k+1} = 0, \tag{8.6}$$

as expected from the symmetry of the distribution, and

$$\mu_{2k} = \frac{(2k)!}{2^k k!} \sigma^{2k}. \tag{8.7}$$

All higher moments are therefore immediately determined once σ^2 is known.

5-9. Quantitative properties of the normal distribution

In fig. 5-7 we present a diagram of the standardized normal distribution $\phi_0(x)$ (eq. (8.3)) and the corresponding distribution function

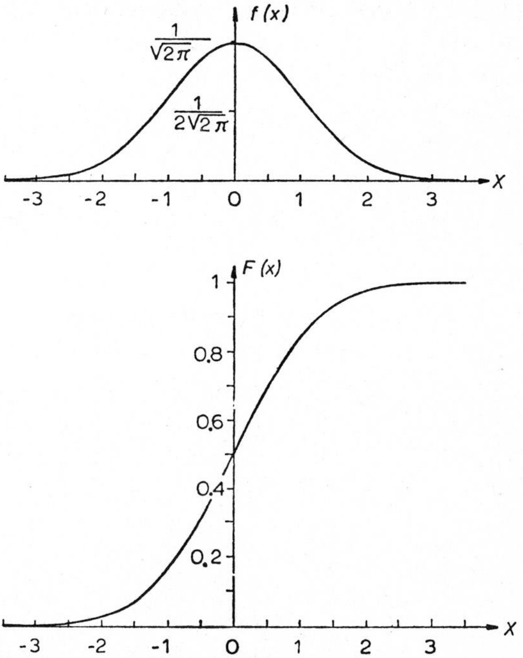

Fig. 5-7. Probability density $f(x)$ and distribution function $F(x)$ of the standardized normal distribution.

$$\psi_0(x) = \int_{-\infty}^{x} \phi_0(\mathrm{x})\,d\mathrm{x} = \frac{1}{\sqrt{(2\pi)}} \int_{-\infty}^{x} \exp(-\tfrac{1}{2}\mathrm{x}^2)\,d\mathrm{x}. \tag{9.1}$$

The probability density ϕ_0 displays the famous Gaussian "bell" shape. We can easily verify that the inflexion points of the function are positioned at $x = \pm 1$ (in the case of the general normal distribution at $x = a \pm b$). The distribution function $\psi_0(x)$ represents the probability that the random variable x takes a value smaller than x

$$\psi_0(x) = P(\mathrm{x} < x). \tag{9.2}$$

For symmetry reasons

$$P(|\mathrm{x}| > x) = 2\psi_0(-|x|) = 2\{1 - \psi_0(|x|)\}, \tag{9.3}$$

or, conversely, the probability of observing x within a band of width $2x$ around (the expectation value) zero is

$$P(|\mathrm{x}| \leqslant x) = 2\psi_0(|x|) - 1. \tag{9.4}$$

The quantities (9.2)–(9.4) can be found in many tables. (References to a number of statistical tables are given in the bibliography.) We have tabulated (9.2) and (9.4) in tables F-2a and F-2b. The above relations can easily be generalized. The distribution function of the general Gaussian distribution is

$$\psi(x) = \frac{1}{\sqrt{(2\pi)}\,b} \int_{-\infty}^{x} \exp\left(-\frac{(\mathrm{x} - a)^2}{b^2}\right) d\mathrm{x}$$

$$= \frac{1}{\sqrt{(2\pi)}} \int_{-\infty}^{(x-a)/b} \exp(-\tfrac{1}{2}u^2)\,du = \psi_0\left(\frac{x - a}{b}\right). \tag{9.5}$$

We want to know the probability of a random variable being observed within an integer multiple of the standard deviation $\sigma = b$ from the mean a

$$P(|\mathrm{x} - a| \leqslant n\sigma) = 2\psi_0\left(\frac{n\sigma}{b}\right) - 1 = 2\psi_0(n) - 1. \tag{9.6}$$

From table F-2b we find

$$\begin{aligned} P(|\mathrm{x} - a| \leqslant \sigma) &= 68.2\%, & P(|\mathrm{x} - a| > \sigma) &= 31.8\%, \\ P(|\mathrm{x} - a| \leqslant 2\sigma) &= 95.4\%, & P(|\mathrm{x} - a| > 2\sigma) &= 4.6\%, \\ P(|\mathrm{x} - a| \leqslant 3\sigma) &= 99.8\%, & P(|\mathrm{x} - a| > 3\sigma) &= 0.2\%. \end{aligned} \tag{9.7}$$

It is interesting to compare this property of the normal distribution with the prediction of Chebychev's inequality that does not assume any knowledge of the distribution function. Relation (3-4.1) yields

$$P(|\mathrm{x} - a| > \sigma) < 1,$$
$$P(|\mathrm{x} - a| > 2\sigma) < 25\%,$$
$$P(|\mathrm{x} - a| > 3\sigma) < 11\%.$$

As already mentioned in ch. 3, §4 the predictions of Chebychev's inequality are rather weak and should be used only if nothing is known about the distribution.

As we shall see later in greater detail one can often assume that measurement errors are distributed normally around the true value a, i.e. that the probability of measuring a value between x and $x + \mathrm{d}x$ is determined by

$$P(x \leqslant \mathrm{x} < x + \mathrm{d}x) = \phi(x)\,\mathrm{d}x.$$

The *standard deviation* σ of the distribution $\phi(x)$, which can be approximately determined by a repetition of measurements, can then be used to characterize the error of measurement. It is an important convention to identify the standard deviation with the measurement error. From relation (9.7) we know that the probability that the result of a single experiment lies within one standard deviation of the true value is only 68.2%. This seems rather low. Therefore it is a habit of many experimenters to multiply the standard deviation by a more or less arbitrary factor, as for example by 3, which increases that probability to 99.8%. This procedure, although apparently prudent, is however very misleading and often harmful. If this factor is not explicitly stated, a comparison of different measurements of the same quantity and especially the calculation of a weighted average (cf. example 7-2) is rendered impossible or liable to be erroneous.

The inverse of the normal distribution is also of great interest. We denote it by

$$x = \Omega(P), \tag{9.8}$$

and obtain for given P a quantity x such that the random number x originating from a standardized normal distribution is smaller than x with probability P:

$$P(\mathrm{x} < \Omega(P)) = P. \tag{9.9}$$

We also define $\Omega'(P)$ such that

$$P(|\mathrm{x}| < \Omega'(P)) = P. \tag{9.10}$$

Obviously

$$P = \int_{-\infty}^{\Omega(P)} \phi_0(x)\,\mathrm{d}x, \tag{9.11}$$

or

$$P = 2\int_{0}^{\Omega'(P)} \phi_0(x)\,\mathrm{d}x, \tag{9.12}$$

respectively. The functions $\Omega(P)$ and $\Omega'(P)$ are tabulated in the tables F-3a and F-3b. From fig. 5-8, in which both functions are sketched, it is apparent that the relation

$$\Omega'(P) = \Omega(\tfrac{1}{2}(1 + P)) \tag{9.13}$$

holds.

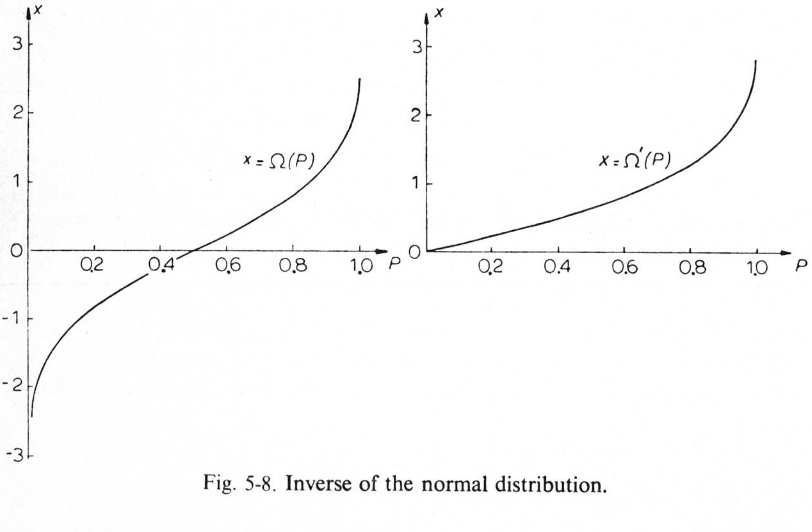

Fig. 5-8. Inverse of the normal distribution.

5-10. Multivariate normal distribution

The joint normal distribution of n variables $\boldsymbol{x} = (x_1, x_2, \ldots, x_n)$ is defined as

$$\phi(\boldsymbol{x}) = k\exp\{-\tfrac{1}{2}(\boldsymbol{x} - \boldsymbol{a})^{\mathrm{T}} B(\boldsymbol{x} - \boldsymbol{a})\}. \tag{10.1}$$

Here k is determined by the normalization of the total probability to one,

$\boldsymbol{a}$ like $\boldsymbol{x}$, is an n-component column vector and B an $(n \times n)$-matrix. Since $\phi(\boldsymbol{x})$ is apparently symmetric about $(\boldsymbol{x} - \boldsymbol{a})$ we have

$$\int_{-\infty}^{\infty} (\boldsymbol{x} - \boldsymbol{a})\,\phi(\boldsymbol{x} - \boldsymbol{a})\,\mathrm{d}\boldsymbol{x} = 0, \qquad (10.2)$$

i.e.

$$E(\boldsymbol{x} - \boldsymbol{a}) = 0$$

or

$$E(\boldsymbol{x}) = \boldsymbol{a}. \qquad (10.3)$$

The vector of expectation values is directly given by $\boldsymbol{a}$. We now differentiate eq. (10.2) with respect to $\boldsymbol{a}$

$$\int_{-\infty}^{\infty} [I - (\boldsymbol{x} - \boldsymbol{a})(\boldsymbol{x} - \boldsymbol{a})^{\mathrm{T}} B]\,\phi(\boldsymbol{x} - \boldsymbol{a})\,\mathrm{d}\boldsymbol{x} = 0,$$

which implies that the expectation value of the square bracket vanishes, i.e.

$$B\,E\{(\boldsymbol{x} - \boldsymbol{a})(\boldsymbol{x} - \boldsymbol{a})^{\mathrm{T}}\} = I,$$

or

$$C = E\{(\boldsymbol{x} - \boldsymbol{a})(\boldsymbol{x} - \boldsymbol{a})^{\mathrm{T}}\} = B^{-1}. \qquad (10.4)$$

Comparison of this result with eq. (4-3.21) shows that C is just the covariance matrix of the variables $\boldsymbol{x} = (x_1, x_2, \ldots, x_n)$.

Let us discuss the normal distribution of 2 variables in greater detail. We have

$$C = B^{-1} = \begin{pmatrix} \sigma_1^2 & \mathrm{cov}\,(x_1, x_2) \\ \mathrm{cov}\,(x_1, x_2) & \sigma_2^2 \end{pmatrix}. \qquad (10.5)$$

By matrix inversion we obtain

$$B = \frac{1}{\sigma_1^2\sigma_2^2 - \mathrm{cov}(x_1, x_2)^2} \begin{pmatrix} \sigma_2^2 & -\mathrm{cov}(x_1, x_2) \\ -\mathrm{cov}(x_1, x_2) & \sigma_1^2 \end{pmatrix}. \qquad (10.6)$$

In the case of vanishing covariance, i.e. independent variables x_1, x_2, this reduces to

$$B_0 = \begin{pmatrix} 1/\sigma_1^2 & 0 \\ 0 & 1/\sigma_2^2 \end{pmatrix}. \tag{10.7}$$

Introducing B_0 into (10.1) we obtain, as expected, a product of 2 normal distributions of a single variable

$$\phi = k \exp\left(-\frac{1}{2} \frac{(x_1 - a_1)^2}{\sigma_1^2} \right) \exp\left(-\frac{1}{2} \frac{(x_2 - a_2)^2}{\sigma_2^2} \right). \tag{10.8}$$

In this case the constant k becomes

$$k_0 = \frac{1}{2\pi \sigma_1 \sigma_2},$$

as can be shown by integration of (10.8) or simply by comparison with eq. (8.1). In the general case of n variables and nonvanishing covariances we have instead

$$k = \frac{1}{(2\pi)^{\frac{1}{2}n} (\det B)^{\frac{1}{2}}}, \tag{10.9}$$

where $\det B$ is the determinant of the matrix B.

In the case where the variables are not independent, i.e. for nonvanishing covariance, the normal distribution of 2 variables is more complicated. For the discussion of this case we use reduced variables

$$u_i = \frac{x_i - a_i}{\sigma_i}, \qquad i = 1, 2,$$

and the correlation coefficient

$$\rho = \frac{\operatorname{cov}(x_1, x_2)}{\sigma_1 \sigma_2} = \operatorname{cov}(u_1, u_2).$$

Eq. (10.1) now takes the simple form

$$\phi(u_1, u_2) = k \exp(-\tfrac{1}{2} \boldsymbol{u}^{\mathrm{T}} B \boldsymbol{u}), \tag{10.10}$$

with

$$B = \frac{1}{1 - \rho^2} \begin{pmatrix} 1 & -\rho \\ -\rho & 1 \end{pmatrix}. \tag{10.11}$$

Lines of constant probability density are determined by requiring the exponent in eq. (10.10) to be constant:

$$\frac{1}{(1-\rho^2)}(u_1^2 + u_2^2 - 2u_1u_2\,\rho) = c. \tag{10.12}$$

We choose momentarily $c = 1$. In the original variables, eq. (10.12) then becomes

$$\frac{(x_1 - a_1)^2}{\sigma_1^2} - 2\rho\,\frac{x_1 - a_1}{\sigma_1}\,\frac{x_2 - a_2}{\sigma_2} + \frac{(x_2 - a_2)^2}{\sigma_2^2} = 1 - \rho^2. \tag{10.13}$$

This is the equation of an ellipse with the centre a_1, a_2. The principal axes of the ellipse have an angle α with respect to the axes x_1 and x_2. This angle, and the semi-diameters p_1 and p_2 along the principal axes, can be derived from eq. (10.13) using the well known properties of conic sections:

$$\tan 2\alpha = \frac{2\rho\;\sigma_1\sigma_2}{\sigma_1^2 - \sigma_2^2}, \tag{10.14}$$

$$p_1^2 = \frac{\sigma_1^2\sigma_2^2(1-\rho^2)}{\sigma_2^2\cos^2\alpha - 2\rho\sigma_1\sigma_2\sin\alpha\cos\alpha + \sigma_1^2\sin^2\alpha}, \tag{10.15}$$

$$p_2^2 = \frac{\sigma_1^2\sigma_2^2(1-\rho^2)}{\sigma_2^2\sin^2\alpha + 2\rho\sigma_1\sigma_2\sin\alpha\cos\alpha + \sigma_1^2\cos^2\alpha}. \tag{10.16}$$

The ellipse with the described properties is called the *ellipse of covariance* of the bivariate normal distribution. For the special case $\sigma_1^2 = 1$, $\sigma_2^2 = 2$ three such ellipses are shown in fig. 5-9 for $\rho = 0.7$, $\rho = 0$ and $\rho = -0.3$, respectively. The covariance ellipse is always inside the rectangle defined by the point (a_1, a_2) and the standard deviations σ_1, σ_2. It touches this rectangle in 4 points. For the extreme cases $\rho = \pm 1$ the ellipse degenerates into a straight line along one of the diagonals of the rectangle.

From (10.13) it is clear that other lines of constant probability are also ellipses, concentric and similar to the ellipse of covariance, and situated inside (outside) it for larger (smaller) probability. The bivariate normal distribution therefore corresponds to a surface in three-dimensional space (fig. 5-10), whose horizontal sections are concentric ellipses. For the largest probability this ellipse degenerates into the point (a_1, a_2). The vertical sections through the centre have the form of a Gaussian distribution whose width is directly proportional to the diameter of the ellipse of covariance along which the section extends.

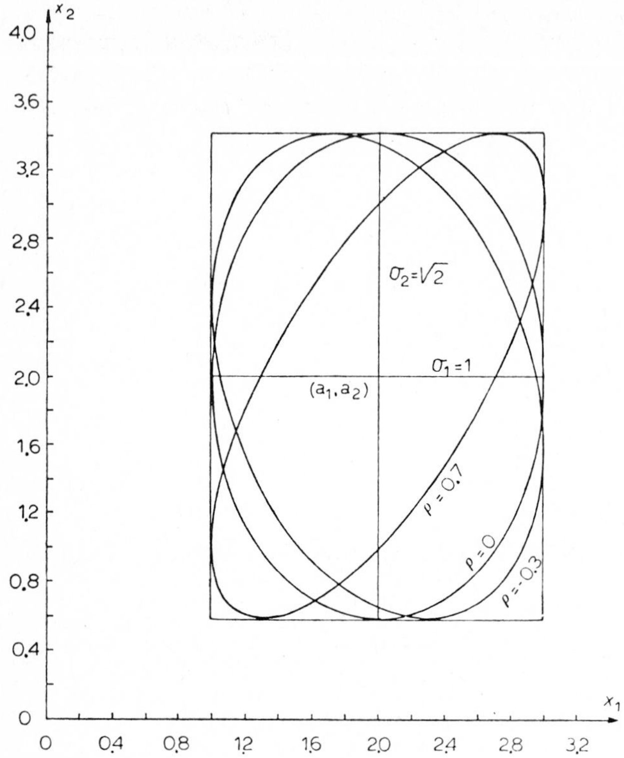

Fig. 5-9. Covariance ellipses.

The probability of observing a pair x_1, x_2 of random variables inside the covariance ellipse is equal to the integral

$$\int_A k\,\phi(\boldsymbol{x})\,\mathrm{d}\boldsymbol{x} = 1 - \mathrm{e}^{-\frac{1}{2}} = \text{const.}, \tag{10.17}$$

where the region of integration A is the surface of the covariance ellipse (10.13). The relation (10.17) can be obtained by applying the transformation of variables $\boldsymbol{y} = T\boldsymbol{x}$ with $T^{-1} = \sqrt{(B^{-1})}$ to the distribution $\phi(\boldsymbol{x})$. The resulting distribution has the properties $\sigma(y_1) = \sigma(y_2) = 1$, $\mathrm{cov}(y_1, y_2) = 0$, i.e. it is of the form (10.8). The region of integration is transformed into the unit circle about (a_1, a_2).

In the discussion of the normal distribution of measurement errors on one variable we found the interval $a - \sigma \leqslant x \leqslant a + \sigma$ as a region of integration comprising values of $f(x)$ above a fixed value. The integral over

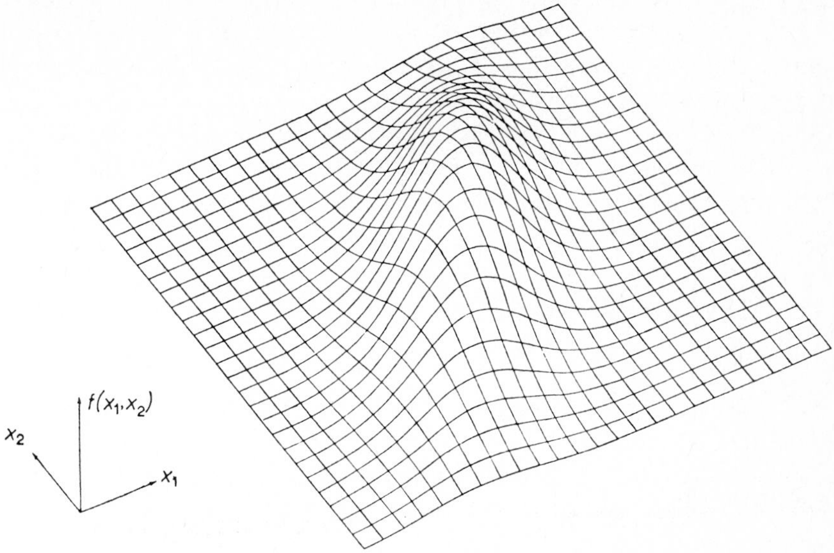

Fig. 5-10. Probability density of a bivariate Gaussian distribution.

this region was independent of σ. In the case of 2 variables the role of this region is taken over by the covariance ellipse, determined by σ_1, σ_2 and ρ,

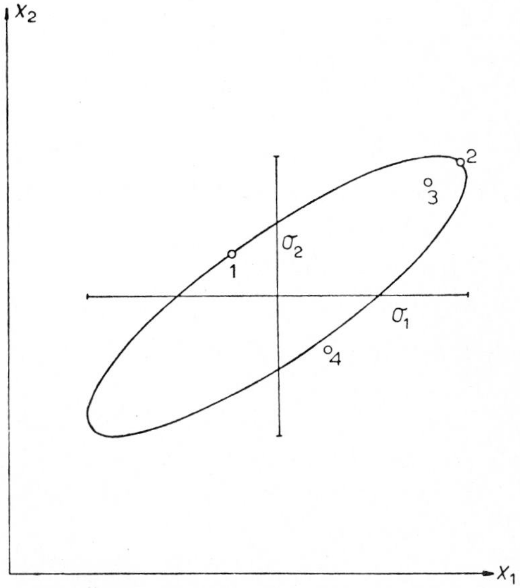

Fig. 5-11. Relative probability of different points for a bivariate normal distribution. $P_1 = P_2 = P_e, \quad P_3 > P_e, \quad P_4 < P_e.$

and not, as is often assumed, by the outer rectangle in fig. 5-9. The significance of the covariance ellipse is also apparent from fig. 5-11. The points 1 and 2, being situated on the covariance ellipse, correspond to equal probability $P(1) = P(2) = P_e$, although the distance of point 1 from the centre is smaller in both coordinate directions. Moreover point 4 is less and point 3 more probable ($P(4) < P_e$, $P(3) > P_e$) even though point 4 deviates less from (a_1, a_2) than point 3.

5-11. The central limit theorem

In deriving the normal distribution we used the following reasoning. We started with a simple experiment with two possible outcomes which we described by a random variable x_i taking one of the values 0 and 1. A finite number n of such experiments lead us to the binomial distribution of the variable Σx_i. Going finally to the limit $n \to \infty$ we found the normal distribution of

$$x = \lim_{n\to\infty} \sum_{i=1}^{n} x_i. \tag{11.1}$$

We are now interested in more general sums of this type. We assume that the x_i are independent and originate from a distribution with mean a and variance σ. Note that we assume no further knowledge about the distribution of the individual x_i. Let the characteristic function of this distribution be denoted by $\varphi(t)$. According to eq. (6.14) the characteristic function of a sum of n variables is then $\{\varphi(t)\}^n$. To simplify the calculation we now assume $a = 0$ (the general case can be reduced to this by a simple coordinate translation $x_i' = x_i - a$).

Because of (6.10) we know the first and second derivatives of $\varphi(t)$ at $t = 0$, i.e.

$$\varphi'(0) = 0, \qquad \varphi''(0) = -\sigma^2,$$

and we can write down the first terms of a Taylor expansion

$$\varphi_{x'}(t) = 1 - \tfrac{1}{2}\sigma^2 t^2 + \ldots .$$

We now introduce a further variable u_i defined by

$$u_i = \frac{x_i'}{\sigma\sqrt{n}} = \frac{x_i - a}{\sigma\sqrt{n}}, \tag{11.2}$$

which for fixed n has simply the effect of contracting the scale. The corresponding characteristic function is

$$\varphi_{u_i}(t) = E\{\exp(itu_i)\} = E\left\{\exp\left(it\frac{x_i - a}{\sigma\sqrt{n}}\right)\right\} = \varphi_{x_i'}\left(\frac{t}{\sigma\sqrt{n}}\right),$$

and therefore

$$\varphi_{u_i}(t) = 1 - \frac{t^2}{2n} + \ldots,$$

where the higher terms are at most of order n^{-2}. We now go to the limit $n \to \infty$ and use

$$u = \lim_{n\to\infty} \sum_{i=1}^{n} u_i = \lim_{n\to\infty} \sum_{i=1}^{n} \frac{x_i - a}{\sigma\sqrt{n}}.$$

Then

$$\varphi_u(t) = \lim_{n\to\infty} \{\varphi_{u_i}(t)\}^n = \lim_{n\to\infty} \left(1 - \frac{t^2}{2n} + \ldots\right)^n,$$

i.e.

$$\varphi_u(t) = \exp(-\tfrac{1}{2}t^2). \tag{11.3}$$

This is just the characteristic function of the standardized normal distribution $\phi_0(u)$. The result can be generalized for the original variable x using eqs. (11.2) and (9.5) and expressed in the central limit theorem:

If the x_i are a set of independent variables each distributed with mean a and variance σ^2 then in the limit $n \to \infty$ their arithmetic mean

$$\bar{x} = \frac{1}{n} \sum_{i=1}^{n} x_i$$

follows a Gaussian distribution with mean a and variance σ^2/n.

It can be shown that sums of random variables follow the Gaussian distribution even under weaker assumptions than those needed for our derivation, in particular in those cases where not all x_i have the same distribution.

5-12. Experimental errors and normal distribution; Herschel's model

In the light of the central limit theorem we can now reconsider Laplace's model of experimental errors as the cumulation of small individual effects (§7). We realize that the original condition, whereby each disturbance has the same magnitude, can now be dropped. A normal distribution of the (disturbed) measurements will still follow. It is interesting to consider the model of John Herschel who on very weak assumptions derived the normal distribution. Small grains are dropped from over the point O onto a plane and the probability of finding one of these grains in an element of area dA is considered. This area element can be considered in terms of cartesian or polar coordinates (fig. 5-12). From the construction of the model there cannot be any systematic deviation of grains from the centre, i.e. for a given r all angles θ are equally probable and therefore the probability density for finding a grain in an element dA about $\boldsymbol{r}$ is only a function of r. We call it $h(r)$.

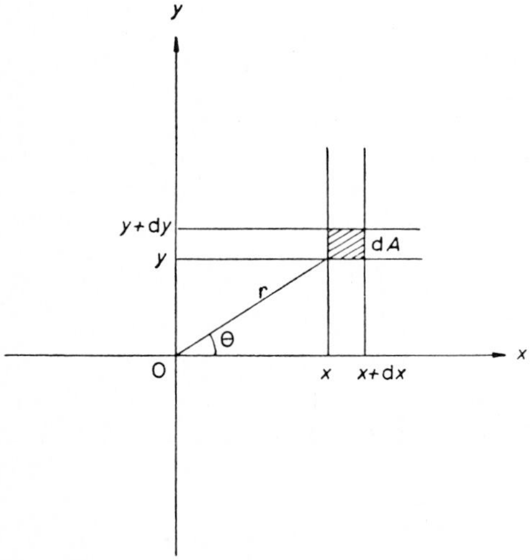

Fig. 5-12. Coordinate systems for the discussion of Herschel's model.

Now we have to make the first assumption, namely that the probability that a grain falls into a strip with boundaries x and $x + \mathrm{d}x$ is independent of the position of the grain with respect to y.

Because of the symmetry in θ the probability densities in x and y then have to be of equal form, i.e. $f(x, y) = f(x)f(y)$. The probability of finding a grain in the element of area dA is therefore

$$h(r)\,\mathrm{d}A = f(x)f(y)\,\mathrm{d}A. \tag{12.1}$$

Taking the logarithm of (12.1) we obtain

$$\ln h(r) = \ln f(x) + \ln f(y),$$

or

$$\ln h(r) = \ln f(r\cos\theta) + \ln f(r\sin\theta). \tag{12.2}$$

Differentiation with respect to θ yields

$$0 = -r\sin\theta\,\frac{f'(r\cos\theta)}{f(r\cos\theta)} + r\cos\theta\,\frac{f'(r\sin\theta)}{f(r\sin\theta)},$$

or, again using x and y,

$$\frac{f'(x)}{xf(x)} = \frac{f'(y)}{yf(y)}. \tag{12.3}$$

Since x and y are completely independent this equality can only hold if both sides are equal to a constant, i.e.

$$f'(x)/f(x) = cx. \tag{12.4}$$

Integration leads to

$$\ln f(x) = \tfrac{1}{2}cx^2 + \text{const.},$$

or

$$f(x) = k\exp(\tfrac{1}{2}cx^2). \tag{12.5}$$

Now since $f(x)$ is a probability density it has to be finite for x approaching infinity, i.e. c has to be negative. We express this by writing $c = -\sigma^{-2}$. Then

$$f(x) = k\exp(-\tfrac{1}{2}x^2/\sigma^2), \tag{12.6}$$

which is the form of a Gaussian distribution. Because of (12.3) we also have

$$f(y) = k\exp(-\tfrac{1}{2}y^2/\sigma^2),$$

and with (12.1)

$$h(r) = k^2\exp\left(-\frac{1}{2}\,\frac{x^2+y^2}{\sigma^2}\right) = k^2\exp\left(-\frac{1}{2}\,\frac{r^2}{\sigma^2}\right). \tag{12.7}$$

From the very simple assumptions of symmetry and independence we have obtained a normal distribution for the deviations of the hits from the target.

This model of Herschel further demonstrates the plausibility of using a Gaussian form for the distribution of measurement errors. The identification of the error distribution with the Gaussian distribution is in fact crucial for many calculations especially for the whole theory of least squares (ch. 9). The normal distribution of errors is however not a law of nature. The causes of experimental errors are too complex if one takes into account all possible experimental arrangements. Symmetry and independence cannot always be easily assured. Therefore the Gaussian distribution of errors has to be verified. This can be done, for example, by performing a χ^2-test on a measured distribution (ch. 8, §6), before performing extensive calculations which are meaningful only if this distribution is Gaussian.

5-13. Convolution of distributions

5-13.1. *Folding integrals*

On various occasions we have already discussed the sums of random variables and, for example, in the derivation of the central limit theorem, found the characteristic function a useful tool in such considerations. Its use however results in a certain loss of directness in the reasoning. In this short section we prefer therefore to discuss the simple problem of the distribution of the sum of 2 random variables without using the characteristic function.

Experimentally one often observes the sum of two distributions. The distribution of emission angles of secondaries in the decay of elementary particles is of interest in determining certain properties of this particle. The observed angle is in this case the sum of two random variables: the decay angle resulting from the random process of radioactive decay and the measurement error. We talk of the two distributions *convoluting** to form a new one.

Let the original quantities be x and y, and their sum defined by

$$u = x + y. \tag{13.1}$$

For our further considerations it is necessary to assume independence of the original variables since only in this case can the joint probability density be expressed as the product of the individual densities

$$f(x, y) = f_x(x) f_y(y). \tag{13.2}$$

* Instead of the term convolution some writers prefer the German word *Faltung* or the French equivalent *composition*.

The distribution function of u, i.e.

$$F(u) = P(u < u) = P(x + y < u), \tag{13.3}$$

is then simply determined by the integration of (13.2) over the shaded area in fig. 5-13:

$$F(u) = \iint_A f_x(x) f_y(y) \, dx dy$$

$$= \int_{-\infty}^{\infty} f_x(x) \, dx \int_{-\infty}^{u-x} f_y(y) \, dy = \int_{-\infty}^{\infty} f_y(y) \, dy \int_{-\infty}^{u-y} f_x(x) \, dx. \tag{13.4}$$

The probability density is then obtained by differentiating $F(u)$

$$f(u) = \frac{dF(u)}{du} = \int_{-\infty}^{\infty} f_x(x) f_y(u - x) \, dx = \int_{-\infty}^{\infty} f_y(y) f_x(u - y) \, dy. \tag{13.5}$$

These expressions are the so-called *folding integrals.*

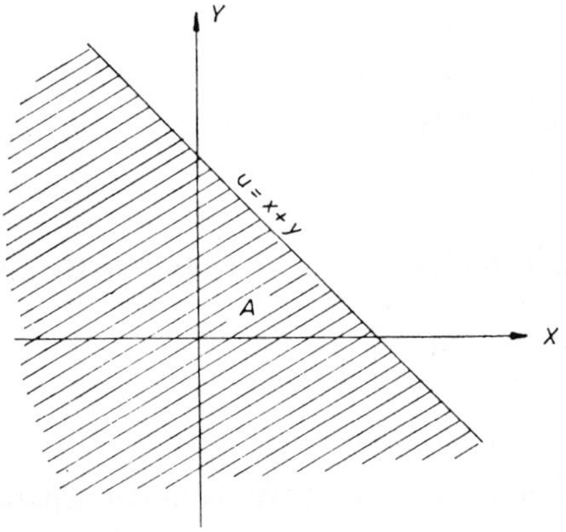

Fig. 5-13. Region of integration of eq. (13.4).

If x or y or both have a restricted range eq. (13.5) is still valid. However the limits of integration become more restricted. We discuss several cases

(a) $0 \leqslant x < \infty, \; -\infty < y < \infty$:

$$f(u) = \int_{-\infty}^{u} f_x(u - y) f_y(y) \, dy. \tag{13.6}$$

(Since $y = u - x$ and for $x_{\min} = 0$ one has $y_{\max} = u$.)

(b) $0 \leqslant x < \infty$, $0 \leqslant y < \infty$:

$$f(u) = \int_0^u f_x(u - y) f_y(y)\,\mathrm{d}y. \tag{13.7}$$

(c) $a \leqslant x < b$, $-\infty < y < \infty$:

$$f(u) = \int_a^b f_x(x) f_y(u - x)\,\mathrm{d}x. \tag{13.8}$$

Case (d) where both x and y vary between finite upper and lower limits is demonstrated using a specific example.

Example 5-9: *Convolution of uniform distributions*
With

$$f_x(x) = \begin{cases} 1, & 0 \leqslant x < 1 \\ 0 & \text{otherwise} \end{cases} \quad \text{and } f_y(y) = \begin{cases} 1, & 0 \leqslant y < 1 \\ 0 & \text{otherwise} \end{cases}$$

and (13.8) we have

$$f(u) = \int_0^1 f_y(u - x)\,\mathrm{d}x.$$

We use the substitution $v = u - x$, $\mathrm{d}v = -\mathrm{d}x$ and obtain

$$f(u) = -\int_u^{u-1} f_y(v)\,\mathrm{d}v = \int_{u-1}^u f_y(v)\,\mathrm{d}v. \tag{13.9}$$

Obviously we have $0 \leqslant u < 2$. We consider separately

(a) $0 \leqslant u < 1$: $$f_1(u) = \int_0^u f_y(v)\,\mathrm{d}v = \int_0^u \mathrm{d}v = u, \tag{13.10a}$$

(b) $1 \leqslant u < 2$: $$f_2(u) = \int_{u-1}^1 f_y(v)\,\mathrm{d}v = \int_{u-1}^1 \mathrm{d}v = 2 - u. \tag{13.10b}$$

Note that the lower (upper) limit of integration does not exceed 0 (1) since $f_y(y)$ vanishes for $y < 0$ and $y > 1$. The result is a triangular distribution (fig. 5-14).

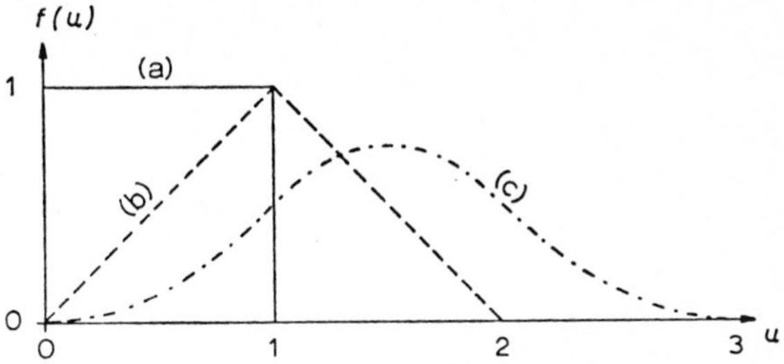

Fig. 5-14. Convolution of uniform distributions
(a) u, (b) $u+u$, (c) $u+u+u$.

If the result is folded again with the uniform distribution, i.e. if u is the sum of 3 independent uniformly distributed variables one obtains

$$f(u) = \begin{cases} \frac{1}{2}u^2, & 0 \leqslant u < 1 \\ \frac{1}{2}(-2u^2+6u-3), & 1 \leqslant u < 2 \\ \frac{1}{2}(u-3)^2, & 2 \leqslant u < 3 \end{cases} \tag{13.11}$$

The proof is left as exercise 5-7 to the reader. The distribution is composed of 3 sections of parabolae (fig. 5-14) and already looks similar to a Gaussian as predicted by the central limit theorem.

5-13.2. *Convolution with the normal distribution*

If a quantity x which is of interest in an experiment is a random variable having the probability density $f(x)$ and is measured with a measurement error y resulting from a Gaussian distribution with zero mean and variance σ^2 the result of the measurement is the sum

$$u = x + y. \tag{13.12}$$

Its probability density is (cf. eq. (13.4))

$$f(u) = \frac{1}{\sqrt{(2\pi)}\sigma} \int_{-\infty}^{\infty} f(x) \exp\left[-(u-x)^2/2\sigma^2\right] \mathrm{d}x. \tag{13.13}$$

By performing many measurements $f(u)$ can be determined experimentally. However $f(x)$ is the function in which the experimenter is interested. Unfortunately eq. (13.13) cannot in general be solved for $f(x)$. This is possible only for a restricted class of functions $f(u)$. Therefore usually the problem is tackled in a different way: From previous experiments or theory one has some knowledge about the form of $f(x)$, e.g. one might assume $f(x)$ to be the uniform distribution without knowing its boundaries a and b. One then performs the convolution (13.13), compares the resulting function $f(u)$ with the experiment and at the same time determines the missing parameters (a and b in our example).

In many cases it is not even possible to perform the integration (13.13) algebraically. Numerical procedures, e.g. the Monte Carlo method then have to be used. Sometimes approximations (cf. example 5-12) give useful results. Because of the importance of convolution with the normal distribution in many experiments we study some examples.

Example 5-10: *Convolution of uniform and normal distributions*

Using eqs. (5.2) and (13.8) and substituting $v = (x - u)/\sigma$ we obtain

$$f(u) = \frac{1}{b-a} \frac{1}{\sqrt{(2\pi)}\sigma} \int_a^b \exp[-(u-x)^2/2\sigma^2] \, dx$$

$$= \frac{1}{b-a} \frac{1}{\sqrt{(2\pi)}} \int_{(a-u)/\sigma}^{(b-u)/\sigma} \exp(-\tfrac{1}{2}v^2) \, dv,$$

$$f(u) = \frac{1}{b-a} \left\{ \psi_0\left(\frac{b-u}{\sigma}\right) - \psi_0\left(\frac{a-u}{\sigma}\right) \right\}, \tag{13.14}$$

the function ψ_0 having been defined in eq. (9.5). Fig. 5-15 shows the result for $a = 0$, $b = 6$, $\sigma = 1$. If $|b - a| \gg \sigma$ (as in the case of fig. 5-15) always one of the two terms in the bracket of eq. (13.14) is zero or 1. The rising flank of the uniform distribution at $u = a$ is then replaced by the distribution function of a normal distribution with

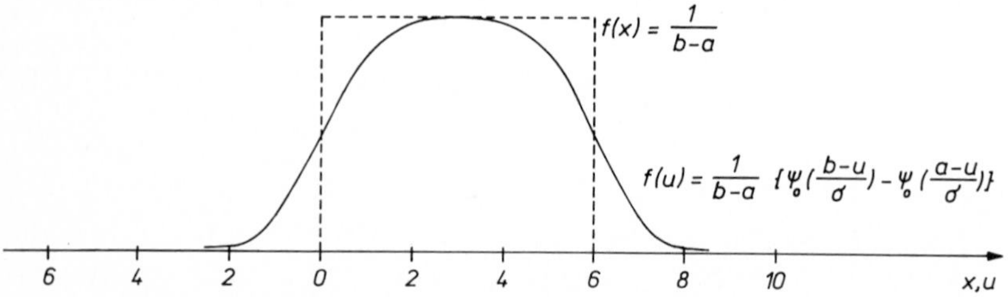

Fig. 5-15. Convolution of uniform and Gaussian distributions.

standard deviation σ (cf. fig. 5-7). The falling flank at $u = b$ is its "mirror image".

Example 5-11: *Convolution of two normal distributions*: "*Quadratic addition of errors*"

If one folds 2 normal distributions of zero mean and variances σ_x^2 and σ_y^2 respectively one obtains from (13.5)

$$f(u) = \frac{1}{\sqrt{(2\pi)}\sigma} \exp\left(-u^2/2\sigma^2\right), \quad \sigma^2 = \sigma_x^2 + \sigma_y^2. \tag{13.15}$$

A proof using the characteristic function was already indicated in § 8. It can also be obtained (cf. exercise 5-7(b)) by computation of the folding integral (13.5). If the distributions $f_x(x)$ and $f_y(y)$ describe two independent sources of measurement errors, the result (13.15) is known as the "*quadratic addition of errors*".

Example 5-12: *Convolution of exponential and normal distributions*

With

$$f(x) = \frac{1}{\tau} \exp\left(-x/\tau\right), \quad x > 0,$$

$$f(y) = \frac{1}{\sqrt{(2\pi)}\sigma} \exp\left(-y^2/2\sigma^2\right)$$

eq. (13.6) takes the form

$$f(u) = \frac{1}{\sqrt{(2\pi)}\sigma\tau} \int_{-\infty}^{u} \exp\left[-(u-y)/\tau\right] \exp\left(y^2/2\sigma^2\right) dy.$$

We can rewrite the exponent

$$-\frac{1}{2\sigma^2\tau}\left[2\sigma^2(u-y) + \tau y^2\right]$$

$$= -\frac{1}{2\sigma^2\tau}\left[2\sigma^2 u - 2\sigma^2 y + \tau y^2 + \frac{\sigma^4}{\tau} - \frac{\sigma^4}{\tau}\right]$$

$$= -\frac{u}{\tau} + \frac{\sigma^2}{2\tau^2} - \frac{1}{2\sigma^2}\left(y - \frac{\sigma^2}{\tau}\right)^2$$

and obtain

$$f(u) = \frac{1}{\sqrt{(2\pi)}\sigma\tau} \exp\left\{\frac{\sigma^2}{2\tau^2} - \frac{u}{\tau}\right\} \int_{-\infty}^{u-\sigma^2/\tau} \exp\left(\frac{v^2}{2\sigma^2}\right) dv.$$

Now we require $\sigma \ll \tau$, i.e. the error of measurement has to be much smaller than the characteristic quantity (width) of the exponential distributions. We also consider only values of u for which $u - \sigma^2/\tau \gg \sigma$, i.e. $u \gg \sigma$. In this case the integral is practically 1, or

$$f(u) \approx \frac{1}{\sqrt{(2\pi)}\sigma\tau} \exp\left\{-\frac{u}{\tau} + \frac{\sigma^2}{2\tau^2}\right\}.$$

In a semilogarithmic plot, i.e. if $\log f(u)$ is plotted against u, the curve $f(u)$ lies above $f(x)$ by the amount $\sigma^2/2\tau^2$ since

$$\ln f(u) = \ln\left(\frac{1}{\sqrt{(2\pi)}\sigma\tau}\right) + \frac{\sigma^2}{2\tau^2} - \frac{u}{\tau} = \ln f(x) + \frac{\sigma^2}{2\tau^2}.$$

This is sketched in fig. 5-16. The result can be easily understood quali-

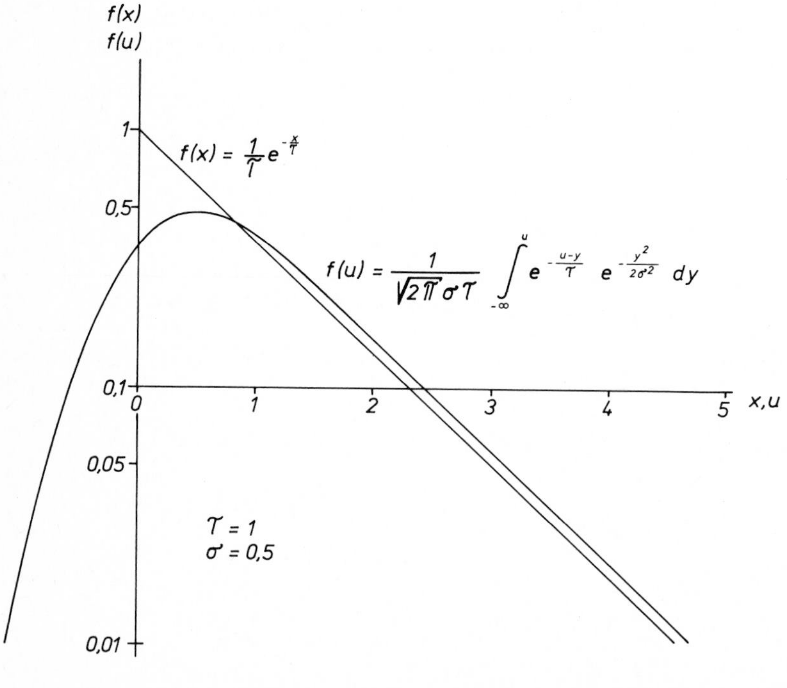

Fig. 5-16. Convolution of exponential and normal distributions.

tatively. For every small interval of x of the exponential distribution the convolution leads to a displacement to the right and left with equal probability. Since the exponential distribution is larger for small x-values at given u, contributions to the convolution $f(u)$ stem with higher probability from the left than from the right. This leads to a general right-shift of $f(u)$ with respect to $f(x)$.

Exercises

5-1: *Binomial distribution.*

(a) Prove the recursion relation

$$W_{k+1}^n = \frac{n-k}{k+1}\frac{p}{q} W_k^n .$$

(b) A certain production process is known to yield a fraction $q = 0.2$ defective objects. So, if 5 objects are produced the expected value of faultless objects is $np = n(1-q) = 5 \cdot 0.8 = 4$. However, what is the probability P_2 and P_3 that only 2 or 3 objects are without defects? Make use of the relation (a) to simplify the calculation.

(c) Determine the value k_m for which the binomial distribution assumes its maximum, i.e. k_m is the mode of the distribution. Hint: Since W_k^n is not a continuous function of k the mode, i.e. the maximum cannot be found by setting a derivative equal to zero. However the finite difference $W_k^n - W_{k-1}^n$ can of course be formed.

(d) In § 1 the binomial probability distribution is found by considering the random variable $\mathsf{x} = \sum_{i=1}^n \mathsf{x}_i$ where x_i is a random variable which can assume only the values 0 and 1 with probabilities $P(\mathsf{x}_i = 1) = p$ and $P(\mathsf{x}_i = 0) = 1 - p = q$ respectively. The binomial distribution

$$f(x) = f(k) = W_k^n = \binom{n}{k} p^k q^{n-k}$$

was then derived by considering in detail the probability of having k times $\mathsf{x}_i = 1$ in a total of n observations of the variable x_i. Derive the binomial distribution in a more formal way by setting up the characteristic function of the variable φ_{x_i} of the variable x_i. By raising it to the nth power the characteristic function of x is found. Hint: use the binomial theorem, eq. (C-6).

5-2: *Poisson distribution*

(a) In a given hospital the expected number of calls on the doctor on duty at night is 3. The number of calls is assumed to be Poisson distributed. What is the probability of the doctor having an undisturbed night?

(b) A random variable x has a probability density $f(x)$. A total of n experiments is made in which x is measured and the number k of times is recorded in which x falls into a small interval $x \leqslant \mathsf{x} < x + \Delta x$. Show that k follows a binomial distribution and determine its parameter p. Show that for $n \to \infty$ but $n/\Delta x =$ const. k follows the Poisson distribution and determine its parameter λ.

(c) For a radioactive nucleus with "mean life" τ which exists at the time $x = 0$, the probability to decay in the time interval $x \leqslant \mathsf{x} < x + \Delta x$ is $f(x)\Delta x = \tau^{-1} e^{-x/\tau} \Delta x$, $x > 0$. Here it is assumed that $\Delta x \ll \tau$. Therefore if a source contains n nuclei at $x = 0$ according to (b) the probability of observing k decays in the interval Δx is Poisson distributed with a given parameter λ. Determine λ as a function of x.

5-3: *Uniform distribution*

Using the method of § 5.3 show how from random numbers x, distributed uniformly between 0 and 1 random numbers y are obtained: (a) following a uniform distribution in the interval $a \leqslant y < b$, (b) following the Cauchy distribution of exercise 3-3.

5-4: *Normal distribution*

The electric resistance R of resistors produced by a certain machine is described by a normal distribution of mean R_0 and standard deviation σ. The production cost for one resistor is C, the sales price $5C$ if $R = R_0 \pm \Delta_1$ and $2C$ if $R = R_0 \pm \Delta_2$. Resistors outside the latter range cannot be sold.

(a) Determine the profit P per produced resistor for $R_m = R_0$, $\Delta_1 = a_1 R_0$, $\Delta_2 = a_2 R_0$, $\sigma = bR_0$. Write the answer using the distribution function ϕ_0.

(b) Use table F-2a to give a numerical value for P if $a_1 = 0.01$, $a_2 = 0.05$, $b = 0.05$.

(c) Show that the probability density eq. (8.1) has points of inflection at $x = a \pm b$.

5-5: *Multivariate normal distribution*

Shooting at a target plane may yield independent normal distributions in x and y with mean 0 and variance σ^2, i.e. $f(x, y)$ is the bivariate normal distribution of exercise 4-2(e). Using the result of that exercise determine (a) the probability $P(R)$ to observe a shot within a given radius R about the origin, (b) the radius R within which a shot is found with a given probability. As a numerical example compute R for $P = 90\%$ and $\sigma = 1$.

5-6: *Central limit theorem*

(a) Write a FORTRAN program which generates random numbers following practically a normal distribution with mean 0 and variance b by forming a sum of N random numbers picked from a uniform distribution. Use the subroutine RAND (program 5-1). Show that the program becomes particularly simple if $N = 12$.

(b) Compare the efficiency of this program with that of program 5-2, i.e. the number of times a "uniformly" random number has to be picked for every "Gaussian" random number.

(c) Prove that for $\lambda \to \infty$ the Poisson distribution approaches the normal distribution with mean λ and variance λ. Hint: Assume λ to be integer. Construct the Poisson distribution with parameter λ as a sum of Poisson distributions with $\lambda = 1$. Then apply the central limit theorem to this sum.

5-7: *Convolution*

(a) Prove eq. (13.11). Begin with eq. (13.9) and use the expressions (13.10) for $f_y(y)$. While in the range $0 \leqslant u < 1$ $[2 \leqslant u < 3]$ only eq. (13.10a) [eq. (13.10b)] applies since always $y < 1$ $(y > 1)$; in the range $1 \leqslant u < 2$ the resulting distribution $f(u)$ has to be composed as a sum of two integrals of the type (13.9) each containing one of the two forms of $f_y(y)$. Special care has to be taken in this case to determine the limits of integration. They are given by the range of u and $f_y(y)$.

(b) Prove eq. (13.15) by performing the integral (13.5) for f_x and f_y Gaussian with mean zero and standard deviation σ_x and σ_y, respectively.

5.8: *Full width at half maximum (FWHM)*

(a) A generalization of the Cauchy distribution (exercises 3-3 and 4-1) is used to describe spectral lines in physics. The function

$$f(x) = \frac{2}{\pi\Gamma}\,\frac{\Gamma^2}{4(x-a)^2+\Gamma^2}$$

is the probability density of the Lorentz distribution (also called the Breit–Wigner distribution). For $a = 0$ and $\frac{1}{2}\Gamma = 1$ it is identical to the Cauchy distribution. Show that Γ describes the "full width at half maximum" (FWHM) of the distribution, i.e. $\Gamma = |x_2 - x_1|$, if $f(x_1) = f(x_2) = \frac{1}{2}f(a)$.

(b) The FWHM is easily obtained graphically from an empirical distribution and often quoted as a measure of the width. Compute it for a normal distribution and compare it to the standard deviation σ.

CHAPTER 6

SAMPLING

In the last chapter we have discussed a number of distributions but we have not specified how they are realized in a particular case. We have only stated the probability that a random variable x will lie within an interval $x < \mathsf{x} \leqslant x + \mathrm{d}x$. Now this probability depends on certain parameters describing its distribution (like λ in the case of the Poisson distribution) which are usually unknown. We therefore have no direct knowledge of the probability distribution and have to approximate it by a *frequency distribution* obtained experimentally. The number of experiments performed for this purpose, called a *sample*, is necessarily finite. To discuss the elements of the theory of sampling we first have to introduce a number of new definitions.

6-1. Random sampling; distribution of a sample; estimates

Each sample is obtained from a set of elements. This set is usually of infinite size. It is composed, e.g., of all conceivable experiments or events of a given type. Such a set is called a *population*. If a sample of n elements is drawn, the sample is said to be of *size n*. Let the distribution of the random variable x in the population be described by the probability density $f(x)$. We are now interested in the values of x taken by the elements of the sample. Let us suppose that a total of l samples of size n are drawn consecutively and the following values of x are observed:

1st sample: $\mathsf{x}_1^{(1)}, \mathsf{x}_2^{(1)}, \ldots, \mathsf{x}_n^{(1)}$,

⋮

jth sample: $\mathsf{x}_1^{(j)}, \mathsf{x}_2^{(j)}, \ldots, \mathsf{x}_n^{(j)}$,

⋮

lth sample: $\mathsf{x}_1^{(l)}, \mathsf{x}_2^{(l)}, \ldots, \mathsf{x}_n^{(l)}$.

We group the result of one sample into an n-dimensional random variable

$$\mathbf{x}^{(j)} = (\mathsf{x}_1^{(j)}, \mathsf{x}_2^{(j)}, \ldots, \mathsf{x}_n^{(j)}), \tag{1.1}$$

which can be considered the position vector in an n-dimensional sample space (ch. 2, §1). It is distributed according to a probability density

$$g(\mathbf{x}) = g(x_1, x_2, \ldots, x_n). \tag{1.2}$$

This probability density has to fulfil two conditions in order to describe the process of *random sampling*.

(a) The various x_i have to be independent, i.e.

$$g(\mathbf{x}) = g_1(x_1)\, g_2(x_2) \ldots g_n(x_n). \tag{1.3}$$

(b) The individual distributions have to be equal and identical to the density distribution of the population

$$g_1(x_1) = g_2(x_2) = \;..\; = g_n(x_n) = f(x). \tag{1.4}$$

Comparing these conditions with (1.2) it is obvious that there is a simple relation between population and sample only if they are fulfilled. If the contrary is not explicitly stated we shall use the word sample to signify the result of random sampling.

It must be stressed that in the actual process of sampling it is often quite difficult to maintain randomness. Because of the wide variation of applications no general recipe can be given to ensure it. In order to obtain reliable results from sampling, we have to take the utmost precautions to fulfil the conditions (1.3) and (1.4). While the independence (1.3) can to a certain extend be controlled by comparing the experimental distributions for the 1st, 2nd, ... elements obtained in a large number of samples, it is not at all easy to be sure that the sample is in fact drawn from a population described by the density $f(x)$. If the elements of the population can be numbered in some way it is sometimes advantageous to use a table of random numbers (table F-8), to select elements for the sample.

We now assume that the n elements of the sample have been ordered according the values of x which may have been marked on an x-axis and we define n_x to be the number of elements for which $\mathsf{x} < x$. The function

$$W_n(x) = n_x/n \tag{1.5}$$

then takes the role of an empirical distribution function. It is a step function increasing by $1/n$, whenever x equals the value x taken by an element of the sample. We call it *distribution of the sample*. Apparently $W_n(x)$ is an approximation for $F(x)$, the distribution function of the population, which it approaches for $n \to \infty$.

A function of a sample (1.1), i.e. a function of the random variable $\mathbf{x}$ that is itself a random variable, is called a *statistic*. A well known example is the *sample mean*

$$\bar{x} = \frac{1}{n}(x_1 + x_2 + \dots + x_n), \tag{1.6}$$

defined as the arithmetic mean of the x_i.

A typical problem of data analysis is the following. The general mathematical form of the probability density of the population is known. Only the numerical value of one (or several) parameters has still to be obtained from a sample. In radioactive decay for example the number of nuclei decaying between $t = 0$ and $t = \tau$ is

$$N_\tau = N_0 \exp(-\lambda\tau),$$

if N_0 nuclei exist at $t = 0$. By means of a finite sample, i.e. by recording the times of a finite number of decays, one tries to obtain the parameter λ. Since the sample is finite the result cannot be exact. We are therefore dealing with the problem of an *estimation of parameters.*

Since the estimated value is obtained by means of sampling it is a statistic, which we call an *estimator*

$$S = S(x_1, x_2, \dots, x_n). \tag{1.7}$$

Sometimes we use the term statistic by itself ("the statistic S serves to estimate the parameter λ"). An estimator is *unbiased* if for any size of the sample its expectation value is equal to the parameter to be estimated

$$E\{S(x_1, x_2. \dots, x_n)\} = \lambda \quad \text{for any } n. \tag{1.8}$$

An estimate is called *consistent* if the result becomes increasingly accurate as the size of the sample increases, i.e. if

$$\lim_{n\to\infty} \sigma(S) = 0. \tag{1.9}$$

Not all estimators are equally suitable. To compare the relative efficiency of two estimators the quotient

$$\eta = \frac{E\{(S_1 - \lambda)^2\}}{E\{(S_2 - \lambda)^2\}} = \frac{\sigma^2(S_1)}{\sigma^2(S_2)}$$

can be used. Sometimes one can state a lower limit for the variance of any estimator of a parameter (ch. 7, §3). If an estimator exists having this minimum variance it should of course be preferred to all others.

In the following sections we shall discuss in some detail sampling from various types of populations.

6-2. Sampling from continuous populations

The case of widest interest in applications concerns sampling from continuous populations described by a probability density $f(x)$. Let us first consider the *sample mean* defined already in eq. (1.6) to be the arithmetic mean of all values of x in the sample

$$\bar{x} = \frac{1}{n}(x_1 + x_2 + \ldots + x_n). \tag{2.1}$$

Being a function of random variables it is itself a random variable. Its expectation value is

$$E(\bar{x}) = \frac{1}{n}\{E(x_1) + E(x_2) + \ldots + E(x_n)\} = \hat{x}, \tag{2.2}$$

i.e. equal to the expectation value of x. Since this is true for any n, the sample mean is an unbiased estimator of the expectation value of x in the population, the *population mean*. The characteristic function of the sample mean (which we shall need later) is according to (2.1)

$$\varphi_{\bar{x}}(t) = \frac{1}{n}\{\varphi_x(t)\}^n = \left\{\varphi_x\left(\frac{t}{n}\right)\right\}^n. \tag{2.3}$$

Next we are interested in the variance of $\bar{x}$:

$$\sigma^2(\bar{x}) = E\{(\bar{x} - E(\bar{x}))^2\} = E\left\{\left(\frac{x_1 + x_2 + \ldots + x_n}{n} - \hat{x}\right)^2\right\}$$

$$= \frac{1}{n^2} E\{((x_1 - \hat{x}) + (x_2 - \hat{x}) + \ldots + (x_n - \hat{x}))^2\}.$$

Since all x_i are independent all mixed terms of the type

$$E\{(x_i - \hat{x})(x_j - \hat{x})\},$$

i.e. the covariances, vanish and we are left with

$$\sigma^2(\bar{x}) = \frac{1}{n}\sigma^2(x). \tag{2.4}$$

The sample mean is therefore a consistent estimate of $\hat{x}$. We now have to find an estimator for the variance of the population. As a first approximation we define the *sample variance* as the arithmetic mean of the quadratic deviations from the sample mean

$$s'^2 = \frac{1}{n}\{(x_1 - \bar{x})^2 + (x_2 - \bar{x})^2 + \dots + (x_n - \bar{x})^2\}. \tag{2.5}$$

This quantity has the expectation value

$$E(s'^2) = \frac{1}{n} E\left\{\sum_{i=1}^{n} (x_i - \bar{x})^2\right\}$$

$$= \frac{1}{n} E\left\{\sum_{i=1}^{n} (x_i - \hat{x} + \hat{x} - \bar{x})^2\right\}$$

$$= \frac{1}{n} \sum_{i=1}^{n} \{E((x_i - \hat{x})^2) - E((\bar{x} - \hat{x})^2)\}$$

$$= \frac{1}{n}\left\{n\sigma^2(x) - n\left(\frac{1}{n}\sigma^2(x)\right)\right\},$$

$$E(s'^2) = \frac{n-1}{n}\sigma^2(x). \tag{2.6}$$

We see that s'^2 is not an unbiased estimator of $\sigma^2(x)$. We can, however, change the definition (2.5) to

$$s^2 = \frac{1}{n-1}\{(x_1 - \bar{x})^2 + (x_2 - \bar{x})^2 + \dots + (x_n - \bar{x})^2\}, \tag{2.7}$$

which is unbiased.

The expression $(n - 1)$ in the denominator, although mathematically justified, seems somewhat strange at first sight. Note however that for $n = 1$ the sample mean is equal to the x value of the only element of the sample ($\bar{x} = x$) and therefore the sample variance if defined according to eq. (2.5) would vanish. The reason for this is that in (2.5) – and also in (2.7) – the sample mean $\bar{x}$ has to be used instead of the population mean $\hat{x}$ since the latter is unknown. Some of the information contained in the sample has first to be used to derive the sample mean. This is lost for the calculation of the sample variance. The effective number of elements in the sample is thus reduced. This fact is taken into account by reducing the denominator of the arithmetic mean (2.5). The argument is repeated quantitatively in §4.

Example 6-1: *Computation of sample mean and variance from given data*

A quantity (e.g. the length of an object) has been measured $n = 7$ times. The results are

10.5, 10.9, 9.2, 9.8, 9.0, 10.4, 10.7.

It simplifies the calculation of eq. (2.1) to note that all values are in the neighbourhood of $a = 10$, i.e. $x_i = a + \delta_i$. Then eq. (2.1) yields

$$\overline{x} = \frac{1}{n}\sum_{i=1}^{n} x_i = \frac{1}{n}\sum_{i=1}^{n} (a + \delta_i) = a + \frac{1}{n}\sum_{i=1}^{n} \delta_i = a + \Delta$$

with

$$\Delta = \frac{1}{n}\sum_{i=1}^{n} \delta_i = \tfrac{1}{7}(0.5 + 0.9 - 0.8 - 0.2 - 1.0 + 0.4 + 0.7)$$

$$= 0.5/7 = 0.07.$$

Therefore $\overline{x} = 10 + \Delta = 10.07$.

To compute the sample variance we use eq. (2.7)

$$s^2 = \frac{1}{n-1}\sum_{i=1}^{n} (x_i - \overline{x})^2$$

$$= \frac{1}{n-1}\sum_{i=1}^{n} (x_i^2 - 2x_i\,\overline{x} + \overline{x}^{\,2})$$

$$= \frac{1}{n-1}\left\{\sum_{i=1}^{n} x_i^2 - n\overline{x}^{\,2}\right\}$$

either in the form of the first or the last line above. The last line contains only 1 instead of n differences. However the numbers to be squared are usually much larger. Therefore this form is suitable only if at least a desk calculator is available. For hand calculations we use the original form

$$s^2 = \tfrac{1}{6}\{0.43^2 + 0.83^2 + 0.87^2 + 0.27^2 + 1.07^2 + 0.33^2 + 0.63^2\}$$

$$= \tfrac{1}{6}\{0.1849 + 0.6889 + 0.7569 + 0.0729 + 1.1449 + 0.1089 + 0.3969\}$$

$$= 3.3543/6 \approx 0.56.$$

The sample standard deviation is

$$s \approx 0.75.$$

6-3. Sampling from partitioned distributions

Often it is advantageous to divide a population G (e.g. all students in Europe) into different subpopulations $G_1, G_2, \ldots, G_t$ (students at the universities

1, 2, ..., t). The quantity x may be described by the probability densities $f_1(x), f_2(x), \ldots, f_t(x)$ in the partial populations. The corresponding distribution functions are

$$F_i(x) = \int_{-\infty}^{x} f_i(x)\,dx = P(\mathsf{x} < x \mid \mathsf{x} \in G_i). \tag{3.1}$$

They have the character of conditional probabilities, since each one is only defined under the condition that the element in question is part of a particular subpopulation. We now make use of the rule of total probability (2-3.4) to obtain the distribution function of the overall population G

$$F(x) = P(\mathsf{x} < x \mid \mathsf{x} \in G) = \sum_{i=1}^{t} P(\mathsf{x} < x \mid \mathsf{x} \in G_i)\, P(\mathsf{x} \in G_i),$$

i.e.

$$F(x) = \sum_{i=1}^{t} P(\mathsf{x} \in G_i)\, F_i(x). \tag{3.2}$$

Correspondingly we have the probability density

$$f(x) = \sum_{i=1}^{t} P(\mathsf{x} \in G_i) f_i(x). \tag{3.3}$$

Abbreviating $P(\mathsf{x} \in G_i)$ by p_i we can now express the population mean of G as

$$\hat{x} = E(\mathsf{x}) = \int_{-\infty}^{\infty} x f(x)\,dx = \int_{-\infty}^{\infty} \sum_{i=1}^{t} x p_i f_i(x)\,dx = \sum_{i=1}^{t} p_i \int_{-\infty}^{\infty} x f_i(x)\,dx,$$

i.e.

$$\hat{x} = \sum_{i=1}^{t} p_i\, \hat{x}_i. \tag{3.4}$$

The expectation value of the population is the mean of the expectation values of the subpopulations, each weighted by its individual probability. For the variance of the population we find that

$$\sigma^2(\mathsf{x}) = \int_{-\infty}^{\infty} (x - \hat{x})^2 f(x)\,dx$$

$$= \int_{-\infty}^{\infty} (x - \hat{x})^2 \sum_{i=1}^{t} p_i f_i(x)\,dx$$

$$= \sum_{i=1}^{t} p_i \int_{-\infty}^{\infty} \{(x - \hat{x}_i) + (\hat{x}_i - \hat{x})\}^2 f_i(x)\,dx.$$

Since the individual x_i are independent there will be no covariance terms, so that

$$\sigma^2(x) = \sum_{i=1}^{t} p_i \left\{ \int_{-\infty}^{\infty} (x - \hat{x}_i)^2 f_i(x)\,dx + (\hat{x}_i - \hat{x})^2 \int_{-\infty}^{\infty} f_i(x)\,dx \right\},$$

$$\sigma^2(x) = \sum_{i=1}^{t} p_i\{\sigma_i^2 + (\hat{x}_i - \hat{x})^2\}. \tag{3.5}$$

This is the weighted mean of a sum of two terms. The first is the variance of the subpopulation, the second the variance of the subpopulation mean about the population mean.

We now draw from each subpopulation a sample of size n_i, with

$$\sum_{i=1}^{t} n_i = n.$$

The arithmetic mean of the total partitioned sample is

$$\bar{x}_p = \frac{1}{n} \sum_{i=1}^{t} \sum_{j=1}^{n_i} x_{ij} = \frac{1}{n} \sum_{i=1}^{t} n_i \bar{x}_i. \tag{3.6}$$

It has the expectation value

$$E(\bar{x}_p) = \frac{1}{n} \sum_{i=1}^{n} n_i \hat{x}_i, \tag{3.7}$$

and the variance

$$\sigma^2(\bar{x}_p) = E\{(\bar{x}_p - E(\bar{x}_p))^2\}$$

$$= E\left\{\left(\sum_{i=1}^{t} \frac{n_i}{n} (\bar{x}_i - \hat{x}_i)\right)^2\right\}$$

$$= \frac{1}{n^2} \sum_{i=1}^{t} n_i^2\, E\{(\bar{x}_i - \hat{x}_i)^2\},$$

$$\sigma^2(\bar{x}_p) = \frac{1}{n^2} \sum_{i=1}^{t} n_i^2 \, \sigma^2(\bar{x}_i). \tag{3.8}$$

With relation (2.4) this becomes

$$\sigma^2(\bar{x}_p) = \frac{1}{n} \sum_{i=1}^{t} \frac{n_i}{n} \, \sigma_i^2. \tag{3.9}$$

The same result can be obtained by applying the law of the propagation of errors (eq. (4-5.7)) to eq. (3.6).

It is intuitively clear that the arithmetic mean $\bar{x}_p$ cannot generally be an estimator for the population mean $\hat{x}$ since it depends on the arbitrary choice of the size n_i of the partial samples. Comparison of (3.7) with (3.4) shows that this is the case only if $p_i = n_i/n$.

The population mean $\hat{x}$ can be estimated by determining first the sample means $\bar{x}_i$ of the subpopulations and then forming the expression

$$\tilde{x} = \sum_{i=1}^{t} p_i \, \bar{x}_i, \tag{3.10}$$

according to eq. (3.4). By error propagation we get for the variance of $\tilde{x}$

$$\sigma^2(\tilde{x}) = \sum_{i=1}^{t} p_i^2 \, \sigma^2(\bar{x}_i) = \sum_{i=1}^{t} \frac{p_i^2}{n_i} \, \sigma_i^2. \tag{3.11}$$

Example 6-2: *Optimal choice of sample size from subpopulations*
In order to minimize $\sigma^2(\tilde{x})$ we cannot simply differentiate (3.11) with respect to all n_i since the n_i are subject to the constraint condition

$$\sum_{i=1}^{t} n_i - n = 0. \tag{3.12}$$

We therefore have to use the method of *Lagrangian multipliers* by adding eq. (3.12) multiplied by a factor μ to eq. (3.11) and setting the partial differentials with respect to the n_i and the Lagrangian multiplier μ equal to zero:

$$L = \sigma^2(\tilde{x}) - \mu(\sum n_i - n) = \sum (p_i^2/n_i)\sigma_i^2 - \mu(\sum n_i - n),$$

$$\frac{\partial L}{\partial n_i} = -\frac{p_i^2 \sigma_i^2}{n_i^2} + \mu = 0, \tag{3.13}$$

$$\frac{\partial L}{\partial \mu} = \sum n_i - n = 0. \tag{3.14}$$

From (3.13) we obtain

$$n_i = p_i \sigma_i / \sqrt{\mu}$$

which together with (3.14) yields

$$1/\sqrt{\mu} = n/\sum p_i \sigma_i$$

and therefore

$$n_i = n p_i \sigma_i / \sum p_i \sigma_i. \tag{3.15}$$

The result (3.15) states that the subpopulation size n_i has to be chosen proportional to the probability p_i that an element of the total population is a member of the subpopulation i and proportional to the standard deviation of that subpopulation.

As a practical example we assume that a scientific publishing company wants to estimate the money spent yearly on scientific books by the two subpopulations (1) students and (2) scientific libraries. Let us further assume that there are 1000 libraries and 10^6 students in the population and that the standard deviation of student expenditure is 100 \$ whereas for the libraries (which are rather different) it is 3×10^5 \$. We then have

$$p_1 \approx 1, \quad p_2 \approx 10^{-3}, \quad \sigma_1 = 100, \quad \sigma_2 = 3 \times 10^5$$

and from (3.15)

$$n_1 = \text{const.} \cdot 100, \quad n_2 = \text{const.} \cdot 300, \quad \text{i.e. } n_2 = 3n_1.$$

Note that the result is not related to the total expenditure of the subpopulations. The p_i and σ_i will in general not be known. They have to be estimated by preliminary sampling.

The discussion of partitioned populations is continued in ch. 11.

6-4. Sampling without replacement from finite populations; mean square deviation; degrees of freedom

We first encountered the concept of sampling while discussing the hypergeometric distribution in ch. 5, §3. We mentioned then that the independence of consecutive drawings was lost because the individual elements were not replaced into the finite (which implies: discrete) population. We are, therefore, no longer dealing with genuine random sampling.

To discuss this further, let the population be composed of N elements $y_1, y_2, \ldots, y_N$. From it, let a sample of size n be drawn having the elements $x_1, x_2, \ldots, x_n$. (In the case of the hypergeometric distributions the y_j and with them the x_i could take only the values 0 and 1.) Since each element of the population has equal probability of being drawn, the population mean is

$$E(y) = \hat{y} = \bar{y} = \frac{1}{N} \sum_{j=1}^{N} y_j. \tag{4.1}$$

Although not a random variable, $\hat{y}$ is an arithmetic mean of the finite number of elements of the population. A definition of a population variance therefore meets with the difficulties discussed at the end of §2. We define according to eq. (2.7)

$$\begin{aligned} \sigma^2(y) &= \frac{1}{N-1} \sum_{j=1}^{N} (y_j - \bar{y})^2 \\ &= \frac{1}{N-1} \left\{ \sum_{j=1}^{N} y_j^2 - \frac{1}{N} \left(\sum_{j=1}^{N} y_j \right)^2 \right\}. \end{aligned} \tag{4.2}$$

Consider now the *sum of squares*

$$\sum_{j=1}^{N} (y_j - \bar{y})^2. \tag{4.3}$$

As we have not restricted the population in any way the y_j can take any value. Therefore also the first term of the sum (4.3) can take any value. The same is true for the 2nd, 3rd, ..., $(N-1)$th terms. The Nth term however is then determined since

$$\sum_{j=1}^{N} (y_j - \bar{y}) = 0. \tag{4.4}$$

This situation is expressed by saying that the *number of degrees of freedom* of the sum of squares (4.3) is $(N-1)$. It can be illustrated geometrically.

We consider the simple case $\bar{y} = 0$ and span an N-dimensional vector space by the y_j. The sum of squares is then the square of the modulus of the position vector in that space. Because of the constraint equation (4.4) the point of that vector can only move in a space of dimension $(N-1)$. The dimension of such a restricted space is called the number of degrees of freedom in mechanics. The case $N = 2$ is sketched in fig. 6-1. Here the position vector is constrained to the straight line $y_2 = -y_1$. A sum of squares (e.g. expression (4.3)) divided by the number of degrees of freedom is called the *mean square* or more explicitly the *mean-square deviation.* This definition does not only apply to populations but also to samples. In the latter case it is used to characterize the variation of individual measurements. Its square root (which has the dimension of the measured quantity, i.e. the mean) has the lengthy name *root-mean-square deviation* frequently abbreviated to *RMS deviation* in the experimental literature.

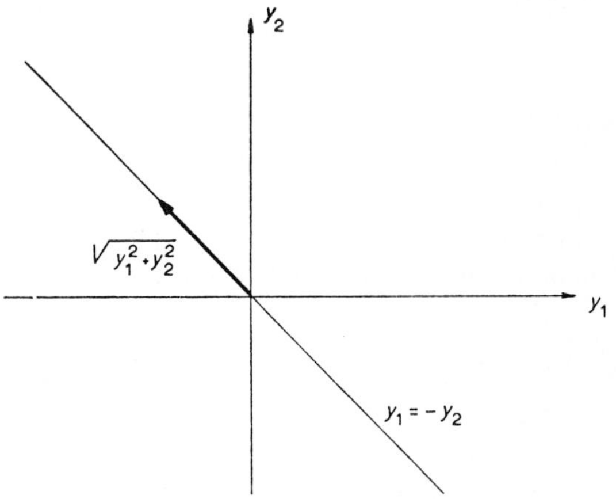

Fig. 6-1. Sample of size 2 yields sum of squares with 1 degree of freedom.

We now turn to the sample $x_1, x_2, \ldots, x_n$. It will facilitate our discussion, if we introduce a Kronecker delta describing the selection process of the sample. We define

$$\delta_i^j = \begin{cases} 1 & \text{if } x_i \text{ is the element } y_j \text{ of the population,} \\ 0 & \text{otherwise.} \end{cases} \tag{4.5}$$

This means

$$x_i = \sum_{j=1}^{N} \delta_i^j y_j. \tag{4.6}$$

Since each y_j has the same probability of being the ith element of the sample, we have

$$P(\delta_i^j = 1) = 1/N. \tag{4.7}$$

Describing a random process δ_i^j is of course a random variable. Its expectation value is obtained readily from eq. (3-3.2) (where $n = 2$, $x_1 = 0$, $x_2 = 1$)

$$E(\delta_i^j) = P(\delta_i^j = 1) = 1/N. \tag{4.8}$$

If one element of the sample is known, say x_i, then for another element the number of possible choices from the population is $(N - 1)$. Therefore

$$P(\delta_i^j \delta_k^l = 1) = \frac{1}{N} \frac{1}{N-1} = E(\delta_i^j \delta_k^l). \tag{4.9}$$

Since the sampling is performed without replacement the same element cannot be sampled twice, i.e. $j \neq l$ or

$$\delta_i^j \delta_k^j = 0. \tag{4.10}$$

Correspondingly we have

$$\delta_i^j \delta_i^l = 0, \tag{4.11}$$

because one element of the sample cannot be represented by two different elements of the population. The expectation value of the first element of the sample x_1 is

$$E(x_1) = E\left\{\sum_{j=1}^{N} \delta_1^j y_j\right\} = \sum_{j=1}^{N} y_j E(\delta_1^j) = \frac{1}{N} \sum_{j=1}^{N} y_j = \bar{y}. \tag{4.12}$$

Now, since there is nothing distinct about x_1, the expectation value of all elements of the sample takes the same value and so does their arithmetic mean

$$E(\bar{x}) = \frac{1}{n} \sum_{i=1}^{n} E(x_i) = \bar{y}, \tag{4.13}$$

which therefore provides an unbiased estimator for the population mean.

To derive the expectation value of the sample variance

$$s_x^2 = \frac{1}{n-1} \sum_{i=1}^{n} (x_i - \bar{x})^2, \tag{4.14}$$

we make use of the identity

$$\sum_{i=1}^{n} x_i^2 = n\,\bar{x}^2 + \sum_{i=1}^{n} (x_i - \bar{x})^2, \tag{4.15}$$

and determine the expectation values of Σx_i^2 and $n\bar{x}^2$. Because of (4.6) we have

$$x_i^2 = \sum_{j=1}^{N} \delta_i^j\, y_j^2,$$

and therefore

$$E(x_1^2) = \frac{1}{N} \sum_{j=1}^{N} y_j^2,$$

which, by making use of (4.2), becomes

$$E(x_1^2) = \frac{1}{N} \{(N-1)\sigma^2(y) + N\bar{y}^2\} = \bar{y}^2 + \left(1 - \frac{1}{N}\right)\sigma^2(y).$$

Again of course x_1 is in no way distinct, i.e.

$$E\left\{\sum_{i=1}^{n} x_i^2\right\} = nE(x_1^2) = n\left\{\bar{y}^2 + \left(1 - \frac{1}{N}\right)\sigma^2(y)\right\}. \tag{4.16}$$

In order to determine the expectation value of $n\bar{x}^2$ we first consider

$$E(x_1\,x_2) = E\Big(\sum_{j\neq l} \delta_1^j \delta_2^l\, y_j\, y_l\Big)$$

$$= \sum_{j\neq l} E\{\delta_1^j \delta_2^l\, y_j\, y_l\}.$$

Because of (4.9) this becomes

$$E(x_1\,x_2) = \frac{1}{N(N-1)} \sum_{j\neq l} y_j\, y_l.$$

The sum extends over all mixed products and contains each combination twice ($y_j\, y_l$ and $y_l\, y_j$). We can write it as

$$E(x_1\,x_2) = \frac{1}{N(N-1)} \left\{\Big(\sum_{k=1}^{N} y_k\Big)^2 - \sum_{k=1}^{N} y_k^2\right\}$$

$$= \frac{1}{N(N-1)} \{(N\bar{y})^2 - ((N-1)\sigma^2(y) + N\bar{y}^2)\}$$

$$= \bar{y}^2 - \frac{\sigma^2(y)}{N}.$$

Using symmetry again this is the result for $E(x_i\, x_k)$ for any i, k with $i \neq k$. Then

$$E\left(\sum_{i \neq k} x_i\, x_k\right) = n(n-1)\left\{\bar{y}^2 - \frac{\sigma^2(y)}{N}\right\}, \tag{4.17}$$

and finally

$$E(n\,\bar{x}^2) = n\,E(\bar{x}^2) = \frac{1}{n}\,E\left\{\left(\sum_{i=1}^{n} x_i\right)^2\right\}$$

$$= \frac{1}{n}\,E\left\{\sum_{i=1}^{n} x_i^2 + \sum_{i \neq k} x_i\, x_k\right\}$$

$$= \frac{1}{n}\,E\left(\sum_{i=1}^{n} x_i^2\right) + \frac{1}{n}\,E\left(\sum_{i \neq k} x_i\, x_k\right).$$

Substituting (4.16) and (4.17) we find that

$$E(n\,\bar{x}^2) = \bar{y}^2 + \left(1 - \frac{1}{N}\right)\sigma^2(y) + (n-1)\,\bar{y} - \frac{n-1}{N}\sigma^2(y)$$

$$= \left(1 - \frac{n}{N}\right)\sigma^2(y) + n\,\bar{y}^2. \tag{4.18}$$

Using (4.18) and (4.16) we can now determine the expectation value of the sample variance (4.14)

$$E(s_x^2) = \frac{1}{n-1}\,E\left\{\sum_{i=1}^{n}(x_i - \bar{x})^2\right\} = \frac{1}{n-1}\left\{E\left(\sum_{i=1}^{n} x_i^2\right) - E(n\,\bar{x}^2)\right\}$$

$$= \frac{1}{n-1}\left\{n\left[\bar{y}^2 + \left(1 - \frac{1}{N}\right)\sigma^2(y)\right] - \left(1 - \frac{n}{N}\right)\sigma^2(y) - n\bar{y}^2\right\}$$

$$= \frac{1}{n-1}\left\{n - \frac{n}{N} - 1 + \frac{n}{N}\right\}\sigma^2(y)$$

$$E(s_x^2) = \sigma^2(y). \tag{4.19}$$

It is an unbiased estimate of the population variance.

Let us consider finally the *variance of the sample mean*

$$\sigma^2(\bar{x}) = E\{(\bar{x} - E(\bar{x}))^2\}.$$

Since $E(\bar{x}) = \bar{y}$ is not a random variable but exactly defined, the variance becomes

$$\sigma^2(\bar{x}) = E(\bar{x}^2) - \bar{y}^2 = \frac{1}{n}\left\{\left(1 - \frac{n}{N}\right)\sigma^2(y) + n\bar{y}^2\right\} - \bar{y}^2,$$

$$\sigma^2(\bar{x}) = \frac{\sigma^2(y)}{n}\left(1 - \frac{n}{N}\right). \tag{4.20}$$

Comparison with the case of an infinite population (eq. (2.4)) reveals an additional factor $(1 - n/N)$. It takes care of the fact that for $n = N$, when the sample mean and population mean become identical, the variance has to vanish.

6-5. Sampling from normal distributions; χ^2-distribution

In this section we return to continuous populations and choose especially a population described by a Gaussian distribution with mean a and variance σ^2. According to eq. (5-8.4) the characteristic function of such a Gaussian distribution is

$$\varphi_x(t) = \exp(ita)\exp(-\tfrac{1}{2}\sigma^2 t^2). \tag{5.1}$$

We now draw a sample of size n from this population. The characteristic function of the sample mean was expressed in terms of the characteristic function of the population in eq. (2.3). We therefore have

$$\varphi_{\bar{x}}(t) = \left\{\exp\left(\mathrm{i}\frac{t}{n}a - \frac{\sigma^2}{2}\left(\frac{t}{n}\right)^2\right)\right\}^n. \tag{5.2}$$

If we consider $(\bar{x} - a) = (\bar{x} - \hat{x})$ instead of x, we obtain

$$\varphi_{\bar{x}-a}(t) = \exp\left(-\frac{\sigma^2 t^2}{2n}\right). \tag{5.3}$$

This is again the characteristic function of a normal distribution. There has only been a change in variance

$$\sigma^2(\bar{x}) = \sigma^2(x)/n. \tag{5.4}$$

In the simple case of the standardized Gaussian distribution ($a = 0, \sigma^2 = 1$) we have

$$\varphi_{\bar{x}}(t) = \exp(-t^2/2n). \tag{5.5}$$

We draw a sample

$$x_1, x_2, \ldots, x_n$$

from this distribution and form the sum of squares of the elements of the sample

$$\chi^2 = x_1^2 + x_2^2 + \ldots + x_n^2. \tag{5.6}$$

We want to show that the quantity* χ^2 follows the distribution function

$$F(\chi^2) = \frac{1}{\Gamma(\lambda)\, 2^\lambda} \int_0^{\chi^2} u^{\lambda-1} e^{-\frac{1}{2}u}\, du, \tag{5.7}$$

with

$$\lambda = \tfrac{1}{2}n. \tag{5.8}$$

Here n is called the *number of degrees of freedom*. Abbreviating

$$\frac{1}{\Gamma(\lambda)\, 2^\lambda} = k, \tag{5.9}$$

we find the probability density

$$f(\chi^2) = k\, (\chi^2)^{\lambda-1} e^{-\frac{1}{2}\chi^2} \tag{5.10}$$

We want to prove (5.7) first for the simple case of one degree of freedom, i.e. $\lambda = \frac{1}{2}$.

We obtain the distribution function

$$F(\chi^2) = P(x^2 < \chi^2) = P(-\sqrt{(\chi^2)} < x < +\sqrt{(\chi^2)})$$

$$= \frac{1}{\sqrt{(2\pi)}} \int_{-\sqrt{(\chi^2)}}^{+\sqrt{(\chi^2)}} e^{-\frac{1}{2}x^2} dx = \frac{2}{\sqrt{(2\pi)}} \int_0^{\sqrt{(\chi^2)}} e^{-\frac{1}{2}x^2} dx.$$

Setting $x^2 = u$, i.e. $du = 2x\,dx$, we have

$$F(\chi^2) = \frac{1}{\sqrt{(2\pi)}} \int_0^{\chi^2} u^{-\frac{1}{2}} e^{-\frac{1}{2}u}\, du, \tag{5.11}$$

* The symbol χ^2 (chi-square) was introduced by K. Pearson. Although it contains an exponent to remind us of its origin as a sum of squares it is treated as a normal variable in our equations.

which is indeed identical to (5.7) for $\lambda = \frac{1}{2}$. To carry out the general proof we first write down the characteristic function of the χ^2-distribution

$$\varphi_{\chi^2}(t) = \int_0^\infty k(\chi^2)^{\lambda-1} \exp(-\tfrac{1}{2}\chi^2 + \mathrm{i}t\chi^2)\,\mathrm{d}\chi^2, \tag{5.12}$$

which, on transforming the integral such that $v = (\frac{1}{2} - \mathrm{i}t)\chi^2$, becomes

$$\varphi_{\chi^2}(t) = 2^\lambda(1 - 2\mathrm{i}t)^{-\lambda} k \int_0^\infty v^{\lambda-1} \mathrm{e}^{-v}\,\mathrm{d}v.$$

According to (D-1) the integral is equal to $\Gamma(\lambda)$. Then

$$\varphi_{\chi^2}(t) = (1 - 2\mathrm{i}t)^{-\lambda}. \tag{5.13}$$

Of course the characteristic function of another χ^2-distribution with λ' is

$$\varphi'_{\chi^2}(t) = (1 - 2\mathrm{i}t)^{-\lambda'}.$$

Since the characteristic function of a sum is the product of the characteristic functions, we arrive at the important theorem:

> The distribution of the sum of two different χ^2-distributions with n_1 and n_2 degrees of freedom respectively, is again a χ^2-distribution with $n = n_1 + n_2$ degrees of freedom.

This theorem applied to eq. (5.11) immediately proves (5.7) because the individual terms of the sum of squares (5.6) are independent and the sum can be interpreted as a sum of n different χ^2-distributions with 1 degree of freedom.

To calculate the expectation value and variance of the χ^2-distribution we use the derivatives of the characteristic function according to (5-6.7)

$$E(\chi^2) = -\mathrm{i}\varphi'(0) = 2\lambda,$$

or

$$E(\chi^2) = n. \tag{5.14}$$

And

$$E\{(\chi^2)^2\} = -\varphi''(0) = 4\lambda^2 + 4\lambda,$$

which leads to

$$\sigma^2(\chi^2) = E\{(\chi^2)^2\} - \{E(\chi^2)\}^2 = 4\lambda,$$

$$\sigma^2(\chi^2) = 2n. \tag{5.15}$$

The mean of the χ^2-distribution is equal to the number of degrees of freedom; the variance is twice as large. In fig. 6-2 the χ^2-distribution is plotted for various values of n. As is also apparent from eq. (5.10), in the case $n = 1$ the distribution has a pole at $\chi^2 = 0$. In the case of $n = 2$, $f(\chi^2)$ is equal to $\frac{1}{2}$ at $\chi^2 = 0$; for $n \geqslant 3$, $f(\chi^2)$ vanishes at this point.

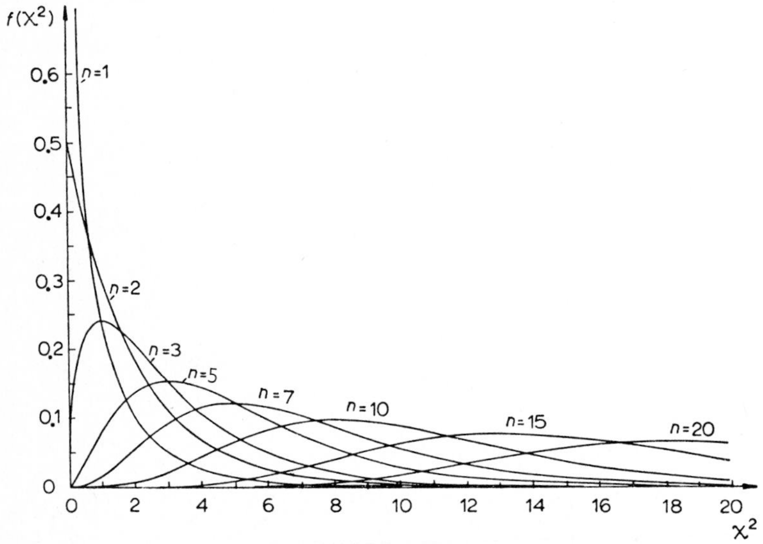

Fig. 6-2. Probability density of χ^2.

For 3 or more degrees of freedom the χ^2-distribution takes a typical skew bell-shape and becomes increasingly symmetric with increasing n. Some values of the distribution are tabulated in table F-4.

The χ^2-distribution is of special importance in many applications where the quantity χ^2 is used as a measure of confidence in a particular result. The smaller the value of χ^2, the more plausible is this result (cf. ch. 8, §6). The distribution function

$$F(\chi^2) = P(\chi^2 < \chi^2) \tag{5.16}$$

gives the probability that the random variable χ^2 does not exceed a given value χ^2. In practice the quantity

$$W(\chi^2) = 1 - F(\chi^2) \tag{5.17}$$

is used as a measure of confidence in a result. $W(\chi^2)$ is sometimes referred to as the confidence level.

It is larger for small χ^2 and decreases with increasing χ^2. The distribution function (5.16) is plotted in fig. 6-3 for some values of n. On the right hand

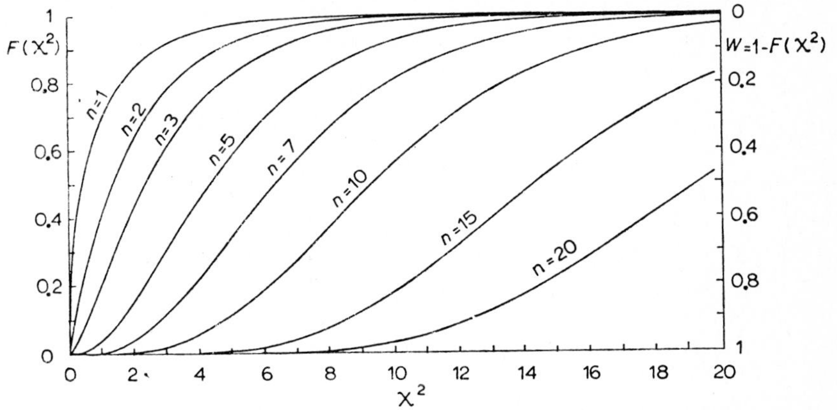

Fig. 6-3. Distribution function of χ^2.

side of the figure the quantity $W(\chi^2)$ is indicated on a second ordinate. The inverse function, describing the fractiles of the χ^2-distribution,

$$\chi_F^2 = \chi^2(F) = \chi^2(1 - W) \tag{5.18}$$

is used expecially in "hypothesis testing" to establish limits of χ^2 inside which a hypothesis can still be accepted (cf. ch. 8, §6). It is tabulated in table F-5.

It remains to be mentioned that in the general case of a normally distributed population with mean a and variance σ^2 it is no longer the sum of squares (5.6) that follows a χ^2-distribution but rather the quantity defined by

$$\chi^2 = \frac{(x_1 - a)^2 + (x_2 - a)^2 + \dots + (x_n - a)^2}{\sigma^2}. \tag{5.19}$$

This is seen readily from eq. (5-9.5).

6-6. χ^2 and empirical variance

As an unbiased, consistent estimator of the variance of a population we had defined in eq. (2.7) the quantity

$$s^2 = \frac{1}{n-1} \sum_{i=1}^{n} (x_i - \bar{x})^2. \tag{6.1}$$

We now discuss the special case where the sample of independent x_i is drawn from a normal distribution with variance σ^2. We intend to show that the random variable

$$\frac{n-1}{\sigma^2} s^2 \tag{6.2}$$

follows a χ^2-distribution with $f = n - 1$ degrees of freedom.

We first perform an orthogonal transformation (ch. 4, §5) of the n variables x_i

$$\begin{aligned}
y_1 &= \frac{1}{\sqrt{1 \cdot 2}} (x_1 - x_2), \\
y_2 &= \frac{1}{\sqrt{2 \cdot 3}} (x_1 + x_2 - 2x_3), \\
y_3 &= \frac{1}{\sqrt{3 \cdot 4}} (x_1 + x_2 + x_3 - 3x_4), \\
&\vdots \\
y_{n-1} &= \frac{1}{\sqrt{(n-1)n}} (x_1 + x_2 + \ldots + x_{n-1} - (n-1)x_n), \\
y_n &= \frac{1}{\sqrt{n}} (x_1 + x_2 + \ldots + x_n) = \sqrt{n}\, \bar{x}.
\end{aligned} \tag{6.3}$$

It can easily be verified that this transformation is indeed orthogonal, i.e. especially that

$$\sum_{i=1}^{n} x_i^2 = \sum_{i=1}^{n} y_i^2. \tag{6.4}$$

Since a sum (or for that matter a difference) of independent normally distributed quantities is again normally distributed (ch. 5, §8), all y_i are normally distributed. The numerical factors in (6.3) ensure that they all have zero mean and unit variance. We can now rewrite eq. (6.1)

$$(n - 1)s^2 = \sum_{i=1}^{n} (x_i - \bar{x})^2 = \sum_{i=1}^{n} x_i^2 - 2\bar{x} \sum_{i=1}^{n} x_i + n\bar{x}^2$$

$$= \sum_{i=1}^{n} x_i^2 - n\bar{x}^2 = \sum_{i=1}^{n} y_i^2 - y_n^2 = \sum_{i=1}^{n-1} y_i^2.$$

This expression is only a sum of $(n - 1)$ independent squares. Comparison with eq. (5.19) now shows that the quantity (6.2) does indeed follow a χ^2-distribution with $(n - 1)$ degrees of freedom.

The squares $(x_i - \bar{x})^2$ are not independent. They are constrained by the equation.

$$\sum_{i=1}^{n} (x_i - \bar{x}) = 0.$$

It can be shown that any additional equation between these squares reduces the number of degrees of freedom again by one. We will frequently make use this general result which we state here without proof.

6-7. Sampling by counting. Small samples

Often elements are drawn from a population and kept in the sample if they possess a given property. Otherwise they are rejected. The sampling process then simply consists of the *counting* of the number of kept events k and the total number of drawings n.

This process is identical to sampling from a binomial distribution. As shown in example 7-5

$$S(p) = \frac{k}{n} \tag{7.1}$$

is the "maximum likelihood" estimator for the parameter p of the binomial distribution. The variance of the distribution of S is

$$\sigma^2(S(p)\cdot) = \frac{p(1 - p)}{n}. \tag{7.2}$$

Using (7.1) it can be estimated from the sample by

$$s^2(S(p)) = \frac{1}{n}\frac{k}{n}\left(1 - \frac{k}{n}\right). \tag{7.3}$$

In practice it is often customary to consider directly the count k rather than the parameter p as the interesting quantity and to define its error as

$$\Delta k = \sqrt{[s^2(S(np))]}. \tag{7.4}$$

Using eq. (3-3.14) we obtain

$$\Delta k = \sqrt{\left[k\left(1 - \frac{k}{n}\right)\right]}. \tag{7.5}$$

The error Δk depends only on the counted number of events and the size of the sample. It is called the *statistical error* of counting.

Especially important is the case of small k, more specifically the case $k \ll n$. This is a limiting case. If we define $\lambda = np$ then according to ch. 5 § 4 the count k can be regarded as a single element of a sample from a Poisson distribution described by the parameter λ. From eqs. (7.1) and (7.5) we get

$$S(\lambda) = S(np) = k, \tag{7.6}$$

$$\Delta k = \sqrt{k}. \tag{7.7}$$

(The same result can be obtained using the result of example 7-4 and setting $N = 1$.)

In order to interpret the statistical error $\Delta k = \sqrt{k}$ we have to inspect the Poisson distribution more closely. Let us begin with the case where k is not too small (say $k > 20$). For large values of λ the Poisson distribution approaches a normal distribution with mean λ and variance $\sigma^2 = \lambda$. This can be seen qualitatively from fig. 5-3 and quantitatively from exercise 5-6.

Therefore for not too small k one can use the well-known properties of the normal distribution for the statement that there is a 68.2 % probability of k falling into the interval $\lambda \pm \Delta k = \lambda \pm \sqrt{k}$. It is called the *confidence interval* corresponding to a *confidence level* of $\alpha = 68.2\,\%$. Conversely the probability of finding k below or above this interval is $(100 - 68.2)/2 = 15.9\,\%$. We now consider 3 extreme cases:

(a) The measured value of k is exactly equal to the parameter λ;

(b) The measured value k lies at the upper limit of the confidence interval $[P(x > k) = \frac{1}{2}(1 - \alpha)]$;

(c) It lies at the lower limit of the confidence interval $[P(x < k) = \frac{1}{2}(1 - \alpha)]$. The situation is illustrated in fig. 6-4. It is obvious from the symmetry of the normal distribution that the limits of the confidence interval in fig. 6-4a correspond exactly to extreme positions of λ in the cases (b) and (c). At a given level of confidence α one therefore simply has to determine the value

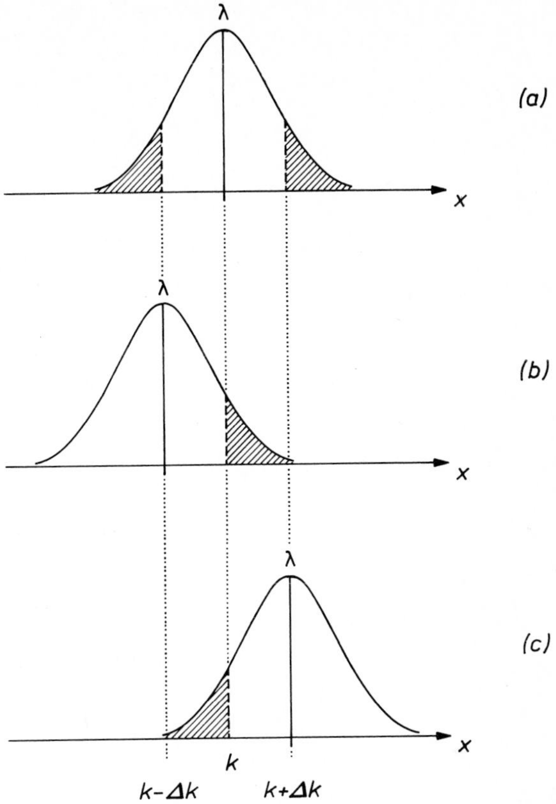

Fig. 6-4. Normal distribution with width $\sigma = \Delta k$ and mean (a) $\lambda = k$, (b) λ, such that $P(x > k) = \frac{1}{2}(1 - \alpha)$, (c) λ, such that $P(x < k) = \frac{1}{2}(1 - \alpha)$.

$(\Delta k)_\alpha$ for a normal distribution in x with mean k and variance k such that the variable x falls into the interval $k \pm (\Delta k)_\alpha = k \pm x_\alpha \sqrt{k}$. For $\alpha = 68.2\,\%$ we find from table F-3b as expected $x_{0.682} = 1$, i.e. $(\Delta k)_{0.682} = \sqrt{k}$. For $\alpha = 90\,\%$ we have $x_{0.90} = 1.65$, i.e. $(\Delta k)_\alpha = 1.65\sqrt{k}$.

For very small values of k, however, the Poisson distribution can no longer be replaced by the normal distribution. One then uses a reasoning analogous to that sketched in figs. 6-4b and 6-4c. Again two extreme cases for the situation of k with respect to λ are considered: They are characterized by the probabilities

(a) $P(x < k) = \frac{1}{2}(1 - \alpha)$.

(b) $P(x > k) = \frac{1}{2}(1 - \alpha)$.

For both conditions values of λ are obtained which serve as upper and lower limits of the confidence interval. For the actual computation of the lower limit one solves

$$P = \tfrac{1}{2}(1 - \alpha) = e^{-\lambda_-} \sum_{n=0}^{k} \lambda_-(n!)^{-1}$$

for the parameter λ_-. Analogously one determines the upper limit λ_+. The corresponding errors Δk_- and Δk_+ are no longer equal. This can be seen from fig. 6-5 where for simplicity the Poisson distribution has been drawn as if k were a continuous variable.

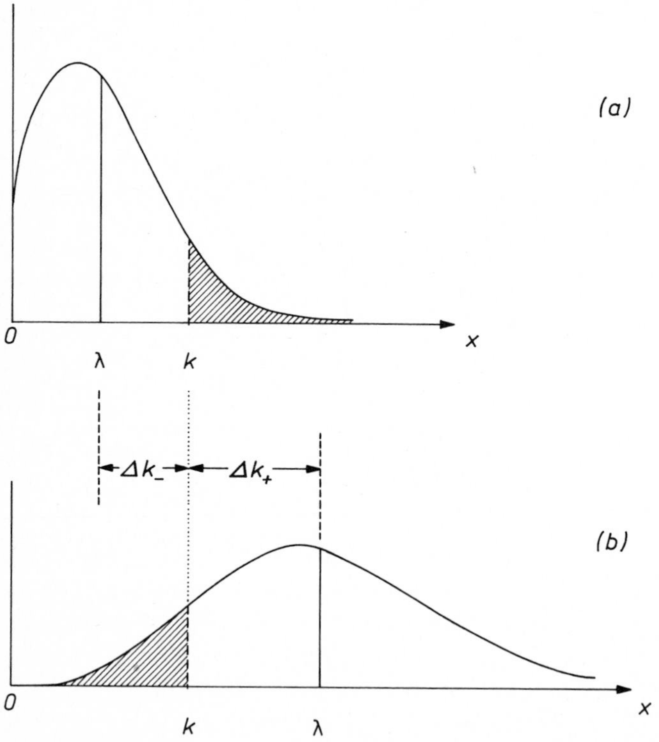

Fig. 6-5. Poisson distribution with mean λ, such that (a) $P(x > k) = \frac{1}{2}(1 - \alpha)$, (b) $P(x < k) = \frac{1}{2}(1 - \alpha)$.

Table 6-1 (reproduced from REGENÉR (1951)) contains the limits of the confidence intervals and the corresponding errors for different values of k for two frequently used confidence levels. For comparison the symmetric errors $(\Delta k)_\alpha$ applicable only for larger k are also given.

TABLE 6-1
Statistical errors for very small samples

Confidence level α for			Lower Poisson error	Lower limit	Count	Upper limit	Upper Poisson error	Normal error
$\lambda > k-\Delta k_-$	$k-\Delta k_- < \lambda < k+\Delta k_+$	$\lambda < k+\Delta k_+$	Δk_-	$k-\Delta k_-$	k	$k+\Delta k_+$	Δk_+	Δk
–	–	0.841	0	0	0	1.84	1.84	0.0
0.841	0.682	0.841	0.83	0.17	1	3.30	2.30	1.0
0.841	0.682	0.841	1.29	0.71	2	4.64	2.64	1.41
0.841	0.682	0.841	1.63	1.37	3	5.92	2.92	1.73
0.841	0.682	0.841	1.91	2.09	4	7.16	3.16	2.00
0.841	0.682	0.841	2.38	3.62	6	9.58	3.58	2.45
0.841	0.682	0.841	3.11	6.89	10	14.26	4.26	3.16
0.841	0.682	0.841	4.47	15.57	20	25.54	5.54	4.72
0.841	0.682	0.841	7.05	42.95	50	58.11	8.11	7.07
0.841	0.682	0.841	$\sqrt{k}$	$k-\sqrt{k}$	large	$k+\sqrt{k}$	$\sqrt{k}$	$\sqrt{k}$
–	–	0.95	0	0	0	3.00	3.00	0
0.95	0.90	0.95	0.95	0.05	1	4.74	3.74	1.65
0.95	0.90	0.95	1.65	0.35	2	6.30	4.30	2.33
0.95	0.90	0.95	2.18	0.82	3	7.75	4.75	2.85
0.95	0.90	0.95	2.63	1.37	4	9.15	5.15	3.30
0.95	0.90	0.95	3.49	2.61	6	11.84	5.84	4.04
0.95	0.90	0.95	4.57	5.43	10	16.96	6.96	5.22
0.95	0.90	0.95	6.74	13.26	20	29.06	9.06	7.38
0.95	0.90	0.95	11.04	38.96	50	63.28	13.28	11.67
0.95	0.90	0.95	$1.65\sqrt{k}$	$k-1.65\sqrt{k}$	large	$k+1.65\sqrt{k}$	$1.65\sqrt{k}$	$1.65\sqrt{k}$

Example 6-3: *Determination of a lower limit of the proton mean life from the observation of no decay events*

As mentioned several times the probability for a radioactive nucleus to decay within a time t is

$$P(t) = \frac{1}{\tau}\int_0^t e^{-x/\tau}\,dx,$$

where τ is the mean life time of the nucleus. For $t \ll \tau$ the relation reduces to

$$P(t) = t/\tau.$$

If n nuclei are observed then

$$k = nP(t) = n \cdot t/\tau$$

are expected to decay within the time t. Usually τ is the quantity of interest. From the observation of k it is estimated to be

$$\tilde{\tau} = \frac{n}{k} t.$$

Of particular interest is the mean life of the proton, one of the main constituents of matter. Several experiments have been performed, where large numbers of protons have been observed for evidence of decay but no decay event was found. Then according to table 6-1 the true value of k does not exceed 3 at 95 % confidence level. Therefore

$$\tau > \frac{n}{3} t$$

at this confidence level. Typical values in actual experiments are $t = 0.3$ years, $n = 10^{28}$, i.e.

$$\tau > 10^{27} \text{ years.}$$

Therefore the proton can be regarded as stable also in the cosmological time scale considering that the age of the universe is of the order of 10^9 years.

6-8. Numerical and graphical analysis of sampled data with computer programs

Now that we have obtained a number of necessary theoretical results in the last sections we can deal with the more practical aspects of the analysis of sampled data. A very important tool is the display of the data in graphical form.

6-8.1. *Scatter diagram and histogram of a one-dimensional sample*

Obviously the simplest way to display a sample

$$x_1, x_2, \ldots, x_N$$

which depends only on a single variable x is to mark all elements of the sample as strokes along an axis measuring x. We call such a plot a *one-*

TABLE 6-2

R-values of a sample of 100 electric resistors with nominally 200 kΩ. The data are displayed graphically in fig. 6-5 and table 6-3

193.199	195.673	195.757	196.051	196.092	196.596	196.679	196.763	196.847	197.267
197.392	197.477	198.189	198.650	198.944	199.070	199.111	199.153	199.237	199.698
199.572	199.614	199.824	199.908	200.118	200.160	200.243	200.285	200.453	200.704
200.746	200.830	200.872	200.914	200.956	200.998	200.998	201.123	201.208	201.333
201.375	201.543	201.543	201.584	201.711	201.878	201.919	202.004	202.004	202.088
202.172	202.172	202.297	202.339	202.381	202.507	202.591	202.716	202.633	202.884
203.051	203.052	203.094	203.094	203.177	203.178	203.219	203.764	203.765	203.848
203.890	203.974	204.184	204.267	204.352	204.352	204.729	205.106	205.148	205.231
205.357	205.400	205.483	206.070	206.112	206.154	206.155	206.615	206.657	206.993
207.243	207.621	208.124	208.375	208.502	208.628	208.670	208.711	210.012	211.394

dimensional scatter-diagram. It contains all the information of the sample. In table 6-2 the values $x_1, x_2, \ldots, x_N$ are shown for a sample of size 100, which were obtained by measuring the electric resistance R of 100 resistors of nominally 200 kΩ. They have been ordered according to x after the sampling. Fig. 6-6a contains the corresponding scatter diagram. From the grouping of strokes in one region and from the width of the grouping sample the mean and variance can be guessed.

Another graphical presentation is usually better suited to visualize the sample by using the second dimension available on paper. The x-axis is used as abscissa and divided into r intervals

$$\xi_1, \xi_2, \ldots, \xi_r$$

of equal width called *bins*. The middle points of the bins may have the x-values

$$x_1, x_2, \ldots, x_r.$$

As ordinate the number of sample elements

$$n_1, n_2, \ldots, n_r$$

falling into the bins $\xi_1, \xi_2, \ldots, \xi_r$ are plotted. The resulting diagram is called a *histogram* of the sample. It can be regarded as an empirical frequency distribution since $h_k = n_k/n$ is an empirical frequency, i.e. a measure for the probability p_k to observe a sample element in the interval ξ_k. Customary ways of plotting are the *bar diagram* in which the n_k are visualized by bars over each x_k (fig. 6-6b1) or a *step diagram* in which the n_k are indicated as horizontal lines of size ξ_k. Neighbouring horizontal lines are connected by vertical lines (fig. 6-6 b2). In this way the area over each interval ξ_k is proportional to n_k. (If the area is plotted intervals of unequal size can also be used.) Usually bar diagrams are preferred in economics. (Sometimes also

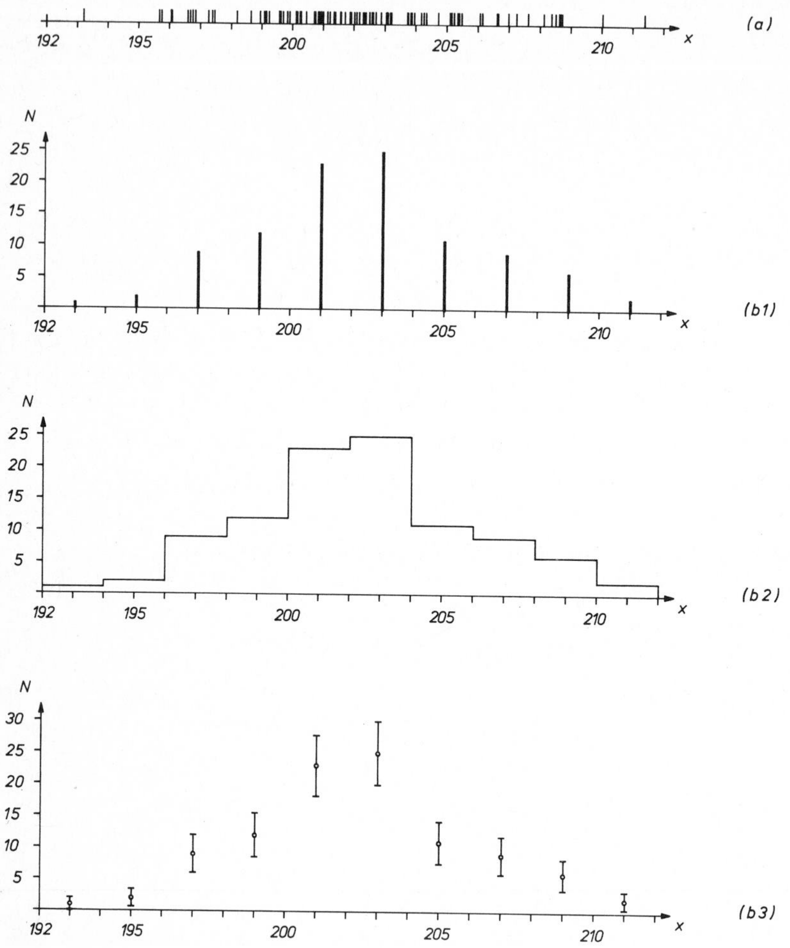

Fig. 6-6. Representation of the data of table 6-2 as (a) one-dimensional scatter diagram, (b) histogram [(b1) bar diagram, (b2) step diagram, (b3) point diagram with error bars].

line diagrams are used in which instead of the bars a series of straight lines are drawn connecting the positions of the bar tops. Contrary to the step diagram, however, the resulting figure does not have an area proportional to the sample size n.) In the physical sciences the step diagram is more customary.

From §7 we know that the statistical error of the numbers n_k can be taken as $\Delta n_k = \sqrt{n_k}$ provided the n_k are not too small. To indicate them the n_k are drawn by points with vertical *error bars* ending at $n_k \pm \sqrt{n_k}$ (fig. 6-6 b3).

It is clear that the relative errors $\Delta n_k/n_k = 1/\sqrt{n_k}$ decrease with increasing n_k, i.e. for fixed sample size n they decrease with increasing bin width. On the other hand increasing the bin width washes out the finer structure of the data with respect to the variable x. The significance of a histogram therefore depends strongly on the proper choice of the bin width, usually found only after several trials.

For larger samples and repetitive trials with different bin size the construction of histograms by hand is obviously rather cumbersome. We therefore discuss a simple FORTRAN program which constructs and prints a histogram from input data. In addition it computes mean and variance of the sample and a corresponding Gauss curve, i.e. a normal distribution with equal mean and variance which is also printed and can be compared with the experimental histogram. The complete program consists of a main program and two subroutines.

Program 6-1: *Main program* HISTO *setting up a histogram from input data*

```
C
C      MAIN PROGRAM HISTO
C      PROGRAM READS DATA, COMPUTES MEAN AND VARIANCE
C      AND PRINTS HISTOGRAM TOGETHER WITH GAUSSCURVE
C
       DIMENSION HIST(50), CURVE(50),TEXT(20)
C
C      SET INITIAL VALUES EQUAL TO ZERO
C
       DO 10 I=1,50
       CURVE(I)=0.
 10    HIST(I)=0.
       SX=0.
       SX2=0.
C
C      READ EXPLANATORY TEXT AND PARAMETERS DESCRIBING DATA AND BINNING
C
       READ(5,1300) TEXT
       READ(5,1000) NDATA,XBEG,DELX,NX
C
C      READ DATA, FILL HISTOGRAM AND COMPUTE SUM AND SUM OF SQUARES
C
       XEND=XBEG+NX*DELX
       DO 40 I=1,NDATA
       READ (5,1100) X
       SX=SX+X
       SX2=SX2+X**2
       IF(X-XBEG)40,20,20
 20    IF(XEND-X) 40,30,30
 30    J=(X-XBEG)/DELX+1
       HIST(J)=HIST(J)+1.
 40    CONTINUE
C
C      COMPUTE MEAN AND VARIANCE OF SAMPLE
C
       XM=SX/NDATA
       VAR=(SX2-NDATA*XM**2)/(NDATA-1)
       SIGMA=SQRT(VAR)
```

```
C
C      COMPUTE GAUSSCURVE WITH THE SAME MEAN AND VARIANCE
C
       CALL GCURVE(XBEG,DELX,NX   ,XM,VAR,NDATA,CURVE)
C
C      WRITE OUTPUT
C
       WRITE(6,1300) TEXT
       WRITE(6,1200) NDATA,XBEG,XEND,DELX,XM,SIGMA
       CALL PRHIST(XBEG,DELX,NX,HIST,CURVE)
 1000  FORMAT(I10,2F10.5,I10)
 1100  FORMAT (F10.5)
 1200  FORMAT(9H NDATA = ,I5/9H XBEG  = ,F7.3,10H, XEND  = ,F7.3,
      *11H, DELTAX = ,F7.3/9H MEAN  = ,F7.3,10H, WIDTH = ,F7.3//)
 1300  FORMAT(20A4)
       END
```

The program HISTO stores results in three arrays of subscripted variables. The variable HIST will contain the histogram, CURVE the values of the Gauss curve and TEXT contains 20 words of alphanumeric text explaining the data. In the first section the contents of HIST and CURVE are set equal to zero as well as two unsubscripted variables SX and SX2 which will later contain the sum and the sum of squares of the input data. Next a punched card with explanatory text is read and its contents stored in TEXT. The second card contains the sample size (NDATA), the lower limit of the first bin (XBEG), the bin width (DELX) and the number of bins (NX). From the last three numbers XEND, the upper limit of the last bin is computed. In a DO-loop the data of the sample are read. For simplicity the x-value of each element of the sample is contained on a different card. Within the loop the sum and the sum of squares of the input data are formed. Also the essential operation of the program is done, i.e. the addition of one into the histogram bin corresponding to the element. First in two IF-statements it is determined whether the element lies in the interval between XBEG and XEND. If so an integer J=(X—BEG)DELX+1 is computed. One easily sees that J is the index of the histogram interval corresponding to the element with measured value. (The number on the right-hand side of the above expression is usually larger than the required index. However only its integer value is stored in J since J is the name of an integer variable.) Accordingly the content of HIST(J) is increased by one. Note that the original input data X are not stored. Therefore histograms of very large samples can be made without requiring a large computer memory. Finally **mean,** variance and width of the sample are computed and stored in XM, VAR and SIGMA, respectively. The subroutine GCURVE computes the corresponding Gauss curve. As the last step the output is written beginning with the explanatory text followed by sample size, description of the histogram, mean and width of the sample and the histogram itself (cf. table 6-3), the latter

TABLE 6-3

Histogram of the data in table 6-2 (XXX) and Gauss curve (*) with mean and variance of the sample. The same data but different bin widths were used for the 3 histograms

```
MEASURED R-VALUES OF RESISTORS WITH NOMINALLY 200 KILOOHMS
NDATA =    100
XBEG  = 180.000, XEND  = 220.000, DELTAX =    4.000
MEAN  = 202.278, WIDTH =    3.544

182.000  0.0
186.000  0.0
190.000  0.0
194.000  3.0  XX*
198.000 21.0  XXXXXXXXXXXXXXXXXXXXX*
202.000 48.0  XXXXXXXXXXXXXXXXXXXXXXXXXXXXXXXXXXXXXXXXXXXXX*XXX
206.000 20.0  XXXXXXXXXXXXXXXXXXXX     *
210.000  8.0  XXX*XXXX
214.000  0.0
218.000  0.0
```

```
MEASURED R-VALUES OF RESISTORS WITH NOMINALLY 200 KILOOHMS
NDATA =    100
XBEG  = 192.000, XEND  = 212.000, DELTAX =    2.000
MEAN  = 202.278, WIDTH =    3.544

193.000  1.0  *
195.000  2.0  XX*
197.000  9.0  XXXXXX*XX
199.000 12.0  XXXXXXXXXXXX  *
201.000 23.0  XXXXXXXXXXXXXXXXXXXX*XX
203.000 25.0  XXXXXXXXXXXXXXXXXXXXX*XXX
205.000 11.0  XXXXXXXXXXX     *
207.000  9.0  XXXXXXXX*
209.000  6.0  XXX*XX
211.000  2.0  *X
```

```
MEASURED R-VALUES OF RESISTORS WITH NOMINALLY 200 KILOOHMS
NDATA =    100
XBEG  = 192.000, XEND  = 212.000, DELTAX =    1.000
MEAN  = 202.278, WIDTH =    3.544

192.500  0.0
193.500  1.0  *
194.500  0.0  *
195.500  2.0  X*
196.500  6.0  XX*XXX
197.500  3.0  XXX *
198.500  3.0  XXX  *
199.500  9.0  XXXXXXX*X
200.500 13.0  XXXXXXXXX*XXX
201.500 10.0  XXXXXXXXXX*
202.500 13.0  XXXXXXXXXX*XX
203.500 12.0  XXXXXXXXXX*X
204.500  5.0  XXXXX   *
205.500  6.0  XXXXXX*
206.500  7.0  XXXXX*X
207.500  2.0  XX *
208.500  6.0  X*XXXX
209.500  0.0  *
210.500  1.0  *
211.500  1.0  X
```

being printed by the subroutine PRHIST. The subroutine GCURVE can easily be replaced if comparison of the histogram with other distributions (e.g. uniform or exponential) is desired. The histogram printing is performed by a subroutine, which is later used also for the representation of two-dimensional samples.

Program 6-2: *Subroutine* GCURVE *computing values of a normal distribution corresponding to an empirically derived histogram*

```
      SUBROUTINE GCURVE (XBEG,DELX,NX   ,XM,VAR,NDATA,CURVE)
      DIMENSION CURVE(1)
      SIGMA= SQRT(VAR)
      DO 10 J=1,NX
      X=XBEG+(J-0.5)*DELX
      C=EXP(-1.*(X-XM)**2/(2.*VAR))/(2.5066*SIGMA)
10    CURVE(J)=C*NDATA*DELX
      RETURN
      END
```

Input parameters are sample size, mean and variance NDATA, XM and VAR, respectively, and the lower limit of the first histogram bin (XBEG), the bin width (DELX) and number of bins (NX). In a DO loop for the middle points of each bin (denoted by X) the probability density C of a Gaussian distribution with mean XM and variance VAR is computed. By multiplication with NDATA and DELX the number of sample elements predicted in this bin for the total sample size NDATA is computed and stored in the array CURVE. It is assumed that C does not vary much within one bin width and therefore the integration over the bin width can be replaced by a simple multiplication.

Program 6-3: *Subroutine* PRHIST *printing a histogram in graphical form*

```
      SUBROUTINE PRHIST (XBEG,DELX,NX,HIST,CURVE)
      DIMENSION HIST(1),CURVE(1),ZLINE(50)
      DATA BLANK, CROSS, AST/1H ,1HX,1H*/
C
C     DETERMINE MAXIMUM BIN CONTENT AND SCALE FACTOR
C
      SCFACT=1.
      HMAX=0.
      DO 2 J=1,NX
      IF(HIST(J)-HMAX) 2,2,1
1     HMAX=HIST(J)
2     CONTINUE
3     IF(HMAX-50.) 5,5,4
4     HMAX=HMAX/2.
      SCFACT=SCFACT/2.
      GO TO 3
C
C     PRINT HISTOGRAM BINWISE
C
5     DO 40 J=1,NX
```

```
      X=XBEG+(J-0.5)*DELX
      DO 10 I=1,50
10    ZLINE(I)=BLANK
      K=HIST(J)*SCFACT+0.5
      IF(K) 25,25,15
15    DO 20 I=1,K
20    ZLINE(I)=CROSS
25    K=CURVE(J)*SCFACT+0.5
      IF(K-50) 30,30,40
30    IF(K) 40,40,35
35    ZLINE(K)=AST
40    WRITE(6,1000) X, HIST(J), ZLINE
      RETURN
1000  FORMAT (1H ,F7.3,F5.1,2H  ,50A1)
      END
```

The program begins with the determination of the scale factor SCFACT which is determined from the maximum bin content HMAX such that if multiplied by SCFACT no histogram bin will contain a number larger than 50 (otherwise the histogram would be larger than the size of the paper used by the computer printer). In the following DO loop (extending down to statement 40), whose index varies from 1 to NX one line of output is printed for each bin. It contains the numbers X (middle point of bin), HIST(J) (contents of bin) and 50 alphanumeric characters contained in the array ZLINE. For each bin first the array ZLINE is filled with blank characters. Next the integer K=HIST(J)*SCFACT+0.5 is computed. It is an integer measure for the contents of bin number J after application of the scale factor. Therefore the first K words of ZLINE are now filled with the character X (denoted by CROSS). Finally a corresponding value of K is computed for CURVE(J) and an asterisk is written into the word number K of ZLINE if K is between 1 and 50. In this way a bar of a length proportional to HIST(J) is printed together with an asterisk at a position proportional to CURVE(J) for every bin of the histogram. Note that no asterisk is printed if CURVE(J) =0. Therefore the subroutine PRHIST can also be used even if no curve is required (cf. program 6-4).

Example 6-4: *Histogram of a sample of electric resistors*

The data of table 6-2 were used to produce histograms of a sample of resistors. They are displayed in table 6-3. Three histograms of the same sample but with different bin widths are shown. The middle one corresponds to fig. 6-6b2.

6-8.2. *Scatter diagram of a two-dimensional sample*

If the elements of a sample depend on two random variables x and y a scatter diagram can be constructed by marking every element as a point in a

Cartesian coordinate system spanned by the variables x and y. Such a *two-dimensional scatter diagram* can give useful qualitative information about the interdependence of the two variables.

A scatter diagram can be printed by the computer if the x- and y-axes are divided into small intervals, so that the x–y-plane is divided into small squares. A symbol (+) is printed in this square if an element of the sample falls into it. If more than 1 element falls into this square the corresponding number is printed instead of the symbol. (Strictly speaking the resulting figure (an example is given in table 6-5) is not a scatter diagram but a two-dimensional histogram in numerical form. However by choosing small intervals it can be made very much like a scatter diagram.) We now discuss a simple program which performs the operations described above.

Program 6-4: Main program SCAT *setting up a two-dimensional scatter diagram from input data*

```
C
C      MAIN PROGRAM SCAT
C      PROGRAM READS DATA OF A TWO-DIMENSIONAL SAMPLE,
C      COMPUTES MEANS, WIDTHS AND CORRELATION COEFFICIENT
C      AND PRINTS A SCATTER DIAGRAM AS WELL AS HISTOGRAMS OF ITS PROJECTIONS
C
       DIMENSION SCAT(50,50), HISTX(50),HISTY(50),CURVE(50),TEXT(20)
C
C      SET INITIAL VALUES EQUAL TO ZERO
C
       DO 10 I=1,50
       DO 10 J=1,50
       HISTX(I)=0.
       HISTY(I)=0.
       CURVE(I)=0.
 10    SCAT(I,J)=0.
       SX=0.
       SX2=0.
       SY=0.
       SY2=0.
       SXY=0.
C
C      READ EXPLANATORY TEXT AND PARAMETERS DESCRIBING DATA AND BINNING
C
       READ(5,1300) TEXT
       READ(5,1000) NDATA,XBEG,DELX,NX,YBEG,DELY,NY
       XEND=XBEG+NX*DELX
       YEND=YBEG+NY*DELY
C
C      READ DATA, FILL SCATTER DIAGRAM AND HISTOGRAMS
C      AND COMPUTE SUMS AND SUMS OF SQUARES
C
       DO 70 I=1,NDATA
       READ(5,1100) X,Y
       SX=SX+X
       SX2=SX2+X**2
       SY=SY+Y
       SY2=SY2+Y**2
       SXY=SXY+X*Y
       IF(X-XBEG)70,30,30
 30    IF(XEND-X)70,40,40
 40    IF(Y-YBEG)70,50,50
 50    IF(YEND-Y)70,60,60
 60    J=(X-XBEG)/DELX+1
       K=(Y-YBEG)/DELY+1
       HISTX(J)=HISTX(J)+1.
       HISTY(K)=HISTY(K)+1.
       SCAT(J,K)=SCAT(J,K)+1.
 70    CONTINUE
```

```
C
C       COMPUTE MEANS, VARIANCES, WIDTHS AND CORRELATION
C
        XM=SX/NDATA
        YM=SY/NDATA
        XVAR=(SX2-NDATA*XM**2)/(NDATA-1)
        YVAR=(SY2-NDATA*YM**2)/(NDATA-1)
        XSIGMA=SQRT(XVAR)
        YSIGMA=SQRT(YVAR)
        COR=(SXY-NDATA*XM*YM)/(NDATA*XSIGMA*YSIGMA)
C
C       WRITE OUTPUT
C
        WRITE(6,1400) TEXT
        WRITE(6,1200) NDATA,XBEG,XEND,DELX,YBEG,YEND,DELY,
       *XM,XSIGMA,YM,YSIGMA,COR
        CALL PRSCAT(XBEG,DELX,NX,YBEG,DELY,NY,SCAT)
        WRITE(6,1500) TEXT
        CALL PRHIST (XBEG,DELX,NX,HISTX,CURVE)
        WRITE(6,1600) TEXT
        CALL PRHIST (YBEG,DELY,NY,HISTY,CURVE)
  1000  FORMAT(I10,2F10.5,I10,2F10.5,I10)
  1100  FORMAT(2F10.5)
  1200  FORMAT(9H NDATA = ,I5/
       *9H XBEG  = ,F7.3,11H, XEND   = ,F7.3,11H, DELTAX = ,F7.3/
       *9H YBEG  = ,F7.3,11H, YEND   = ,F7.3,11H, DELTAY = ,F7.3/
       *9H XMEAN = ,F7.3, 11H, XWIDTH = ,F7.3/
       *9H YMEAN = ,F7.3, 11H, YWIDTH = ,F7.3/
       *15H CORRELATION = , F7.3//)
  1300  FORMAT(20A4)
  1400  FORMAT(//20A4/)
  1500  FORMAT(//20A4/22H  PROJECTION ON X-AXIS/)
  1600  FORMAT(//20A4/22H  PROJECTION ON Y-AXIS/)
        END
```

The program SCAT sets up the scatter diagram in the two-dimensional array SCAT(I,J). I and J are the indices counting the intervals in x and y, respectively. As projections onto the x- and y-axis histograms are also stored in HISTX and HISTY. An explanatory text is stored in the array TEXT. An array CURVE is also foreseen. It could be used if curves should be printed for comparison with one or both histograms. In the initial phase the arrays SCAT, HISTX, HISTY and CURVE are filled with zeros as well as the five variables SX, SY, SX2, SY2 and SXY which are used later for the computation of sample means, variances and the correlation coefficient. As a next step a card with text and a second card specifying the sample size (NDATA), the lower limit for x and y (XBEG, YBEG), the interval sizes (DELX, DELY) and the number of intervals (NX, NY) are read. From the latter 6 quantities the upper limits (XEND, YEND) are computed. Now the data for the sample are read. Each element is contained on a separate card. The values of X and Y are expected to be punched in columns 1–10 and 11–20, respectively. With each element the variables SX, SY, SX2, SY2 and SXY are updated and the appropriate words of HISTX, HISTY and SCAT are increased by one if the element falls within the ranges XBEG < X < XEND, YBEG < Y < YEND. Finally sample means, variances and the correlation coefficient are computed and printed under a heading identical to the input text followed by the sample size. The scatter diagram is printed by the subroutine PRSCAT and the histograms of the projections are printed by PRHIST. Before each histogram a heading is printed repeating the input text and a line PROJECTION ON X (or Y) AXIS.

Program 6-5: *Subroutine* PRSCAT *printing a two-dimensional scatter diagram in graphical form*

```
      SUBROUTINE PRSCAT(XBEG,DELX,NX,YBEG,DELY,NY,SCAT)
      DIMENSION SCAT(50,50),XLINE(50),ZLINE(51),ZNUMBR(9),XSCALE(10)
      DATA BLANK,AST,ZNUMBR/1H ,1H*,1H+,1H2,1H3,1H4,1H5,1H6,1H7,1H8,1H9/
      DATA HOR,VERT /1H-,1HI/
C
C     WRITE SCALE ON X-AXIS
C
      DO 5 I=1,50
    5 XLINE(I)=BLANK
      DO 10 I=1,NX
   10 XLINE(I)=HOR
      NMARKS=(NX-1)/10+1
      DO 15 I=1,NMARKS
      XLINE((I-1)*10+1)=VERT
   15 XSCALE(I)=XBEG+((I-1)*10.+0.5)*DELX
      WRITE (6,1400)
      WRITE (6,1000) (XSCALE(I),I=1,NMARKS)
      WRITE (6,1200) XLINE
C
C     PRINT SCATTER PLOT LINEWISE TOGETHER WITH SCALE ON Y-AXIS
C
      DO 70 M=1,NY
      K=NY-M+1
      Y=YBEG+(K-0.5)*DELY
      DO 20 I=1,51
   20 ZLINE(I)=BLANK
      DO 60 J=1,NX
      IF(SCAT(J,K)-1.) 60,30,30
   30 IF(SCAT(J,K)-9.)50,50,40
   40 ZLINE(J)=AST
      GO TO 60
   50 N=SCAT(J,K)+0.5
      ZLINE(J)=ZNUMBR(N)
   60 CONTINUE
      ZLINE(NX+1)=VERT
   70 WRITE(6,1100) Y, ZLINE
      WRITE (6,1300) XLINE
      WRITE (6,1000) (XSCALE(I),I=1,NMARKS)
      RETURN
 1000 FORMAT(4X,6F10.3)
 1100 FORMAT(1H ,F7.3,1X,1HI,51A1)
 1200 FORMAT(4X,1HY,5X,50A1)
 1300 FORMAT(10X,50A1)
 1400 FORMAT(16X,1HX)
      END
```

The lay out of the scatter diagram is such that x is plotted along the horizontal and y along the vertical axis (cf. table 6-5). Each printed line therefore belongs to a fixed interval in y. Since the paper is printed starting with the top and ending with the bottom line, the first line corresponds to the highest value of y. Before the diagram is started the x-axis is printed. It consists of horizontal lines (–) interrupted by vertical lines (I) at every tenth interval. They are stored in the array XLINE. The value of x corresponding to the middle points of the intervals marked by I are printed as a scale on top of the x-axis.

In the following DO-loop which extends down to statement 70 one line of output is prepared for every interval in y, beginning with the highest value of y. The index K gives the interval number in y, the variable Y the middle point of the interval. For every value of K the array ZLINE, in which the line of output is prepared, is filled with blank characters. In the inner DO-loop extending to statement 60 the index J is varied from 1 to NX, thus taking on all possible interval numbers in X. ZLINE(J) is left blank if SCAT(J, K) is zero, a special character (+) is entered into ZLINE(J) if SCAT(J, K) equal 1, a character corresponding to one of the numbers 2 to 8 is used if SCAT(J, K) contains the corresponding number. If it is larger the symbol * is used. Finally ZLINE(NX+1) is filled with a vertical line (I) to serve as a right-hand margin of the plot. In statement 70 the array ZLINE is printed. It is preceded by a number indicating the middle point of the interval in y and a vertical line for the left-hand margin. After all lines of the plot the x-axis and below it the scale in x are repeated.

Example 6-5: *Two-dimensional scatter diagram of dividend versus share price for industrial firms*

Table 6-4 is a printout of the first 10 punched cards containing the dividend in 1967 (in columns 1–10), the price per share (on Dec. 31st 1967 in columns 11–20) and the name of the firm (beginning in column 25) for all West German industrial firms with a stock capital over 10 million marks. The cards are used as input data of the program SCAT. The output is reproduced in table 6-5. As expected it shows a strong correlation between dividend and share price. It can be seen however that for increasing share price the dividend does not grow

TABLE 6-4

Dividend, price per 100 DM nominal share and name of West German industrial firms

12.	133.	ACKERMANN-GOEGGINGEN
08.	417.	ADLERWERKE KLEYER
17.	346.	AGROB AG FUER GROB U. FEINKERAMIK
25.	765.	AG.F.ENERGIEWIRTSCHAFT
16.	355.	AG F. LICHT- U. KRAFTVERS.,MCHN.
20.	315.	AG.F. IND.U.VERKEHRSW.
08.	138.	AG. WESER
16.	295.	AEG ALLG.ELEKTR.-GES.
20.	479.	ANDREAE-NORIS ZAHN
10.	201.	ANKERWERKE

```
DIVIDEND (X-AXIS) VS. SHARE PRICE (Y-AXIS)

NDATA =     226
XBEG  =     0.000, XEND    =  30.000, DELTAX =    1.000
YBEG  =     0.000, YEND    = 800.000, DELTAY =   20.000
XMEAN =    12.372, XWIDTH =    6.621
YMEAN =   271.733, YWIDTH =  237.894
CORRELATION =    0.796

                  X
        0.500     10.500    20.500
  Y     I---------I---------I---------
790.000 I                              I
770.000 I                         +    I
750.000 I+                             I
730.000 I                              I
710.000 I                              I
690.000 I                  +           I
670.000 I              +               I
650.000 I                              I
630.000 I                              I
610.000 I                        +     I
590.000 I                        +     I
570.000 I           +                  I
550.000 I                              I
530.000 I              2    +          I
510.000 I                              I
490.000 I            + + +3            I
470.000 I              +  ++++         I
450.000 I         +    + +  +          I
430.000 I              + +   2         I
410.000 I        +      +    2         I
390.000 I              ++2+4+          I
370.000 I              23+++           I
350.000 I              ++++2+        + I
330.000 I            + 224             I
310.000 I           + +332++ 2         I
290.000 I            ++++2++           I
270.000 I            + +23 3           I
250.000 I      + 2 + +22+4 + +         I
230.000 I       ++   5+22+             I
210.000 I      2   3+2 +               I
190.000 I        2 +22+                I
170.000 I          333+2   +           I
150.000 I+   +  +  2++                 I
130.000 I+   + 2 +26++                 I
110.000 I      4+3++                   I
 90.000 I3   ++33 +3                   I
 70.000 I4   25+ +                     I
 50.000 I2   +                         I
 30.000 I+                             I
 10.000 I+                             I
        I---------I---------I---------
        0.500     10.500    20.500
```

```
DIVIDEND (X-AXIS) VS. SHARE PRICE (Y-AXIS)
PROJECTION ON Y-AXIS

 10.000  1.0  X
 30.000  1.0  X
 50.000  3.0  XXX
 70.000 13.0  XXXXXXXXXXXXX
 90.000 15.0  XXXXXXXXXXXXXXX
110.000 10.0  XXXXXXXXXX
130.000 15.0  XXXXXXXXXXXXXXX
150.000  7.0  XXXXXXX
170.000 13.0  XXXXXXXXXXXXX
190.000  8.0  XXXXXXXX
210.000  9.0  XXXXXXXXX
230.000 13.0  XXXXXXXXXXXXX
250.000 16.0  XXXXXXXXXXXXXXXX
270.000 10.0  XXXXXXXXXX
290.000  8.0  XXXXXXXX
310.000 14.0  XXXXXXXXXXXXXX
330.000  9.0  XXXXXXXXX
350.000  8.0  XXXXXXXX
370.000  8.0  XXXXXXXX
390.000 10.0  XXXXXXXXXX
410.000  4.0  XXXX
430.000  4.0  XXXX
450.000  4.0  XXXX
470.000  5.0  XXXXX
490.000  6.0  XXXXXX
510.000  0.0
530.000  3.0  XXX
550.000  0.0
570.000  1.0  X
590.000  1.0  X
610.000  1.0  X
630.000  0.0
650.000  0.0
670.000  1.0  X
690.000  1.0  X
710.000  0.0
730.000  0.0
750.000  1.0  X
770.000  1.0  X
790.000  0.0
```

```
DIVIDEND (X-AXIS) VS. SHARE PRICE (Y-AXIS)
PROJECTION ON X-AXIS

 0.500 14.0  XXXXXXXXXXXXXX
 1.500  0.0
 2.500  0.0
 3.500  0.0
 4.500  6.0  XXXXXX
 5.500  6.0  XXXXXX
 6.500 13.0  XXXXXXXXXXXXX
 7.500  6.0  XXXXXX
 8.500 11.0  XXXXXXXXXXX
 9.500  5.0  XXXXX
10.500 20.0  XXXXXXXXXXXXXXXXXXXX
11.500 10.0  XXXXXXXXXX
12.500 19.0  XXXXXXXXXXXXXXXXXXX
13.500  7.0  XXXXXXX
14.500 25.0  XXXXXXXXXXXXXXXXXXXXXXXXX
15.500 17.0  XXXXXXXXXXXXXXXXX
16.500 23.0  XXXXXXXXXXXXXXXXXXXXXXX
17.500  9.0  XXXXXXXXX
18.500 16.0  XXXXXXXXXXXXXXXX
19.500  5.0  XXXXX
20.500  8.0  XXXXXXXX
21.500  0.0
22.500  0.0
23.500  0.0
24.500  2.0  XX
25.500  1.0  X
26.500  0.0
27.500  0.0
28.500  1.0  X
29.500  0.0
```

TABLE 6-5
Scatter diagram from the data of table 6-4

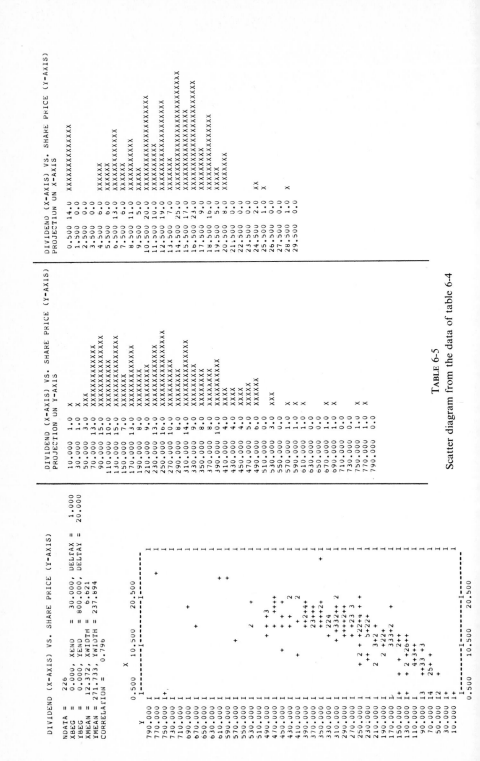

proportionally. Properties other than immediate profit alone seem to lead people to pay high prices for shares. The projections are reproduced on the right-hand side rather than below the scatter diagram. The projection on the x-axis shows the non-statistical nature of the dividend. In Germany it is given in percent of the nominal share price and is nearly always an integer number. It can be seen that even numbers are strongly preferred.

Exercises

6-1: *Efficiency of estimators*
x_1, x_2, x_3 are elements of a sample drawn from a continuous population of unknown mean $\hat{x}$ but known variance σ^2.
(a) Show that

$$S_1 = \tfrac{1}{4}x_1 + \tfrac{1}{4}x_2 + \tfrac{1}{2}x_3,$$

$$S_2 = \tfrac{1}{5}x_1 + \tfrac{2}{5}x_2 + \tfrac{2}{5}x_3,$$

$$S_3 = \tfrac{1}{6}x_1 + \tfrac{1}{3}x_2 + \tfrac{1}{2}x_3$$

are unbiased estimators of $\hat{x}$. (Hint: it is easy to show generally that $S = \sum_{i=1}^{n} a_i x_i$ is unbiased if $\sum_{i=1}^{n} a_i = 1$.)
(b) Construct the variances $\sigma^2(S_1)$, $\sigma^2(S_2)$, $\sigma^2(S_3)$ using eq. (4-5.7) and assuming that the elements x_1, x_2, x_3 are independent.
(c) Show that the arithmetic mean $\bar{x} = \tfrac{1}{3}x_1 + \tfrac{1}{3}x_2 + \tfrac{1}{3}x_3$ has smallest variance of all estimators of the type $S = \sum_{i=1}^{3} a_i x_i$ with the constraint $\sum_{i=1}^{3} a_i = 1$. (Hint: minimize the variance of $S = a_1 x_1 + a_2 x_2 + (1 - a_1 - a_2)x_3$ with respect to a_1 and a_2.) Compute this variance and compare it with the variances obtained in (b).

6-2: *Sample mean and variance*
Compute sample mean $\bar{x}$, sample variance s^2 and an estimate for the variance of the sample mean $s_{\bar{x}}^2 = (1/n)s^2$ for the following sample

18, 21, 23, 19, 20, 21, 20, 19, 20, 17,

using the method of example 6-1.

6-3: *Sampling from a partitioned population*
An opinion poll is to be performed on the outcome of an election. In our (highly constructed) example the population is divided into 3 subpopulations $i = 1, 2, 3$ and a preliminary sample of size 10 is drawn from each subpopulation. Each element of a sample can have the value 0 (vote for party A) or 1 (vote for party B).

The samples are

$i = 1$ $(p_i = 0.1)$: $x_{ij} = 0, 0, 0, 0, 1, 0, 1, 0, 0, 0,$
$i = 2$ $(p_i = 0.7)$: $x_{ij} = 0, 0, 1, 1, 0, 1, 0, 1, 1, 0,$
$i = 3$ $(p_i = 0.2)$: $x_{ij} = 0, 1, 1, 1, 1, 0, 1, 1, 1, 0.$

(a) Use these samples to give an estimate $\tilde{x}$ for the result of the election and its variance $s_{\tilde{x}}^2$. Does the result favour one party significantly?

(b) Use the samples to determine the ratios n_i/n for larger subsamples of size n_i to the total sample size n so that $\tilde{x}$ will have smallest variance (cf. example 6-2).

6-4: *χ^2-distribution*
(a) Determine the skewness $\gamma = \mu_3/\sigma^3$ of the χ^2-distribution using eq. (5-6.7). Start by expressing μ_3 in terms of λ_3, $\hat{x}$ and $E(x^2)$.
(b) Show that $\gamma \to 0$ for $n \to \infty$.

6-5: *Counting error in small sample*
Determine the counting errors Δk_+ and Δk_- for $k = 2$ from table 6-1 at the confidence level $\alpha = 90\%$. Compute their relative deviations $\Delta k_+/\sqrt{k}$, $\Delta k_-/\sqrt{k}$ and $(\Delta k_+ + \Delta k_-)/2\Delta k$ from the "normal" error Δk.

6-6: *Histogram*
Construct a histogram of the following 30 measurements:
26.02, 27.13, 24.78, 26.19, 22.57, 25.16, 24.39, 22.73, 25.35, 26.13, 23.15, 26.29, 24.89, 23.76, 26.04, 22.03, 27.48, 25.42, 24.27, 26.58, 27.17, 24.22, 21.86, 26.53, 27.09, 24.73, 26.12, 28.04, 22.78, 25.36.
Choose a bin width of $\Delta x = 1$. (Hint: draw each measurement as a cross of width Δx and height 1. In this way no ordering of the measurements is necessary since the next cross in the same bin is drawn above the preceding cross. Thus the bars of the histogram grow as the process of plotting proceeds.)

CHAPTER 7

THE METHOD OF "MAXIMUM LIKELIHOOD"

7-1. Likelihood quotient; likelihood function

In the last chapter we introduced the concept of parameter estimation. We have also described the desirable properties of estimators though without specifying how such estimators can be constructed in a particular case. Only for the important quantities expectation value and variance we have derived estimators. We now tackle the general problem.

To specify explicitly the set of p parameters

$$\boldsymbol{\lambda} = (\lambda_1, \lambda_2, \ldots, \lambda_p)$$

that characterize a distribution, we write the probability of the random variables

$$\mathbf{x} = (x_1, x_2, \ldots, x_n)$$

in the form

$$f = f(\mathbf{x}; \boldsymbol{\lambda}). \tag{1.1}$$

A single experiment, i.e. a single measurement of the quantities $\mathbf{x}$ or, in the language of sampling, a sample of size 1 may have lead to the result $\mathbf{x}^{(j)}$. We can assign to the experiment the number

$$\mathrm{d}P^{(j)} = f(\mathbf{x}^{(j)}; \boldsymbol{\lambda})\,\mathrm{d}\mathbf{x}. \tag{1.2}$$

The number $\mathrm{d}P^{(j)}$ has the character of an *a posteriori probability*. It states – after the result of the experiment is known – how probable it was to find just this result, i.e. a value of $\mathbf{x}^{(j)}$ with $x_i^{(j)} < x_i^{(j)} \leqslant x_i + \mathrm{d}x_i$ $(i = 1, 2, \ldots, n)$. For a sample of N (independent) elements or experiments the corresponding probability becomes, according to eq. (2-3.6)

$$\mathrm{d}P = \prod_{j=1}^{N} f(\mathbf{x}^{(j)}; \boldsymbol{\lambda})\,\mathrm{d}\mathbf{x}. \tag{1.3}$$

It is a function of the parameters $\boldsymbol{\lambda}$. There are cases where we know that the population could be characterized by one of only two possible sets of parameters λ_1 or λ_2 which are distinct from one another. This is the case, for instance, in nuclear physics where the parity of a state is either "even" or "odd". We can then form the quotient

$$Q = \frac{\prod_{j=1}^{N} f(x^{(j)}; \lambda_1)}{\prod_{j=1}^{N} f(x^{(j)}; \lambda_2)}. \tag{1.4}$$

The result of the sampling process can be expressed by saying "the set of parameters λ_1 is Q times more probable than the set λ_2". Expression (1.4) is called the *likelihood quotient*, a product of the form

$$L = \prod_{j=1}^{N} f(x^{(j)}; \lambda) \tag{1.5}$$

the *likelihood function*. It is important to stress the difference between an *a posteriori* probability like (1.3) and the normal *a priori* probability. Apart from the fact that the former is a random variable whereas the latter is – although often unknown – a well-defined quantity, the use of a "probability after the event" seems somewhat artificial and has led in the past to many, partly philosophical, discussions. Under no circumstances should the likelihood function (1.5) be confused with a probability density. To keep these entirely different concepts distinct, the term *likelihood* is reserved for *a posteriori* probabilities.

Example 7-1: *Likelihood quotient*

An asymmetric coin is tossed a few times. From the result we try to decide whether it belongs to class A or class B which exhibit the following probability behaviour:

	class A	class B
heads	$\frac{1}{3}$	$\frac{2}{3}$
tails	$\frac{2}{3}$	$\frac{1}{3}$

A sample of 5 tosses yields "heads" once and "tails" 4 times. Then

$L_A = \frac{1}{3}(\frac{2}{3})^4, \qquad L_B = \frac{2}{3}(\frac{1}{3})^4$

and

$Q = L_A/L_B = 8.$

It is thus highly probable that the coin in question belongs to class A.

7-2. The concept of maximum likelihood

It is now apparent how we should generalize the foregoing considerations on likelihood quotients. We shall have the highest confidence in that set of parameters for which the likelihood function (1.5) takes on a maximum value. In fig. 7-1 the situation is sketched for several possible likelihood functions determined by one parameter. Frequently we have the desirable case of a symmetric bell-shaped likelihood function. Here $\tilde{\lambda}$, the value of λ for which the likelihood function $L(\lambda)$ takes its maximum, can undoubtedly be assumed to be the best estimate of λ. The square root of the variance of the distribution about $\tilde{\lambda}$ can be taken as the error of the estimate. In the case of an asymmetric distribution $\tilde{\lambda}$ is still the best estimate, but the variance, although still meaningful mathematically, is an unsatisfactory parameter for the description of the likelihood function near its maximum. It is preferable in this case to present the distribution $L(\lambda)$ itself along with $\tilde{\lambda}$. It can happen in some cases that $L(\lambda)$ has several maxima. We shall then prefer the one with the largest value of the likelihood function. The computation can however become very complex especially if there are several parameters λ. We need to be particularly cautious if there are several maxima of nearly equal magnitude and the number of observations is small.

To determine the position of the maximum we can simply use the standard methods of differential calculus, i.e. find the first derivative of the likelihood function with respect to λ and set it equal to zero. The derivative of a product

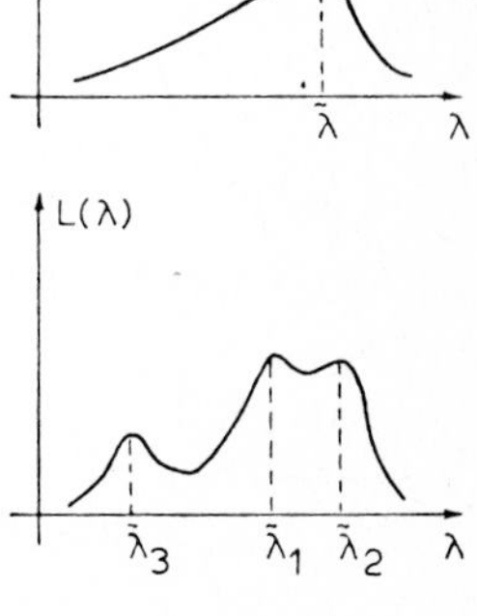

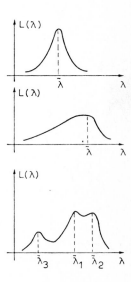

Fig. 7-1. Likelihood functions.

with many factors is however quite cumbersome to handle. We therefore first construct the logarithm of L

$$l = \ln L = \sum_{j=1}^{N} \ln f(x^{(j)}; \boldsymbol{\lambda}). \tag{2.1}$$

This function is also frequently referred to as the likelihood function. To distinguish between the definitions (1.5) and (2.1) we shall usually refer to (2.1) explicitly as the *logarithmic likelihood function*. Obviously the positions of the maxima of (1.5) and (2.1) are identical. It is therefore sufficient to differentiate (2.1) in order to estimate the most likely value of $\boldsymbol{\lambda}$.

We first consider the case of a single parameter, i.e. $\boldsymbol{\lambda} = \lambda$. The problem then reduces to the solution of the so-called *likelihood equation*

$$l' = \mathrm{d}l/\mathrm{d}\lambda = 0. \tag{2.2}$$

Using (2.1) we have

$$l' = \sum_{j=1}^{N} \frac{\mathrm{d}}{\mathrm{d}\lambda} \ln f(x^{(j)}; \lambda) = \sum_{j=1}^{N} \frac{f'}{f} = \sum_{j=1}^{N} \varphi(x^{(j)}; \lambda). \tag{2.3}$$

Here

$$\varphi(x^{(j)}; \lambda) = \frac{\dfrac{\mathrm{d}}{\mathrm{d}\lambda} f(x^{(j)}; \lambda)}{f(x^{(j)}; \lambda)} \tag{2.4}$$

is the usual abbreviation for the *logarithmic derivative* of the function f with respect to λ. In the general case of p parameters the likelihood equation (2.2) is replaced by a system of p simultaneous equations

$$\frac{\partial l}{\partial \lambda_i} = 0, \qquad i = 1, 2, \ldots, p. \tag{2.5}$$

Example 7-2: Repeated measurements of different accuracy

If a quantity is measured repeatedly by different apparatuses or methods the errors of measurement can be different. We shall assume that the errors are normally distributed. An individual measurement therefore corresponds to the drawing of a sample of size one from a Gaussian distribution with mean λ and variance σ_j^2. The *a posteriori* probability for the measured value $x^{(j)}$ is then

$$f(x^{(j)}; \lambda)\,\mathrm{d}x = \frac{1}{\sqrt{(2\pi)}\,\sigma_j} \exp\left(-\frac{(x^{(j)} - \lambda)^2}{2\sigma_j^2}\right) \mathrm{d}x.$$

If a total of N measurements is performed they yield the likelihood function

$$L = \prod_{j=1}^{N} \frac{1}{\sqrt{(2\pi)}\,\sigma_j} \exp\left(-\frac{(x^{(j)} - \lambda)}{2\sigma_j^2}\right) \tag{2.6}$$

with the logarithm

$$l = -\tfrac{1}{2} \sum_{j=1}^{N} \frac{(x^{(j)} - \lambda)^2}{\sigma_j^2} + \text{const.} \tag{2.7}$$

The likelihood equation becomes

$$\frac{\mathrm{d}l}{\mathrm{d}\lambda} = \sum_{j=1}^{N} \frac{x^{(j)} - \lambda}{\sigma_j^2} = 0.$$

Its solution is

$$\tilde{\lambda} = \frac{\displaystyle\sum_{j=1}^{N} \frac{x^{(j)}}{\sigma_j^2}}{\displaystyle\sum_{j=1}^{N} \frac{1}{\sigma_j^2}}. \tag{2.8}$$

The maximum likelihood result is the mean of the individual measurements, each weighted by its inverse variance.

Example 7-3: Estimator for the parameter N of the hypergeometric distribution

In example 5-2 we have discussed the problem of estimating a zoological population using marking and recapture. According to eq. (5-3.10) the probability of catching exactly n fish, of which k are marked, from a pond containing a total (unknown) of N fish, of which K were marked, is given by

$$L(k; n, K, N) = \frac{\binom{K}{k}\binom{N-K}{n-k}}{\binom{N}{n}}.$$

We have to find that value of N for which the function L is maximized. We study the quotient

$$\frac{L(k;\, n,\, k,\, N)}{L(k;\, n,\, k,\, N-1)} = \frac{(N-n)(N-k)}{(N-n-k+k)N} \begin{cases} > 1, & \text{if } Nk < nK, \\ < 1, & \text{if } Nk > nK. \end{cases}$$

Therefore L is largest if N is the integer nearest to nK/k.

7-3. Information inequality; minimum variance and sufficient estimates

We now return to the problem of constructing estimators with desirable properties. In ch. 6, §1 we have defined an estimator S to be unbiased if the *bias* vanished for any sample

$$B(\lambda) = E(S) - \lambda = 0. \tag{3.1}$$

For a "good" estimator it is, however, not sufficient to be unbiased. In addition we require that the variance

$$\sigma^2(S)$$

be as small as possible. Frequently it is necessary to find a compromise between the requirements of minimum bias and minimum variance, because there exists a relation between the two quantities described by the *information inequality**. That such a relation exists can be seen from the observation that it is always possible to construct an estimator with $\sigma^2(S) = 0$ simply by choosing any constant for S. We consider an estimator $S(x^{(1)}, x^{(2)}, \ldots, x^{(N)})$ to be, as before, a function of the sample $x^{(1)}, x^{(2)}, \ldots, x^{(N)}$. The joint probability density of the (random) sample is according to (6-1.3) and (6-1.4)

$$f(x^{(1)}, x^{(2)}, \ldots, x^{(N)}; \lambda) = f(x^{(1)}; \lambda) f(x^{(2)}; \lambda) \cdots f(x^{(N)}; \lambda).$$

The expectation value of S then becomes

$$E(S) = \int S(x^{(1)}, x^{(2)}, \ldots, x^{(N)}) f(x^{(1)}; \lambda) f(x^{(2)}; \lambda) \cdots f(x^{(N)}; \lambda) \times \mathrm{d}x^{(1)} \mathrm{d}x^{(2)} \mathrm{d}x^{(N)}. \tag{3.2}$$

According to (3.1) also

$$E(S) = B(\lambda) + \lambda.$$

We now assume that we can differentiate under the integral with respect to λ. Doing this we obtain

* This inequality was found independently by H. Cramèr, M. Fréchet and C. R. Rao and other authors. It is also called the Cramèr–Rao-inequality or Fréchet-inequality.

$$1 + B'(\lambda) = \int S \left(\sum_{j=1}^{N} \frac{f'(x^{(j)};\lambda)}{f(x^{(j)};\lambda)} \right) f(x^{(1)};\lambda) f(x^{(2)};\lambda) \cdots f(x^{(N)};\lambda) \times \mathrm{d}x^{(1)} \mathrm{d}x^{(2)} \cdots \mathrm{d}x^{(N)},$$

which is equivalent to

$$1 + B'(\lambda) = E\left\{ S \sum_{j=1}^{N} \frac{f'(x^{(j)};\lambda)}{f(x^{(j)};\lambda)} \right\} = E\left\{ S \sum_{j=1}^{N} \varphi(x^{(j)};\lambda) \right\}.$$

From eq. (2.3) we have

$$l' = \sum_{j=1}^{N} \varphi(x^{(j)};\lambda).$$

Therefore

$$1 + B'(\lambda) = E\{Sl'\}. \tag{3.3}$$

Of course we have

$$\int f(x^{(1)};\lambda) f(x^{(2)};\lambda) \cdots f(x^{(N)};\lambda)\, \mathrm{d}x^{(1)} \mathrm{d}x^{(2)} \cdots \mathrm{d}x^{(N)} = 1.$$

Differentiating this equation as well we obtain

$$\int \sum_{j=1}^{N} \frac{f'(x^{(j)};\lambda)}{f(x^{(j)};\lambda)} f(x^{(1)};\lambda) \cdots f(x^{(N)};\lambda)\, \mathrm{d}x^{(1)} \cdots \mathrm{d}x^{(N)} = E(l') = 0.$$

Multiplying this equation by $E(S)$ and subtracting the result from (3.3) we get

$$1 + B'(\lambda) = E\{Sl'\} - E(S)\,E(l') = E\{[S - E(S)]l'\}. \tag{3.4}$$

To discuss the significance of this expression we need an inequality of the Cauchy–Schwarz type in the following form:

> If x and y are random variables and if x^2 and y^2 have finite expectation values, then
>
> $$\{E(xy)\}^2 \leqslant E(x^2)\,E(y^2). \tag{3.5}$$

To prove this inequality we consider the expression

$$E((ax + y)^2) = a^2E(x^2) + 2aE(xy) + E(y^2) \geqslant 0, \tag{3.6}$$

which is a non-negative number for all values of a. If, for the moment, we consider this expression to be zero then it is a quadratic equation for a with the solution

$$a_{1,2} = \frac{E(xy)}{E(x^2)} \pm \sqrt{\left\{\left(\frac{E(xy)}{E(x^2)}\right)^2 - \frac{E(y^2)}{E(x^2)}\right\}}. \tag{3.7}$$

The inequality (3.6) holds if the discriminant in eq. (3.7) vanishes or becomes negative

$$\frac{\{E(xy)\}^2}{\{E(x^2)\}^2} - \frac{E(y^2)}{E(x^2)} \leqslant 0.$$

This is equivalent to (3.5). We now apply (3.5) to (3.4) and obtain

$$\{1 + B'(\lambda)\}^2 \leqslant E\{[S - E(S)]^2\}\, E(l'^2). \tag{3.8}$$

We use (2.3) to rewrite $E(l'^2)$:

$$E(l'^2) = E\left\{\left(\sum_{j=1}^{N} \varphi(x^{(j)}; \lambda)\right)^2\right\}$$

$$= E\left\{\sum_{j=1}^{N} (\varphi(x^{(j)}; \lambda))^2\right\} + E\left\{\sum_{i \neq j} \varphi(x^{(i)}; \lambda)\varphi(x^{(j)}; \lambda)\right\}.$$

All mixed product terms on the right-hand side vanish since

$$E\{\varphi(x^{(i)}; \lambda)\,\varphi(x^{(j)}; \lambda)\} = E\{\varphi(x^{(i)}; \lambda)\}\, E\{\varphi(x^{(j)}; \lambda)\},$$

$$E\{\varphi(x; \lambda)\} = \int_{-\infty}^{\infty} \frac{f'(x; \lambda)}{f(x; \lambda)} f(x; \lambda)\mathrm{d}x = \int f'(x; \lambda)\mathrm{d}x,$$

and

$$\int_{-\infty}^{\infty} f(x; \lambda)\,\mathrm{d}x = 1,$$

which by differentiation with respect to λ under the integral yields

$$\int_{-\infty}^{\infty} f'(x; \lambda)\,\mathrm{d}x = 0.$$

We therefore have simply

$$E(l'^2) = E\left\{\sum_{j=1}^{N} (\varphi(x^{(j)};\lambda))^2\right\} = E\left\{\sum_{j=1}^{N} \left(\frac{f'(x^{(j)};\lambda)}{f(x^{(j)};\lambda)}\right)^2\right\}.$$

Since the terms of the sum are independent, the expectation value of the sum is a sum of expectation values. The individual expectation values do not depend on the particular elements of the sample. Therefore

$$I(\lambda) = E(l'^2) = NE\left\{\left(\frac{f'(x;\lambda)}{f(x;\lambda)}\right)^2\right\}.$$

This expression is called the *information of the sample with respect to* λ. It is a non-negative number that vanishes if the likelihood function does not depend on the parameter λ.

It is sometimes useful to write down the information in a different form. To derive it we differentiate the expression

$$E\left(\frac{f'(x;\lambda)}{f(x;\lambda)}\right) = \int_{-\infty}^{\infty} \frac{f'(x;\lambda)}{f(x;\lambda)} f(x;\lambda)\,\mathrm{d}x = 0$$

another time with respect to λ and obtain

$$0 = \int_{-\infty}^{\infty} \left\{\frac{f'^2}{f} + f\left(\frac{f'}{f}\right)'\right\}\mathrm{d}x = \int_{-\infty}^{\infty} \left\{\left(\frac{f'}{f}\right)^2 + \left(\frac{f'}{f}\right)'\right\} f\,\mathrm{d}x$$

$$= E\left\{\left(\frac{f'}{f}\right)^2\right\} + E\left\{\left(\frac{f'}{f}\right)'\right\}.$$

The information can therefore be written

$$I(\lambda) = NE\left\{\left(\frac{f'(x;\lambda)}{f(x;\lambda)}\right)^2\right\} = -NE\left\{\left(\frac{f'(x;\lambda)}{f(x;\lambda)}\right)'\right\}$$

or

$$I(\lambda) = E(l'^2) = -E(l''). \tag{3.9}$$

We can now write (3.8) as

$$\{1 + B'(\lambda)\}^2 \leqslant \sigma^2(S)\, I(\lambda),$$

or

$$\sigma^2(S) \geqslant \frac{\{1 + B'(\lambda)\}^2}{I(\lambda)}. \tag{3.10}$$

This is the *information inequality*. It provides a relation between the bias and variance of an estimate and the information contained in a sample. Note that no special choice of estimator has been made. The right-hand side of the inequality (3.10) is therefore a lower bound for the variance of a given estimator. It is called the *minimum-variance bound* or the Cramèr-Rao bound. In cases in which the bias does not depend on λ, in particular in cases of vanishing bias, the inequality (3.10) reduces to

$$\sigma^2(S) \geqslant 1/I(\lambda). \tag{3.11}$$

This relation justifies the name information: the larger the information contained in a sample, the smaller the variance of the estimate.

We would now like to ask: Under what circumstances is the minimum-variance bound reached? More explicitly, when does the equality sign in relation (3.10) hold? In relation (3.6) this is the case if $(ax + y)$ vanishes, because only then is $E\{(ax + y)^2\} = 0$ for any a, x and y. Applied to (3.8) this implies that

$$l' + a(S - E(S)) = 0,$$

or

$$l' = A(\lambda)(S - E(S)). \tag{3.12}$$

Here A is any number which must not depend on the sample $x^{(1)}, x^{(2)}, \ldots, x^{(N)}$ although it may be a function of λ. By integration we find that

$$l = \int l' \,\mathrm{d}\lambda = B(\lambda)S + C(\lambda) + D, \tag{3.13}$$

and finally that

$$L = d\exp\{B(\lambda)S + C(\lambda)\}. \tag{3.14}$$

The quantities d and D do not depend on λ.

We see therefore that estimators which are accompanied by likelihood functions of the special type (3.14) attain the minimum variance given in (3.10). They are called *minimum variance estimators*.

In the case of an unbiased minimum variance estimator we have from (3.11) that

$$\sigma^2(S) = \frac{1}{I(\lambda)} = \frac{1}{E(l'^2)}. \tag{3.15}$$

Substituting (3.12)

$$\sigma^2(S) = \frac{1}{(A(\lambda))^2 E\{(S - \lambda)^2\}} = \frac{1}{(A(\lambda))^2 \sigma^2(S)}$$

or

$$\sigma^2(S) = \frac{1}{A(\lambda)}. \tag{3.16}$$

If instead of (3.14) only the weaker condition

$$L = g(S, \lambda)\, c(x^{(1)}, x^{(2)}, \ldots, x^{(N)}) \tag{3.17}$$

holds, the estimator S is called *sufficient* for λ. It can be shown [cf. KENDALL and STUART, Vol. 2, 1967] that no other estimator can contribute information to the knowledge of λ that is not already contained in S, if the condition (3.17) is fulfilled. Hence the name sufficient estimator (or statistic).

Example 7-4: Estimator for the parameter of the Poisson distribution

From eq. (5-4.1) we have

$$f(k) = \frac{\lambda^k}{k!} \mathrm{e}^{-\lambda}.$$

The likelihood function for a sample $k^{(1)}, k^{(2)}, \ldots, k^{(N)}$ is

$$l = \sum_{j=1}^{N} \{k^{(j)} \ln \lambda - \ln(k^{(j)}!) - \lambda\},$$

and its derivative with respect to λ is

$$\frac{\mathrm{d}l}{\mathrm{d}\lambda} = l' = \sum_{j=1}^{N} \left\{\frac{k^{(j)}}{\lambda} - 1\right\} = \frac{1}{\lambda} \sum_{j=1}^{N} \{k^{(j)} - \lambda\},$$

$$l' = \frac{N}{\lambda}(\bar{k} - \lambda). \tag{3.18}$$

Comparison with (3.12) and (3.16) shows that the arithmetic mean $\bar{k}$ is an unbiased minimum variance estimator with variance λ/N.

Example 7-5: Estimator for the parameter of the binomial distribution

With $p = \lambda$, $q = 1 - \lambda$ the likelihood function is given directly by eq. (5-1.3)

$$L(k, \lambda) = \binom{n}{k} \lambda^k (1 - \lambda)^{n-k}.$$

Then

$$l = \ln L = k \ln \lambda + (n - k) \ln (1 - \lambda) + \ln \binom{n}{k},$$

$$l' = \frac{k}{\lambda} - \frac{n-k}{1-\lambda} = \frac{n}{\lambda(1-\lambda)} \left(\frac{k}{n} - \lambda \right).$$

Comparison with (3.12) and (3.16) reveals the arithmetic mean k/n as minimum variance estimator with variance $\lambda(1 - \lambda)/n$.

Example 7-6: *Law of combination of errors from repeated measurements*

We return to the problem of example 7-2, i.e. repeated measurements of the same quantity with different accuracy or, in the language of sampling, sampling from normal distributions with equal mean λ and different but known variances σ_j. From (2.7) we have

$$\frac{dl}{d\lambda} = l' = \sum_{j=1}^{N} \frac{x^{(j)} - \lambda}{\sigma_j^2}.$$

This can be rewritten as

$$l' = \sum \frac{x^{(j)}}{\sigma_j^2} - \sum \frac{\lambda}{\sigma_j^2}$$

$$= \sum_{j=1}^{N} \frac{1}{\sigma_j^2} \left\{ \frac{\sum \frac{x^{(j)}}{\sigma_j^2}}{\sum \frac{1}{\sigma_j^2}} - \lambda \right\}.$$

As in example 7-2 we recognize

$$S = \tilde{\lambda} = \frac{\sum \frac{x^{(j)}}{\sigma_j^2}}{\sum \frac{1}{\sigma_j^2}} \tag{3.19}$$

as unbiased estimator of λ. Comparison with (3.12) now reveals that it also possesses minimum variance. From (3.16) we can read off this variance

$$\sigma^2(\tilde{\lambda}) = \left(\sum_{j=1}^{N} \frac{1}{\sigma_j^2} \right)^{-1}. \tag{3.20}$$

Eq. (3.20) is usually known as the *law of combination of errors*. It could also have been obtained by applying the law of propagation of errors (4-5.7) to eq. (3.19). Identifying $\sigma(\tilde{\lambda})$ with the error of the estimate $\tilde{\lambda}$ and σ_j with the error of the jth measurement we can write it in its usual form

$$\Delta\tilde{\lambda} = \left(\frac{1}{(\Delta x_1)^2} + \frac{1}{(\Delta x_2)^2} + \ldots + \frac{1}{(\Delta x_n)^2} \right)^{-\frac{1}{2}}. \tag{3.21}$$

If all measurements are of equal accuracy $\sigma = \sigma_j$, eqs. (3.19) and (3.20) reduce to the relations

$$\tilde{\lambda} = \bar{x}, \qquad \sigma^2(\tilde{\lambda}) = \sigma^2/n,$$

which were already found in ch. 6, §2.

7-4. Asymptotic properties of the likelihood function and of maximum likelihood estimators

We now want to show heuristically some important properties of the likelihood function and maximum likelihood estimators for very large samples, i.e. for the limit $N \to \infty$. The estimator $S = \tilde{\lambda}$ was defined as a solution of the likelihood equation

$$l'(\lambda) = \sum_{j=1}^{N} \left(\frac{f'(x^{(j)}; \lambda)}{f(x^{(j)}; \lambda)} \right)_{\tilde{\lambda}} = 0. \tag{4.1}$$

Assuming that the derivative $l'(\lambda)$ can be differentiated further with respect to λ we develop it into a Taylor series about $\lambda = \tilde{\lambda}$

$$l'(\lambda) = l'(\tilde{\lambda}) + (\lambda - \tilde{\lambda})l''(\tilde{\lambda}) + \ldots . \tag{4.2}$$

The first term on the right-hand side vanishes because of (4.1). In the second term we can write explicitly

$$l''(\tilde{\lambda}) = \sum_{j=1}^{N} \left(\frac{f'(x^{(j)}; \lambda)}{f(x^{(j)}; \lambda)} \right)'_{\tilde{\lambda}}.$$

This expression has the form of a sample mean. For very large N it can be replaced by the corresponding expectation value (cf. ch. 6, §2)

$$l''(\tilde{\lambda}) = NE\left\{ \left(\frac{f'(x; \lambda)}{f(x; \lambda)} \right)'_{\tilde{\lambda}} \right\}. \tag{4.3}$$

Using eq. (3.9) we can write

$$l''(\tilde{\lambda}) = E(l''(\tilde{\lambda})) = -E(l'^2(\tilde{\lambda})) = -I(\tilde{\lambda}) = -1/b^2. \tag{4.4}$$

We have thus replaced the expression $l''(\tilde{\lambda})$, which is a function of the particular sample $x^{(1)}, x^{(2)}, \ldots, x^{(10)}$, by the number $-1/b^2$ which depends only on the density f and the estimator $\tilde{\lambda}$. Neglecting the higher order terms, eq. (4.2) can now be written as

$$l'(\lambda) = -\frac{1}{b^2}(\lambda - \tilde{\lambda}). \tag{4.5}$$

By integration we obtain

$$l(\lambda) = -\frac{1}{2b^2}(\lambda - \tilde{\lambda})^2 + c,$$

or

$$L(\lambda) = k\exp\{-(\lambda - \tilde{\lambda})^2/2b^2\}. \tag{4.6}$$

Here c and k are constants. The likelihood function $L(\lambda)$ has the form of a normal distribution with mean $\tilde{\lambda}$ and variance b^2.

We can now compare eq. (4.5) with eqs. (3.12) and (3.16). Since we are estimating the parameter we have $S = \tilde{\lambda}$ and $E(S) = \lambda$. Therefore $\tilde{\lambda}$ is an unbiased minimum variance estimator with variance

$$\sigma^2(\tilde{\lambda}) = b^2 = \frac{1}{I(\tilde{\lambda})} = \frac{1}{E(l'^2(\tilde{\lambda}))} = -\frac{1}{E(l''(\tilde{\lambda}))}. \tag{4.7}$$

Since the estimator $\tilde{\lambda}$ has this property only in the limit $N \to \infty$, we refer to it as being *asymptotically unbiased.* This is equivalent to saying that the maximum likelihood estimator is consistent (cf. ch. 6, §1). Similarly the likelihood function is called *asymptotically normal.*

In §2 we have found that the likelihood function $L(\lambda)$ is a measure of the probability that the true value λ_0 of the parameter λ is equal to λ. The result of an estimation is often presented in the abbreviated form

$$\lambda = \tilde{\lambda} \pm \sigma(\tilde{\lambda}) = \tilde{\lambda} \pm \Delta\lambda.$$

Since the likelihood function is asymptotically normal this should be interpreted as follows (cf. ch. 5, §9): The probability that the true value λ_0 lies in the interval

$$\tilde{\lambda} - \Delta\tilde{\lambda} < \lambda_0 < \tilde{\lambda} + \Delta\tilde{\lambda}$$

is 68.2%.

In practice the relations derived above are used in large, but finite, samples. Unfortunately no general rules can be given as to when a sample is large enough to allow this procedure. It is evident that for finite N, eq. (4.3) is an approximation whose accuracy depends not only on N but also on the particular probability density $f(x;\lambda)$.

7-5. Solution of the likelihood equation by iteration

Frequently the likelihood equation cannot be solved analytically. In simple cases graphical methods can be used to obtain the maximum of the likelihood functions L or l. We can also use various computational methods to reach a solution by successive approximation. We use some relations derived in the previous paragraph to describe one such method. We begin with the likelihood equation

$$l' = \sum_{j=1}^{N} \frac{f'(x^{(j)};\lambda)}{f(x^{(j)};\lambda)} = 0, \tag{5.1}$$

and assume that some first approximation λ_0 of the parameter to be estimated is available (by previous experiment or guess). Calling the second approximation λ_1, we can write

$$\lambda_1 = \lambda_0 + \delta\lambda_0. \tag{5.2}$$

The correction $\delta\lambda_0$ can be derived from an expansion

$$l'(\lambda_1) = l'(\lambda_0) + \delta\lambda_0\, l''(\lambda_0) + \dots . \tag{5.3}$$

Neglecting higher terms we have

$$\delta\lambda_0 = \frac{l'(\lambda_0)}{-\,l''(\lambda_0)}. \tag{5.4}$$

In the above it has been assumed that the left-hand side of eq. (5.3) vanishes, i.e. that the likelihood equation holds for the second approximation.

Now just as in §4 we may replace

$$-\,l''(\lambda_0) \approx -\,E(l''(\lambda_0)) = E(l'^2(\lambda_0)) = I(\lambda_0). \tag{5.5}$$

Although for finite samples the equality in (5.5) is not exact, we may still perform this substitution since λ_1 is also only an approximation. The process can be repeated in iterations:

$$\lambda_1 = \lambda_0 + \delta\lambda_0 = \lambda_0 + \frac{l'(\lambda_0)}{I(\lambda_0)},$$

$$\lambda_2 = \lambda_1 + \delta\lambda_1 = \lambda_1 + \frac{l'(\lambda_1)}{I(\lambda_1)},$$

$$\vdots$$

The approximation λ_i will be regarded as sufficiently accurate if $|\delta\lambda_{i-1}|$ falls below some chosen limit.

7-6. Simultaneous estimation of several parameters

We have already encountered (eq. (2.5)) a system of equations that allows the simultaneous estimation of p parameters $\boldsymbol{\lambda} = (\lambda_1, \lambda_2, \ldots, \lambda_p)$. We find that it is not the determination of the parameters themselves, but rather the estimation of their errors which gives rise to greater complications. In addition the covariances between different parameters have to be taken into consideration.

We extend the discussion of §4 to consider the asymptotic properties of the likelihood function in the case of several parameters. The logarithmic likelihood function

$$l(x^{(1)}, x^{(2)}, \ldots, x^{(N)}; \boldsymbol{\lambda}) = \sum_{j=1}^{N} \ln f(x^{(j)}; \boldsymbol{\lambda}) \tag{6.1}$$

can be developed into a Taylor series expansion about the point

$$\tilde{\boldsymbol{\lambda}} = (\tilde{\lambda}_1, \tilde{\lambda}_2, \ldots, \tilde{\lambda}_p), \tag{6.2}$$

$$l(\boldsymbol{\lambda}) = l(\tilde{\boldsymbol{\lambda}}) + \sum_{k=1}^{p} \left(\frac{\partial l}{\partial \lambda_k}\right)_{\tilde{\boldsymbol{\lambda}}} (\lambda_k - \tilde{\lambda}_k) - \tfrac{1}{2} \sum_{l=1}^{p} \sum_{m=1}^{p} \left(\frac{\partial^2 l}{\partial \lambda_l\, \partial \lambda_m}\right)_{\tilde{\boldsymbol{\lambda}}} (\lambda_l - \tilde{\lambda}_l)(\lambda_m - \tilde{\lambda}_m) + \ldots \tag{6.3}$$

Since (from the definition of $\tilde{\boldsymbol{\lambda}}$)

$$\left(\frac{\partial l}{\partial \lambda_k}\right)_{\tilde{\boldsymbol{\lambda}}} = 0; \qquad k = 1, 2, \ldots, p,$$

for all k, the expansion simplifies to

$$l(\boldsymbol{\lambda}) = l(\tilde{\boldsymbol{\lambda}}) - \tfrac{1}{2}(\boldsymbol{\lambda} - \tilde{\boldsymbol{\lambda}})^{\mathrm{T}} A (\boldsymbol{\lambda} - \tilde{\boldsymbol{\lambda}}) + \ldots, \tag{6.4}$$

with

$$A = \begin{pmatrix} \dfrac{\partial^2 l}{\partial \lambda_1^2} & \dfrac{\partial^2 l}{\partial \lambda_1 \partial \lambda_2} & \cdots & \dfrac{\partial^2 l}{\partial \lambda_1 \partial \lambda_p} \\ \dfrac{\partial^2 l}{\partial \lambda_1 \partial \lambda_2} & \dfrac{\partial^2 l}{\partial \lambda_2^2} & \cdots & \dfrac{\partial^2 l}{\partial \lambda_2 \partial \lambda_p} \\ \vdots & & & \\ \dfrac{\partial^2 l}{\partial \lambda_1 \partial \lambda_p} & \dfrac{\partial^2 l}{\partial \lambda_2 \partial \lambda_p} & \cdots & \dfrac{\partial^2 l}{\partial \lambda_p^2} \end{pmatrix}_{\lambda = \tilde{\lambda}} \tag{6.5}$$

In the limit $N \to \infty$ we may replace the elements of A, which still depend on the particular sample, by the corresponding expectation values

$$B = E(A) = \begin{pmatrix} E\left(\dfrac{\partial^2 l}{\partial \lambda_1^2}\right) & E\left(\dfrac{\partial^2 l}{\partial \lambda_1 \partial \lambda_2}\right) & \cdots & E\left(\dfrac{\partial^2 l}{\partial \lambda_1 \partial \lambda_p}\right) \\ E\left(\dfrac{\partial^2 l}{\partial \lambda_1 \partial \lambda_2}\right) & E\left(\dfrac{\partial^2 l}{\partial \lambda_2^2}\right) & \cdots & E\left(\dfrac{\partial^2 l}{\partial \lambda_2 \partial \lambda_p}\right) \\ \vdots & & & \\ E\left(\dfrac{\partial^2 l}{\partial \lambda_1 \partial \lambda_p}\right) & E\left(\dfrac{\partial^2 l}{\partial \lambda_2 \partial \lambda_p}\right) & \cdots & E\left(\dfrac{\partial^2 l}{\partial \lambda_p^2}\right) \end{pmatrix}_{\lambda = \tilde{\lambda}} . \tag{6.6}$$

Neglecting higher-order terms, we can now write down the non-logarithmic likelihood function

$$L = k \exp\{\tfrac{1}{2}(\lambda - \tilde{\lambda})^{\mathrm{T}} B(\lambda - \tilde{\lambda})\}. \tag{6.7}$$

Comparison with eq. (5-10.1) shows that this is a p-dimensional normal distribution with mean $\tilde{\lambda}$ and covariance matrix

$$C = B^{-1}. \tag{6.8}$$

The variances of the maximum likelihood estimates $\tilde{\lambda}_1, \tilde{\lambda}_2, \ldots, \tilde{\lambda}_p$ are given by the diagonal elements of the matrix (6.8). The off-diagonal elements are the covariances between all possible pairs of estimates

$$\sigma^2(\tilde{\lambda}_i) = c_{ii}, \tag{6.9}$$

$$\operatorname{cov}(\tilde{\lambda}_j, \tilde{\lambda}_k) = c_{jk}. \tag{6.10}$$

One can define the correlation coefficient between the estimates $\tilde{\lambda}_j, \tilde{\lambda}_k$:

$$\rho(\tilde{\lambda}_j, \tilde{\lambda}_k) = \frac{\operatorname{cov}(\tilde{\lambda}_j, \tilde{\lambda}_k)}{\sigma(\tilde{\lambda}_j)\sigma(\tilde{\lambda}_k)}. \tag{6.11}$$

As in the case of a single parameter, the square roots of the variances are called the errors or standard deviations of the estimates

$$\Delta\tilde{\lambda}_i = \sigma(\tilde{\lambda}_i) = \sqrt{c_{ii}}. \qquad (6.12)$$

In §4 we have seen that by presenting the maximum likelihood estimate and its error a region was defined which contained the true value with 68.2% probability. Since in the multiparameter case the likelihood function is a multivariate Gaussian distribution, such a region is not defined by the errors alone, but the complete covariance matrix must be taken into account. For the particular case of two parameters this region is the covariance ellipse, introduced in ch. 5, §10.

Example 7-7: Estimators for mean and variance of the normal distribution

From a sample of size N we want to determine the mean λ_1 and variance λ_2 of a normal distribution. This problem occurs for example in the study of the range of α-particles in matter. Because of the statistical nature of energy loss by a large number of individual independent collisions, the range of particles is normally distributed around the mean. By measuring the ranges $x^{(j)}$ of N individual particles the mean λ_1 and the "straggling-constant" $\lambda_2 = \sigma$ can be estimated. We get the likelihood functions

$$L = \prod_{j=1}^{N} \frac{1}{\sqrt{(2\pi)}\,\lambda_2} \exp\left(-\frac{(x^{(j)} - \lambda_1)^2}{2\lambda_2^2}\right),$$

and

$$l = -\tfrac{1}{2}\sum_{j=1}^{N}\frac{(x^{(j)}-\lambda_1)^2}{\lambda_2^2} - N\ln\lambda_2 - \text{const.}$$

The system of likelihood equations becomes

$$\frac{\partial l}{\partial \lambda_1} = \sum_{j=1}^{N}\frac{x^{(j)} - \lambda_1}{\lambda_2^2} = 0,$$

$$\frac{\partial l}{\partial \lambda_2} = \frac{1}{\lambda_2^3}\sum_{j=1}^{N}(x^{(j)} - \lambda_1)^2 - \frac{N}{\lambda_2} = 0.$$

Its solution is

$$\tilde{\lambda}_1 = \frac{1}{N}\sum_{j=1}^{N} x^{(j)},$$

$$\tilde{\lambda}_2 = \sqrt{\left(\frac{\sum_{j=1}^{N}(x^{(j)} - \tilde{\lambda}_1)^2}{N}\right)}.$$

The maximum likelihood method yields for the expectation value the arithmetic mean of the individual measurements; for the variance it gives the quantity s'^2 (eq. (6-2.5)) which is somewhat biased and not s, the unbiased estimator (eq. (6-2.7))

We now determine the matrix B. The second derivatives are

$$\frac{\partial^2 l}{\partial \lambda_1^2} = -\frac{N}{\lambda_2^2},$$

$$\frac{\partial^2 l}{\partial \lambda_1 \partial \lambda_2} = -\frac{2\sum(x^{(j)} - \lambda_1)}{\lambda_2^2},$$

$$\frac{\partial^2 l}{\partial \lambda_2^2} = -\frac{3\sum(x^{(j)} - \lambda_1)^2}{\lambda_2^4} + \frac{N}{\lambda_2^2}.$$

Following the method of eq. (6.6) we substitute λ_1, λ_2 by $\tilde{\lambda}_1, \tilde{\lambda}_2$ and find that

$$B = \begin{pmatrix} N/\tilde{\lambda}_2^2 & 0 \\ 0 & 2\,N/\tilde{\lambda}_2^2 \end{pmatrix}$$

or the covariance matrix

$$C = B^{-1} = \begin{pmatrix} \tilde{\lambda}_2^2/N & 0 \\ 0 & \tilde{\lambda}_2^2/2\,N \end{pmatrix}.$$

We interprete the diagonal elements as the errors of the corresponding parameters, i.e.

$$\Delta\tilde{\lambda}_1 = \tilde{\lambda}_2/\sqrt{N}$$

which, incidentally, is exactly eq. (6-2.4) and

$$\Delta\tilde{\lambda}_2 = \tilde{\lambda}_2/\sqrt{(2N)}.$$

There is no correlation between λ_1 and λ_2

Example 7-8: Estimators for the parameters of a bivariate normal distribution

From ch. 5, § 10 we have

$$f(x_1, x_2) = \frac{1}{2\pi\,\sigma_1\sigma_2\sqrt{(1-\rho^2)}}$$

$$\times \exp\left[-\frac{1}{2(1-\rho^2)}\left\{\frac{(x_1-a_1)^2}{\sigma_1^2} + \frac{(x_2-a_2)^2}{\sigma_2^2} - 2\rho\,\frac{(x_1-a_1)(x_2-a_2)}{\sigma_1\sigma_2}\right\}\right].$$

By solving the set of simultaneous likelihood equations for the five parameters $a_1, a_2, \sigma_1^2, \sigma_2^2, \rho$ it can be shown that

$$\bar{x}_1 = \frac{1}{N} \sum_{j=1}^{N} x_1^{(j)}, \qquad \bar{x}_2 = \frac{1}{N} \sum_{j=1}^{N} x_2^{(j)},$$

$$s_1'^2 = \frac{1}{N} \sum_{j=1}^{N} (x_1^{(j)} - \bar{x}_1)^2, \quad s_2'^2 = \frac{1}{N} \sum_{j=1}^{N} (x_2^{(j)} - \bar{x}_2)^2,$$

$$r = \frac{\sum_{j=1}^{N} (x_1^{(j)} - \bar{x}_1)(x_2^{(j)} - \bar{x}_2)}{N\, s_1'\, s_2'}, \tag{6.13}$$

are their maximum likelihood estimators. As in example 7-7 the estimators $s_1'^2$ and $s_2'^2$ of the variances are biased. This is also true for the expression (6.13), the *sample correlation coefficient*. Like all maximum likelihood estimators r is consistent, i.e. it yields a good estimate of ρ for very large samples. For $N \to \infty$ the distribution of r approaches a Gaussian distribution with mean ρ and variance

$$\sigma^2(r) = (1 - \rho^2)^2/N. \tag{6.14}$$

For finite samples the distribution is asymmetric. It is therefore important to ensure a sufficiently large sample size when using (6.13). As a rule of thumb it is usually recommended that N be at least 500.

7-7. Uniqueness of the method; confidence interval

Usually any process or experiment can be described by one of several equivalent quantities (parameters), e.g. half life, mean life or decay constant; velocity, momentum or energy; volume or mass etc., that are related to one another by other known quantities or constants. Of course the result of an estimation must not depend on the choice of a particular quantity (e.g. $\alpha(\lambda)$ instead of λ).

If we choose λ we have the likelihood equation

$$\frac{\partial l}{\partial \lambda} = 0. \tag{7.1}$$

If we had chosen $\alpha = \alpha(\lambda)$ instead, it would have been

$$\frac{\partial l}{\partial \alpha} = 0. \tag{7.2}$$

But since

$$\frac{\partial l}{\partial \lambda} = \frac{\partial l}{\partial \alpha}\frac{\partial \alpha}{\partial \lambda},$$

by the validity of (7.1), eq. (7.2) is already satisfied. The choice of the particular parameter is irrelevant. The maximum likelihood method yields *unique solutions.*

The situation is much less satisfactory in what concerns error considerations. We now introduce a new measure for the uncertainty of an estimate, called the *confidence interval.* We want to know the probability of the true value of the parameter λ lying between two fixed values λ' and λ'' (fig. 7-2)

$$P(\lambda' < \lambda < \lambda'') = \frac{\int_{\lambda'}^{\lambda''} L(\lambda)\mathrm{d}\lambda}{\int_{-\infty}^{\infty} L(\lambda)\mathrm{d}\lambda}. \tag{7.3}$$

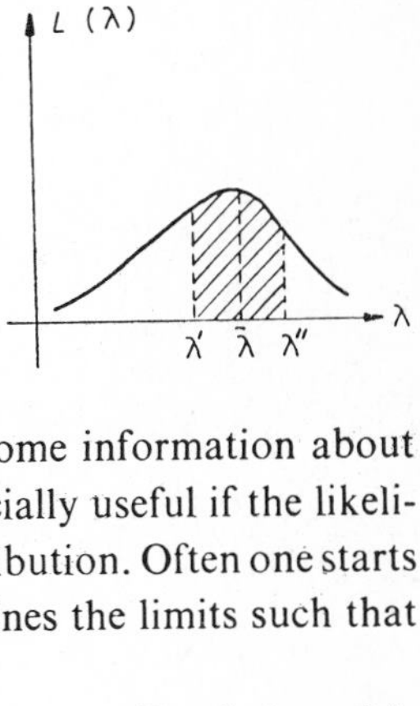

Fig. 7-2. Confidence interval.

Such a definition has the advantage of containing some information about the actual form of the likelihood function. It is especially useful if the likelihood function is not very similar to the Gaussian distribution. Often one starts with a fixed probability (e.g. $P = 95\%$) and determines the limits such that eq. (7.3) is fulfilled.

Unfortunately the probability (7.3) depends on the specific choice of λ. If we choose $\alpha(\lambda)$, we obtain (cf. ch.4, §4)

$$P(\lambda' < \lambda < \lambda'') = \frac{\int_{\lambda'}^{\lambda''} L(\lambda)\mathrm{d}\lambda}{\int_{-\infty}^{\infty} L(\lambda)\mathrm{d}\lambda} = \frac{\int_{\alpha'}^{\alpha''} L\,\frac{\partial \lambda}{\partial \alpha}\,\mathrm{d}\alpha}{\int_{-\infty}^{\infty} L\,\mathrm{d}\lambda} \neq \frac{\int_{\alpha'}^{\alpha''} L\,\mathrm{d}\alpha}{\int_{-\infty}^{\infty} L\,\mathrm{d}\alpha}.$$

7-8. Bartlett's *S*-function

To avoid some of the above difficulties, BARTLETT [1953] introduced the following function

$$S(\lambda) = \frac{1}{C}\frac{\partial l}{\partial \lambda}, \tag{8.1}$$

with

$$C^2 = -\int_{\lambda_{\min}}^{\lambda_{\max}} \frac{\partial^2 l}{\partial \lambda^2} L(\lambda)\, \mathrm{d}\lambda. \tag{8.2}$$

The integral extends over the whole region of variability of λ or – in other words – over the region for which $L(\lambda) \neq 0$. The function S now has zero mean and unit variance independent of the choice of λ:

$$E(S) = \frac{1}{C}\int_{\lambda_{\min}}^{\lambda_{\max}} \frac{\partial l}{\partial \lambda} L\, \mathrm{d}\lambda = \frac{1}{C}\int_{\lambda_{\min}}^{\lambda_{\max}} \frac{\partial L}{\partial \lambda}\, \mathrm{d}\lambda = \frac{1}{C}\int_{\lambda_{\min}}^{\lambda_{\max}} \mathrm{d}L$$

$$= \frac{1}{C}\{L(\lambda_{\max}) - L(\lambda_{\min})\}.$$

Since $L(\lambda)$ vanishes at the boundaries of integration this becomes

$$E(S) = 0. \tag{8.3}$$

The variance then becomes

$$\sigma^2(S) = E\{S^2 - (E(S))^2\} = E\{S^2\} = \frac{1}{C^2}\int_{\lambda_{\min}}^{\lambda_{\max}} \left(\frac{\partial l}{\partial \lambda}\right)^2 L\, \mathrm{d}\lambda$$

$$= \frac{1}{C^2}\int \frac{1}{L^2}\left(\frac{\partial L}{\partial \lambda}\right)^2 L\, \mathrm{d}\lambda = \frac{\int \left(\frac{\partial L}{\partial \lambda}\right)^2 \frac{1}{L}\, \mathrm{d}\lambda}{-\int \frac{\partial}{\partial \lambda}\left(\frac{1}{L}\frac{\partial L}{\partial \lambda}\right) L\, \mathrm{d}\lambda}$$

$$= \frac{\int \left(\frac{\partial L}{\partial \lambda}\right)^2 \frac{1}{L}\, \mathrm{d}\lambda}{-\int \frac{\partial^2 L}{\partial \lambda^2}\, \mathrm{d}\lambda + \int \left(\frac{\partial L}{\partial \lambda}\right)^2 \frac{1}{L}\, \mathrm{d}\lambda}.$$

Outside the range of possible values of λ, i.e. for $\lambda < \lambda_{min}$ and $\lambda > \lambda_{max}$ the likelihood function $L(\lambda)$ is identically zero. Then the derivative $\partial L/\partial\lambda$ also vanishes. Therefore

$$-\int_{\lambda_{min}}^{\lambda_{max}} \frac{\partial^2 L}{\partial\lambda^2}\,\mathrm{d}\lambda = \left(\frac{\partial L}{\partial\lambda}\right)_{\lambda_{min}} - \left(\frac{\partial L}{\partial\lambda}\right)_{\lambda_{max}} = 0,$$

and we simply get

$$\sigma^2(S) = 1. \tag{8.4}$$

We have so far obtained the mean and variance of the function $S(\lambda)$. But we have not yet illustrated how it would look in a particular case. Let us assume now that the likelihood function is a Gaussian with respect to λ with mean $\tilde{\lambda}$ and variance ε^2. Using eqs. (4.5) and (4.6) we easily get

$$C^2 = \frac{1}{\varepsilon^2}, \qquad S(\lambda) = \frac{\tilde{\lambda} - \lambda}{\varepsilon}.$$

The Bartlett S-function is in this case a straight line that passes through zero at $\lambda = \tilde{\lambda}$ and assumes the values ± 1, ± 2, ... when λ deviates by 1, 2, ... standard deviations from $\tilde{\lambda}$. After having found the S-function one therefore solves the equations

$$S(\lambda_-) = +1, \qquad S(\lambda_+) = -1,$$

and determines

$$\Delta_- = \tilde{\lambda} - \lambda_-, \qquad \Delta_+ = \lambda_+ - \tilde{\lambda}.$$

If these are approximately equal we shall consider the error to be symmetrical. Otherwise the result of the analysis is stated in the form

$$\lambda = \tilde{\lambda}\,{}^{+\Delta_+}_{-\Delta_-}.$$

Example 7-9: Determination of the Λ-hyperon's mean life time and its error using Bartlett's S-function

The Λ-hyperon is an elementary particle that decays with mean life time τ according to the law of radioactive decay.
The probability of observing a decay between the times t and $t+\mathrm{d}t$ after production is described by the probability density

$$f(t) = \frac{1}{\tau}\exp(-t/\tau).$$

If the particle moves in some direction with a speed v and we denote the speed of light by c then, because of the relativistic time dilatation, we have instead

$$f(t) = \frac{1}{\tau} \exp\left[- \frac{1}{\tau} t \sqrt{\{1 - (v/c)^2\}} \right].$$

The Λ-particles can be produced by exposing a hydrogen bubble chamber to a beam of K^- mesons. The production point is visible in the chamber by the fact that a K^- track suddenly ends. The Λ-particle, being electrically uncharged, does not ionize, but its decay particles form a characteristic pattern. Their momenta can be measured and added to give the Λ momentum and velocity ($v^{(j)}$). From the distance between production and decay point then the decay time $t^{(j)}$ can be determined. Each individual decay yields the *a posteriori* probability

$$f_j = \frac{1}{\tau} \exp\left[- \frac{1}{\tau} t^{(j)} \sqrt{\{1 - (v^{(j)}/c)^2\}} \right],$$

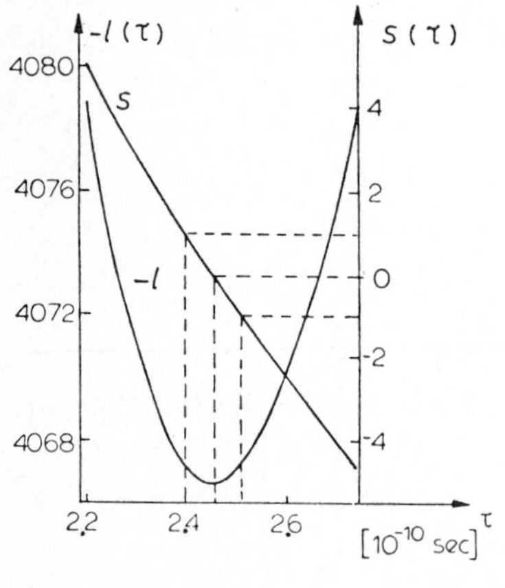

Fig. 7-3. Logarithmic likelihood function and Bartlett's S-function for example 7-9.

which must be further modified to make certain experimental corrections (for unobservable decays etc.)

By forming a product or sum over all observed events we finally obtain the likelihood functions L or $l = \ln L$ and the Bartlett S-function. In fig. 7-3 the latter two functions are displayed obtained from a total of 2213 decays in an experiment of ENGELMANN et al. [1966]. By reading off the maximum of the likelihood function (here

$-l$ is plotted) and the values of τ for $S = \pm 1$ we obtain the result

$$\tau_\Lambda = \left(2.452 \begin{matrix} +\ 0.056 \\ -\ 0.054 \end{matrix}\right) \times 10^{-10} \text{ sec.}$$

Exercises

7-1: *Maximum likelihood estimators*

(a) A random variable x is known to follow the uniform distribution $f(x) = 1/b$, $0 < x < b$. The parameter b has to be estimated from a sample. Show that the maximum likelihood estimator of b is $S = \tilde{b} = x_{\max}$. (Hint: the result cannot be obtained by differentiation but simply by inspection of the likelihood function.)

(b) Set up the likelihood equations for the two parameters a and Γ of the Lorentz distribution (exercise 5-8a). Show that there exists no unique solution. It is easy to see, however, that for $x^{(j)} \ll \Gamma$ the arithmetic mean $\bar{x}$ is the estimator of a.

7-2: *Information*

(a) Construct the information $I(\lambda)$ of a sample of size N drawn from a normal distribution of known variance σ^2 but unknown mean $\lambda = a$.

(b) Construct the information $I(\lambda)$ of a sample of size N drawn from a normal distribution of known mean a but unknown variance $\lambda = \sigma^2$. Show that the maximum likelihood estimator of σ^2 is

$$S = \frac{1}{N} \sum_{j=1}^{N} (x^{(j)} - a)^2$$

and that this estimator is unbiased, i.e. $B(S) = E(S) - \lambda = 0$.

7-3: *Variance of an estimator*

(a) Use the information inequality to obtain a lower limit for the variance of an estimator of the mean in exercise 7-2(a). Show that the limit is equal to the minimum variance found in exercise 6-1(c).

(b) Use the information inequality to obtain a lower limit for the variance of S in exercise 7-2(b).

(c) Use eq. (3.16) to show that S is a minimum variance estimator whose variance is equal to the lower limit obtained in (b).

CHAPTER 8

TESTING OF STATISTICAL HYPOTHESES

Frequently the problem of statistical analysis is not the determination of completely unknown parameters. One is likely to have a prior expectation of the values of the parameters based on previous results or on some model or theory. The purpose of the sample (or measurement) is to test this *hypothesis*. In production control, for example, we assume that certain critical quantities are distributed normally about their nominal value. A sample of products is drawn to test this hypothesis. We want to illustrate the procedure of statistical tests using this example. For simplicity we consider the hypothesis that the variable x of a population is described by the standardized Gaussian distribution. To test it we draw a sample of size 10.

The analysis of the sample may have given the arithmetic mean $\bar{x} = 0.154$. If we assume the hypothesis to be true, the random variable $\bar{x}$ is normally distributed with mean 0 and variance $1/n$. We now ask: what is the probability of observing a value $|\bar{x}| \geqslant 0.154$ in such a distribution? According to (5-9.5) and table F-2 we find

$$P(|\bar{x}| \geqslant 0.154) = 2\{1 - \psi_0(0.154\sqrt{10})\} = 0.62.$$

Therefore even if our hypothesis is true, there is a probability of 62% that a sample of size 10 yields a mean that deviates by 0.154 or more from the population mean.

We now find ourselves in difficulties trying to answer the simple question, "Is the hypothesis true or false?". A clear cut answer cannot be given directly. This might have been expected because all results in mathematical statistics are probability results. We can somewhat artificially get around this situation by introducing the concept of the *level of significance*. The procedure then is the following. Before analyzing the sample we fix a certain (usually small) probability α. We then ask: Assuming the hypothesis to be true, is the probability of finding a sample with the observed properties larger or smaller than α? In our case the question would be, is $P(|\bar{x}| \geqslant 0.154) < \alpha$? If the probability is indeed smaller than α we would conclude that it is very unlikely for a population with the assumed distribution to yield such a sample, and therefore *reject* the hypothesis. The inverse reasoning is not possible. If the probability in question exceeds α one cannot say that "the hypothesis is true" but only that "it is not inconsistent with the result of the sample".

The choice of the level of significance, α, depends on the problem under investigation. In the quality control of pencils we might, for instance, be satisfied with 1 %. On the other hand, if we want to determine the premiums of an insurance company, such that the probability of a crash of the company is less than α, we might regard 0.01 % as still too high. In the analysis of scientific data, usually values of 5 %, 1 % or 0.1 % are taken. We can calculate from table F-3 the limits for $|\bar{x}|$ corresponding to these levels

$$\begin{aligned} 0.05 &= 2\{1 - \psi_0(1.96)\} = 2\{1 - \psi_0(0.62\sqrt{10})\}, \\ 0.01 &= 2\{1 - \psi_0(2.58)\} = 2\{1 - \psi_0(0.82\sqrt{10})\}, \\ 0.001 &= 2\{1 - \psi_0(3.29)\} = 2\{1 - \psi_0(1.04\sqrt{10})\}. \end{aligned}$$

Using these values of α we would require $|\bar{x}|$ to exceed 0.62, 0.82 or 1.04 respectively before rejecting the hypothesis.

In some cases the sign of the quantity in question (in our example $\bar{x}$) is relevant. A baker producing overweight loaves of bread loses profit; loaves which are underweight might cause him to lose his licence. We might therefore test the deviation in one direction only and ask if

$$P(\bar{x} \geqslant x'_\alpha) < \alpha.$$

This is called a "one-tailed test" in contrast to the "two-tailed test" already mentioned (fig. 8-1).

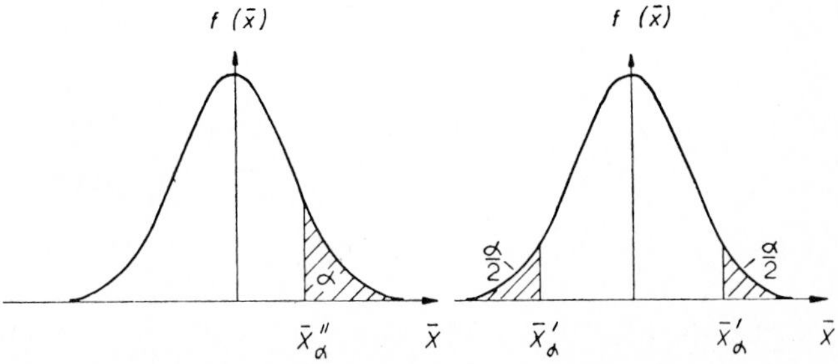

Fig. 8-1. One-tailed and two-tailed tests.

In general some quantity other than the mean might be of interest. We then proceed as follows: First we define a function of the sample especially suitable to perform the test on. This is called the *test statistic T*. We then fix a level of significance α and determine a subregion U within the region of variation of T such that

$$P_H(T \in U) = \alpha.$$

The index H signifies that this probability has been calculated under the assumption that the hypothesis H is true. We now draw a sample. It may, for example, yield the particular value T' of the test statistic. If T' falls inside the critical region U the hypothesis H is rejected.

In the following sections we first discuss a few specific tests and then return to a more rigorous treatment of the theory of statistical tests.

8-1. *F*-test on equality of variances

In the development of measuring or production techniques it is often necessary to compare the variances of populations that have equal means. As an example: a quantity is measured with two different instruments that have no systematic errors; do the measurements have the same variance, i.e. are the instruments of equal quality?

In order to test this hypothesis we assume that the populations are normally distributed and draw samples of size N_1 and N_2, respectively.

For each sample we form the empirical variance (6-2.7) and consider the *variance ratio*

$$F = s_1^2/s_2^2. \tag{1.1}$$

If our hypothesis of equal population variances is true this ratio must be near to unity. From ch. 6, §6 we know that for each sample we can construct a statistic that follows a χ^2-distribution. We have

$$X_1^2 = \frac{(N_1 - 1)s_1^2}{\sigma_1^2} = \frac{f_1\, s_1^2}{\sigma_1^2},$$

$$X_2^2 = \frac{(N_2 - 1)s_2^2}{\sigma_2^2} = \frac{f_2\, s_2^2}{\sigma_2^2}.$$

The two distributions have $f_1 = (N_1 - 1)$ and $f_2 = (N_2 - 1)$ degrees of freedom respectively. Assuming our hypothesis to be true, i.e. $\sigma_1^2 = \sigma_2^2$ the quotient F becomes

$$F = \frac{f_2}{f_1}\,\frac{X_1^2}{X_2^2}.$$

The probability density of a χ^2-distribution with f degrees of freedom is (eq. (6-5.10))

$$f(\chi^2) = \frac{1}{\Gamma(\frac{1}{2}f)\,2^{\frac{1}{2}f}}\,(\chi^2)^{\frac{1}{2}(f-2)}\,e^{-\frac{1}{2}\chi^2}.$$

We can now calculate the probability*

$$W(Q) = P\left(\frac{X_1^2}{X_2^2} < Q\right).$$

that the quotient X_1^2/X_2^2 becomes less than Q

$$W(Q) = \frac{1}{\Gamma(\frac{1}{2}f_1)\,\Gamma(\frac{1}{2}f_2)\,2^{\frac{1}{2}(f_1+f_2)}} \int\int_{\substack{x>0\\ y>0\\ x/y>Q}} x^{\frac{1}{2}f_1-1}\,e^{-\frac{1}{2}x}\,y^{\frac{1}{2}f_2-1}\,e^{-\frac{1}{2}y}\,dx\,dy.$$

The integration leads to

$$W(Q) = \frac{\Gamma(\frac{1}{2}f)}{\Gamma(\frac{1}{2}f_1)\,\Gamma(\frac{1}{2}f_2)} \int_0^Q t^{\frac{1}{2}f_1-1}\,(t+1)^{-\frac{1}{2}f}\,dt, \tag{1.2}$$

with $f = f_1 + f_2$. We can now go back to the original quotient (1.1) and set

$$F = Qf_2/f_1.$$

We then have

$$W(F) = P\left(\frac{s_1^2}{s_2^2} < F\right).$$

This is a distribution function. The corresponding distribution is called Fisher's F-distribution**. It depends on the parameters f_1 and f_2.

The probability density of the F-distribution is

$$f(F) = \left(\frac{f_1}{f_2}\right)^{\frac{1}{2}f_1} \frac{\Gamma(\frac{1}{2}(f_1+f_2))}{\Gamma(\frac{1}{2}f_1)\,\Gamma(\frac{1}{2}f_2)}\,F^{\frac{1}{2}f_1-1}\left(1+\frac{f_1}{f_2}F\right)^{-\frac{1}{2}(f_1+f_2)}. \tag{1.3}$$

It is plotted in fig. 8-2 for specific values of f_1 and f_2. The distribution

* We use the symbol W here to denote a distribution function in order to avoid confusion with the quotient (1.1).

** Other names sometimes used are v^2-distribution, ω^2-distribution, Snedecor distribution.

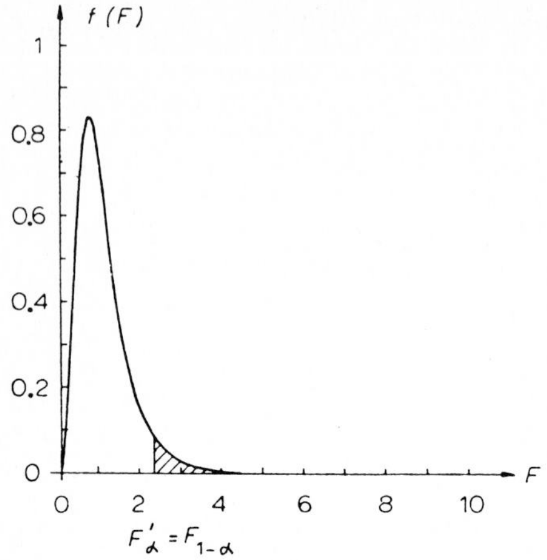

Fig. 8-2. Probability density of F-distribution for $f_1 = 10, f_2 = 20$. The bound $F_{1-\alpha}$ is drawn for $\alpha = 0.05$.

resembles the χ^2-distribution; it is different from zero only for positive values of F and extends a tail for $F \to \infty$. Therefore it cannot be symmetric. It can quite easily be shown that for $f_2 > 2$ the expectation value is

$$E(F) = f_2/(f_2 - 2).$$

We can now determine a bound F'_α by requiring (fig. 8-2)

$$P\left(\frac{s_1^2}{s_2^2} > F'_\alpha\right) = \alpha. \tag{1.4}$$

It is clear from this relation that the bound F'_α is equal to the *fractile* (cf. ch. 3, §3) $F_{1-\alpha}$ of the F-distribution

$$P\left(\frac{s_1^2}{s_2^2} < F'_\alpha\right) = P\left(\frac{s_1^2}{s_2^2} < F_{1-\alpha}\right) = 1 - \alpha. \tag{1.5}$$

If this bound is exceeded we say that $\sigma_1^2 > \sigma_2^2$ at the significance level α. In table F-6, $F_{1-\alpha}$ is tabulated for different pairs of (f_1, f_2). Generally of course we want to use the two-tailed test, i.e. test if the quotient F falls between two bounds F''_α and F'''_α determined by

$$P\left(\frac{s_1^2}{s_2^2} > F''_\alpha\right) = \tfrac{1}{2}\alpha, \qquad P\left(\frac{s_1^2}{s_2^2} < F'''_\alpha\right) = \tfrac{1}{2}\alpha. \tag{1.6}$$

Since F is a quotient the requirement

$$s_1^2/s_2^2 < F_\alpha'''(f_1, f_2)$$

is equivalent to

$$s_2^2/s_1^2 > F_\alpha''(f_2, f_1).$$

Here the first argument always denotes the number of degrees of freedom corresponding to the numerator, the second the one corresponding to the denominator. The requirement (1.6) can therefore be replaced by

$$P\left(\frac{s_1^2}{s_2^2} > F_\alpha''(f_1, f_2)\right) = \tfrac{1}{2}\alpha, \qquad P\left(\frac{s_2^2}{s_1^2} > F_\alpha''(f_2, f_1)\right) = \tfrac{1}{2}\alpha. \qquad (1.7)$$

Comparison of (1.7) and (1.4) shows

$$F_\alpha''(f_1, f_2) = F_{1-\frac{1}{2}\alpha}(f_1, f_2).$$

Because of this table F-6 can be used for both one-tailed and two-tailed *F*-tests.

From table F-6 we find that $F_{1-\frac{1}{2}\alpha} > 1$ for all reasonable values of α, i.e. $\alpha \leqslant 0.1$. Then we need only perform the test

$$s_L^2/s_S^2 > F_{1-\frac{1}{2}\alpha}(f_L, f_S). \qquad (1.8)$$

TABLE 8-1

F-test on equality of variance; data of example 8-1

Number of measurement	Measurements performed on: Instrument 1 [μ]	Instrument 2 [μ]
1	100	97
2	101	102
3	103	103
4	98	96
5	97	100
6	98	101
7	102	100
8	101	
9	99	
10	101	
Mean	100	99.8
Degrees of freedom	9	6
s^2	34/9 = 3.7	39/6 = 6.5

$F = s_L^2/s_S^2 = 6.5/3.7 = 1.8$

Here the indices L and S indicate the larger and smaller values of the sample variances respectively, i.e. $s_L^2 > s_S^2$. If the inequality (1.8) is satisfied then the hypothesis of equal variances has to be rejected.

> *Example* 8-1: *F-test on the hypothesis of equal variance of two series of measurements*
>
> A standard length (100 μ object micrometer) is measured repeatedly by two different microscopes. The measurements and calculations are summarized in table 8-1. From table F-6 we find for the two-tailed F-test at the 10 % level
>
> $$F''_{0.1}(6, 9) = F_{0.95}(6, 9) = 3.37.$$
>
> Therefore the hypothesis of equal variance cannot be rejected.

8-2. Student's test; comparison of means

We consider a random variable x which follows the normal distribution. We draw a sample of size N with mean $\bar{x}$. The variance of this mean is determined by eq. (6-2.4)

$$\sigma^2(\bar{x}) = \sigma^2(x)/N. \tag{2.1}$$

For sufficiently large samples the sample mean $\bar{x}$ is normally distributed with expectation value $\hat{x}$ and variance $\sigma^2(\bar{x})$ because of the central limit theorem, i.e.

$$y = (\bar{x} - \hat{x})/\sigma(\bar{x}) \tag{2.2}$$

follows the standardized Gaussian distribution. Usually, however, $\sigma(\bar{x})$ is not known. Instead we have only the estimate

$$s_x^2 = \frac{1}{N-1}\sum_{j=1}^{N}(x_j - \bar{x})^2 \tag{2.3}$$

for $\sigma^2(x)$. Using eq. (2.1) we can derive an estimate for $\sigma^2(\bar{x})$

$$s_{\bar{x}}^2 = \frac{1}{N(N-1)}\sum_{j=1}^{N}(x_j - \bar{x})^2. \tag{2.4}$$

We now ask: How much does (2.2) deviate from the standardized Gaussian distribution if $\sigma(\bar{x})$ is replaced by $s_{\bar{x}}$? By a simple coordinate translation we can always achieve $\hat{x} = 0$. We therefore consider the distribution of

$$t = \bar{x}/s_{\bar{x}} = \bar{x}\sqrt{N}/s_x. \tag{2.5}$$

Since $(N - 1)s_x^2 = fs_x^2$ follows a χ^2-distribution with $f = N - 1$ degrees of freedom, we can write

$$\mathrm{t} = \bar{x}\sqrt{N}\sqrt{f}/\chi.$$

The distribution function of t is described by the probability

$$F(t) = P(\mathrm{t} < t) = P\left(\frac{\bar{x}\sqrt{N}\sqrt{f}}{\chi} < t\right). \tag{2.6}$$

After a somewhat lengthy calculation we find that

$$F(t) = \frac{\Gamma(\frac{1}{2}(f + 1))}{\Gamma(\frac{1}{2}f)\sqrt{\pi}\sqrt{f}} \int_{-\infty}^{t} \left(1 + \frac{t^2}{f}\right)^{-\frac{1}{2}(f+1)} \mathrm{d}t. \tag{2.7}$$

The corresponding probability density is

$$f(t) = \frac{\Gamma(\frac{1}{2}(f + 1))}{\Gamma(\frac{1}{2}f)\sqrt{\pi}\sqrt{f}} \left(1 + \frac{t^2}{f}\right)^{-\frac{1}{2}(f+1)} \tag{2.8}$$

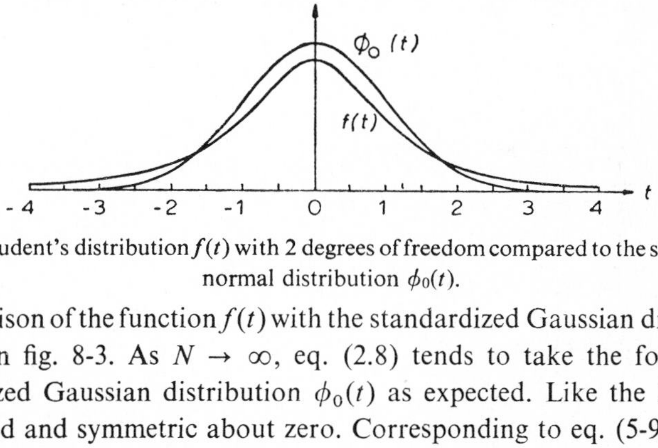

Fig. 8-3. Student's distribution $f(t)$ with 2 degrees of freedom compared to the standardized normal distribution $\phi_0(t)$.

A comparison of the function $f(t)$ with the standardized Gaussian distribution is given in fig. 8-3. As $N \to \infty$, eq. (2.8) tends to take the form of the standardized Gaussian distribution $\phi_0(t)$ as expected. Like the latter it is bell shaped and symmetric about zero. Corresponding to eq. (5-9.4)

$$P(|\mathrm{t}| \leqslant t) = 2\,F(|t|) - 1. \tag{2.9}$$

We can now again determine bounds $\pm\, t'_\alpha$ corresponding to a given level of significance α by requiring

$$\int_0^{t'_\alpha} f(t)\,\mathrm{d}t = \tfrac{1}{2}(1 - \alpha), \tag{2.10}$$

i.e.

$$t'_\alpha = t_{1-\frac{1}{2}\alpha}.$$

The fractiles $t_{1-\frac{1}{2}\alpha}$ are tabulated in table F-7 for different values of α and f.

We can now describe the method of applying the Student's test*:

> A hypothesis predicts a certain expectation value λ_0. A sample of size N is drawn and yields the mean $\bar{x}$ and the empirical variance s_x^2. If the inequality
>
> $$|t| = \frac{|\bar{x} - \lambda_0|\sqrt{N}}{s_x} > t'_\alpha = t_{1-\frac{1}{2}\alpha} \tag{2.11}$$
>
> is fulfilled the hypothesis must be rejected.

This is a two-tailed test. If deviations in only one direction are important, the test with the same level of significance is

$$t = \frac{(\bar{x} \pm \lambda_0)\sqrt{N}}{s_x} > t'_{2\alpha} = t_{1-\alpha}. \tag{2.12}$$

We now want to generalize the test and apply it to the comparison of two sample means. From two populations X and Y samples of size N_1 and N_2, respectively, have been drawn. We seek a test statistic for the hypothesis of equal population means, i.e.

$$\hat{x} = \hat{y}.$$

Because of the central limit theorem the sample means are approximately normally distributed. Their variances are

$$\sigma^2(\bar{x}) = \frac{1}{N_1}\sigma^2(x), \qquad \sigma^2(\bar{y}) = \frac{1}{N_2}\sigma^2(y). \tag{2.13}$$

They are estimated by

$$s_{\bar{x}}^2 = \frac{1}{N_1(N_1 - 1)} \sum_{j=1}^{N_1} (x_j - \bar{x})^2,$$

$$s_{\bar{y}}^2 = \frac{1}{N_2(N_2 - 1)} \sum_{j=1}^{N_2} (y_j - \bar{y})^2. \tag{2.14}$$

According to ch. 5, §8 the difference

$$\Delta = \bar{x} - \bar{y} \tag{2.15}$$

is also nearly normally distributed with

* The t distribution was introduced and applied to the problem of statistical tests by W. S. Gosset who published under the pseudonym "Student".

$$s_{\Delta}^2 = s_{\bar{x}}^2 + s_{\bar{y}}^2. \tag{2.16}$$

If our hypothesis of equal population means is true then of course $\hat{\Delta} = 0$. The quotient

$$\Delta/\sigma(\Delta) \tag{2.17}$$

would then follow the standardized Gaussian distribution. Using eq. (5-9.4) we could therefore immediately write down the probability for the hypothesis to be true if $\sigma(\Delta)$ were known. Instead only s_{Δ} can be obtained from the samples. The corresponding quotient

$$\Delta/s_{\Delta} \tag{2.18}$$

will generally be somewhat larger.

Usually the hypothesis $\hat{x} = \hat{y}$ tacitly implies that $\bar{x}$ and $\bar{y}$ have been obtained from the same population. Therefore $\sigma^2(x)$ and $\sigma^2(y)$ are identical and as the best estimate for s_x^2 and s_y^2 we can use the weighted average of both. The weights are $(N_1 - 1)$ and $(N_2 - 1)$ respectively

$$s^2 = \frac{(N_1 - 1)s_x^2 + (N_2 - 1)s_y^2}{(N_1 - 1) + (N_2 - 1)}. \tag{2.19}$$

We construct

$$S_{\bar{x}}^2 = \frac{s^2}{N_1}, \qquad S_{\bar{y}}^2 = \frac{s^2}{N_2},$$

and

$$s_{\Delta}^2 = S_{\bar{x}}^2 + S_{\bar{y}}^2 = \frac{N_1 + N_2}{N_1 N_2} s^2. \tag{2.20}$$

It can be proved [cf. KENDALL and STUART, Vol. 2, 1967] that the quotient (2.18) follows the Student distribution with $f = N_1 + N_2 - 2$ degrees of freedom.

The equality of means can now be tested by *Student's difference test*. From the results of the two samples the quantity (2.18) is computed. Its absolute value is compared with the fractile of the Student distribution with $f = N_1 + N_2 - 2$ degrees of freedom corresponding to the level of significance α. If

$$|t| = \frac{|\Delta|}{s_{\Delta}} = \frac{|\bar{x} - \bar{y}|}{s_{\Delta}} \geqslant t'_{\alpha} = t_{1-\frac{1}{2}\alpha}, \tag{2.21}$$

the hypothesis of equal means has to be rejected. We can then assume that $\hat{x} > \hat{y}$ or $\hat{x} < \hat{y}$ depending on the sign of $\Delta = \bar{x} - \bar{y}$.

TABLE 8-2
Student's difference test on equality of means; data of example 8-2

x	y
21	16
24	20
18	22
19	19
25	18
17	19
18	19
22	
21	
23	
18	
13	
16	
23	
22	
24	
$N_1 = 16$	$N_2 = 7$
$\overline{x} = 20.3$	$\overline{y} = 19.0$
$s_x^2 = 171.8/15$	$s_y^2 = 20/6$

Example 8-2: *Student's test on the hypothesis of equal means of two series of measurements*

In table 8-2 measurements are presented (in arbitrary units) on the concentration of neuramin acid in erythrocytes of patients suffering from a blood disease (column x) and for a test group of persons of normal health (y). From the sample means and variances we find

$$|\Delta| = |\bar{x} - \bar{y}| = 1.3,$$

$$s^2 = \frac{15\,s_x^2 + 6\,s_y^2}{21} = 9.15,$$

$$s_\Delta^2 = \frac{23}{112}s^2 = 1.88.$$

For $\alpha = 5\%$ and $f = 21$ we find that $t_{1-\frac{1}{2}\alpha} = 2.08$. We therefore conclude that the experimental material is insufficient to reveal any relation between the disease and the concentration.

8-3. Some aspects of a general theory of tests

So far we have given formulae for certain tests on a rather imprecise basis. We have not, for instance, explicitly stated the reasons for our particular choice of the critical regions. In this section the basic ideas of a more rigorous theory of tests are sketched. Of course an extensive treatment of this subject is far beyond the scope of this book. The interested reader is referred to the bibliography.

Each sample of size N leads to N points in the sample space defined in ch. 2, §1. We consider here for simplicity the case of one continuous random variable x where the sampling process results in N points $(x^{(1)}, x^{(2)}, \dots, x^{(N)})$ on the x-axis. In the case of r random variables there would be N points in an r-dimensional space. The result of sampling can also be represented by one single point in a space of dimension Nr. Thus, for instance, a sample of size 2 with one variable can be visualized as a point in a plane spanned by $x^{(1)}, x^{(2)}$. We will abbreviate such a space by E. Any *hypothesis* H consists of an assumption on the probability density

$$f(x; \lambda_1, \lambda_2, \dots, \lambda_p) = f(x; \boldsymbol{\lambda}) \tag{3.1}$$

of the variable. The hypothesis is called *simple* if the function f is completely defined by it, i.e. if the hypothesis specifies all parameters λ_i; it is *composite* if the general mathematical form of f is known but if the exact value of at least one parameter is not determined. A simple hypothesis may, for example, predict a standardized Gaussian distribution for f, whereas a composite hypothesis might only predict a Gaussian distribution with zero mean but unknown variance. The hypothesis H_0 to be tested is called a *null hypothesis*. We may occasionally write explicitly

$$H_0(\boldsymbol{\lambda} = \boldsymbol{\lambda}_0) = H_0(\lambda_1 = \lambda_{10}, \lambda_2 = \lambda_{20}, \dots, \lambda_p = \lambda_{p0}). \tag{3.2}$$

Any other possible hypothesis is called an *alternative hypothesis*, e.g.

$$H_1(\boldsymbol{\lambda} = \boldsymbol{\lambda}_1) = H_1(\lambda_1 = \lambda_{11}, \lambda_2 = \lambda_{21}, \dots, \lambda_p = \lambda_{p1}). \tag{3.3}$$

Often we have to test a simple null hypothesis (3.2) against one composite alternative hypothesis

$$H_1(\boldsymbol{\lambda} \neq \boldsymbol{\lambda}_0) = H_1(\lambda_1 \neq \lambda_{10}, \lambda_2 \neq \lambda_{20}, \dots, \lambda_p \neq \lambda_{p0}). \tag{3.4}$$

Since the null hypothesis predicts the probability distribution in sample space it also determines the probability of observing a point $X = (x^{(1)}, x^{(2)}, \ldots, x^{(N)})$ in any region of space E. We can define a *critical region* S_c at the significance level α by requiring

$$P(X \in S_c \,|\, H_0) = \alpha, \tag{3.5}$$

i.e. that the probability of observing a point X inside S_c, under the condition that H_0 is true, be equal to α. If X is in fact found to fall inside S_c the hypothesis H_0 is rejected. Note that the requirement (3.5) does not by any means uniquely determine the critical region S_c.

Although the use of E-space is conceptually elegant, it is inconvenient for actually performing a test. We therefore form a *test statistic*

$$T = T(X) = T(x^{(1)}, x^{(2)}, \ldots, x^{(N)}), \tag{3.6}$$

and determine the region U of the variable T that corresponds to the critical region S_c, i.e. one performs a mapping

$$X \to T(X), \qquad S_c(X) \to U(X). \tag{3.7}$$

The null hypothesis is rejected if $T \in U$.

Of course the null hypothesis might be true although it was rejected because $X \in S_c$. The probability of such an error, called an *error of the first kind,* is just equal to α. There is another possibility of making a wrong decision. This is not to reject the hypothesis, because X is not contained in S_c although in fact it is wrong and an alternative hypothesis H_1 is true. This is an *error of the second kind.* Its probability

$$P(X \bar{\in} S_c \,|\, H_1) = \beta \tag{3.8}$$

of course depends on the particular alternative hypothesis H_1. We have now found a means of determining S_c. It is clear that a test will be particularly useful, if for a given probability α the critical region is determined in such a way that the probability β for an error of the second kind is minimized. The critical region, and therefore the test itself, naturally depends on the alternative hypotheses which are anticipated.

Once the critical region S_c is fixed, we can consider the probability of rejecting the null hypothesis as a function of the "true" hypothesis or the parameters describing it. By analogy with eq. (3.5) this is

$$M(S_c, \lambda) = P(X \in S_c \,|\, H) = P(X \in S_c \,|\, \lambda). \tag{3.9}$$

This probability is a function of S_c and the parameters λ, the *power function* of the test. The inverse probability, namely the probability of accepting* the null hypothesis as a function of the parameters of the true hypothesis

$$L(S_c, \boldsymbol{\lambda}) = 1 - M(S_c, \boldsymbol{\lambda}) \tag{3.10}$$

is called the *acceptance probability* or the *operation characteristic* of the test. Obviously we have

$$\begin{aligned} M(S_c, \boldsymbol{\lambda}_0) &= \alpha, & M(S_c, \boldsymbol{\lambda}_1) &= 1 - \beta, \\ L(S_c, \boldsymbol{\lambda}_0) &= 1 - \alpha, & L(S_c, \boldsymbol{\lambda}_1) &= \beta. \end{aligned} \tag{3.11}$$

The *most powerful test* of the simple hypothesis H_0 relative to the simple alternative hypothesis H_1 is now naturally defined as the one for which

$$M(S_c, \boldsymbol{\lambda}_1) = 1 - \beta = \text{maximum}. \tag{3.12}$$

A *uniformly most powerful* test may exist. It is most powerful with respect to any possible alternative hypothesis, especially also composite alternatives.

A test is *unbiased* if its power is larger than or equal to α for every alternative hypothesis

$$M(S_c, \boldsymbol{\lambda}_1) \geqslant \alpha. \tag{3.13}$$

This definition is reasonable since the probability of rejecting the null hypothesis should be smallest in that case when this hypothesis is true. An *unbiased most powerful test* is the most powerful among all unbiased tests. *Unbiased uniformly most powerful tests* are defined correspondingly. A method of constructing a test with such desirable properties is discussed in the next section. Before turning to this problem we illustrate the definitions just introduced with an example.

Example 8-3: *Test on the hypothesis that a normal population of given variance σ^2 has the mean $\lambda = \lambda_0$*

As a test statistic we use the arithmetic mean $\bar{x} = (1/n)(x_1 + x_2 + \ldots + x_n)$ (we will see in example 8-4 that this is the best statistic for our purpose). From ch. 6, § 2 we know that $\bar{x}$ is normally distributed with mean λ and variance σ^2/n, i.e. the probability density of $\bar{x}$ in the case $\lambda = \lambda_0$ is

$$f(\bar{x}; \lambda_0) = \frac{\sqrt{n}}{\sqrt{(2\pi)}\,\sigma} \exp\left(-\frac{n}{2\sigma^2}(\bar{x} - \lambda_0)^2\right). \tag{3.14}$$

* We use the word "accept" for brevity although it should be "have no reason to reject".

It is drawn in fig. 8-4 together with 4 different critical regions U corresponding to the same level of significance α. They are

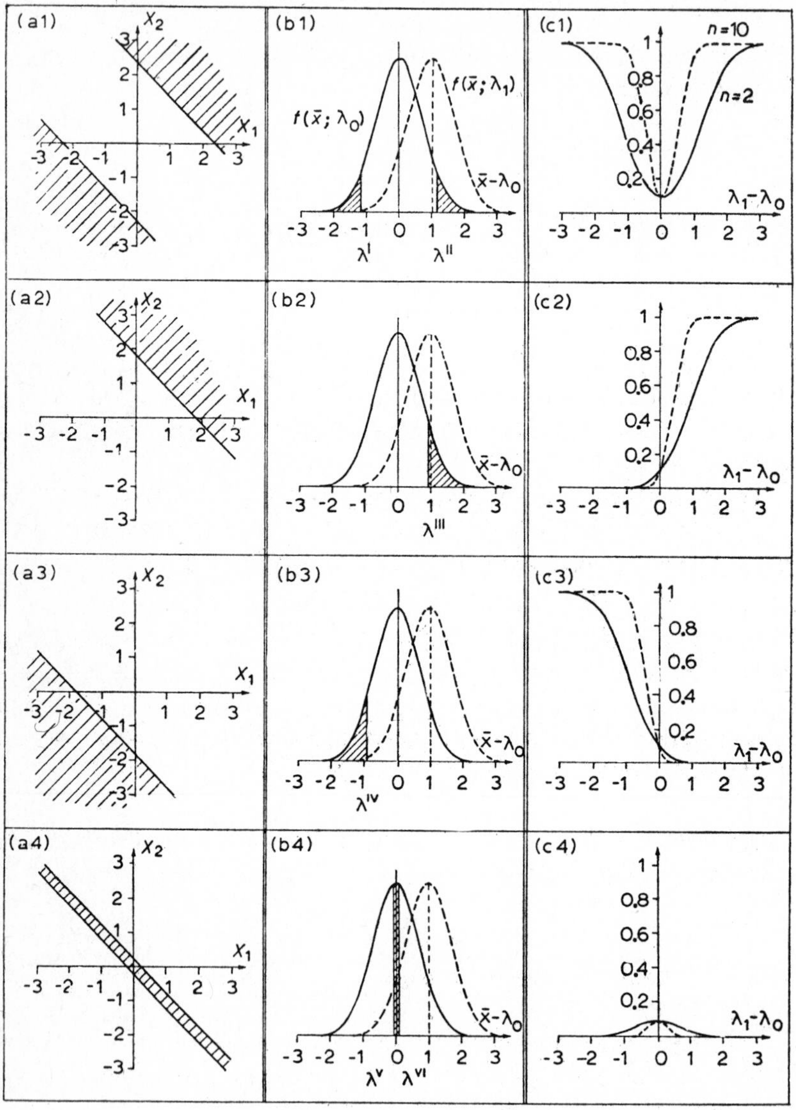

Fig. 8-4. Critical region in E-space (a), critical region of test statistic (b) and power function (c) for the test of example 8-3.

$$U_1:\quad \bar{x} < \lambda' \quad \text{and} \quad \bar{x} > \lambda'' \qquad \text{with} \int_{-\infty}^{\lambda'} f(\bar{x})\,\mathrm{d}\bar{x} = \int_{\lambda''}^{\infty} f(\bar{x})\,\mathrm{d}\bar{x} = \tfrac{1}{2}\alpha;$$

$$U_2:\quad \bar{x} > \lambda''' \qquad \text{with} \int_{\lambda'''}^{\infty} f(\bar{x})\,\mathrm{d}\bar{x} = \alpha;$$

$$U_3:\quad \bar{x} < \lambda^{\mathrm{IV}} \qquad \text{with} \int_{-\infty}^{\lambda^{\mathrm{IV}}} f(\bar{x})\,\mathrm{d}\bar{x} = \alpha;$$

$$U_4:\quad \lambda^{\mathrm{V}} \leqslant \bar{x} < \lambda^{\mathrm{IV}} \qquad \text{with} \int_{\lambda^{\mathrm{V}}}^{\lambda_0} f(\bar{x})\,\mathrm{d}\bar{x} = \int_{\lambda_0}^{\lambda^{\mathrm{IV}}} f(\bar{x})\,\mathrm{d}\bar{x} = \tfrac{1}{2}\alpha.$$

To obtain the power function corresponding to each of these regions we have now to vary λ. The probability density of $\bar{x}$ for any value of λ is by analogy with eq. (3.14)

$$f(\bar{x};\lambda) = \frac{\sqrt{n}}{\sqrt{(2\pi)}\,\sigma} \exp\left[-\frac{n}{2\sigma^2}(\bar{x}-\lambda)\right]. \tag{3.15}$$

The broken curve in fig. 8-4b represents this density for $\lambda = \lambda_1 = \lambda_0 + 1$. The power function (3.9) is now simply

$$P(\bar{x} \in U \mid \lambda) = \int_U f(\bar{x};\lambda)\,\mathrm{d}\bar{x}. \tag{3.16}$$

The power functions thus obtained for the critical regions U_1, U_2, U_3, U_4 are drawn in fig. 8-4c for $n = 2$ and $n = 10$ (solid and broken curves, respectively).

We can now compare the efficiency of the 4 tests we have constructed. From fig. 8-4c we see directly that U_1 corresponds to an unbiased test since clearly the condition (3.13) is fulfilled.

On the other hand the region U_2 is more powerful if the alternative hypothesis is $H_1(\lambda_1 > \lambda_0)$ whereas it is very poor for $H_1(\lambda_1 < \lambda_0)$. The contrary is true for region U_3. Region U_4 finally yields a test whose rejection power is largest if the null hypothesis is true. This is of course not desirable. The test was only constructed for the purpose of demonstration. Comparing the first three tests we see that none of them is more powerful than the other two for all values of λ_1. We have, therefore, not succeeded in finding a uniformly most powerful test. In example 8-4, which continues the discussion of the present example, we shall find that no such test exists for our problem.

8-4. Neyman–Pearson theorem and applications

In the last section we have introduced the space E in which a sample is described by a point X. The probability of finding X inside the critical region S_c subject to the condition that the null hypothesis holds was defined in (3.5)

$$P(X \in S_c \mid H_0) = \alpha. \tag{4.1}$$

We now define the conditional probability density in E-space

$$f(X \mid H_0).$$

Obviously we have

$$\int_{S_c} f(X \mid H_0)\,\mathrm{d}X = P(X \in S_c \mid H_0) = \alpha. \tag{4.2}$$

The following is the theorem of Neyman and Pearson.

A test of the simple hypothesis H_0 relative to a simple hypothesis H_1 is most powerful if the critical region S_c in E-space is such that

$$\frac{f(X \mid H_0)}{f(X \mid H_1)} \begin{cases} \leqslant c \text{ for each } X \in S_c, \\ \geqslant c \text{ for each } X \overline{\in} S_c. \end{cases} \tag{4.3}$$

Here c is a constant which depends on the level of significance.

We prove the theorem by considering another region S besides S_c. It may partly overlap with S_c (fig. 8-5). We choose the size of S such that it could serve as a critical region with the same level of significance, i.e.

$$\int_{S} f(X \mid H_0)\,\mathrm{d}X = \int_{S_c} f(X \mid H_0)\,\mathrm{d}X = \alpha.$$

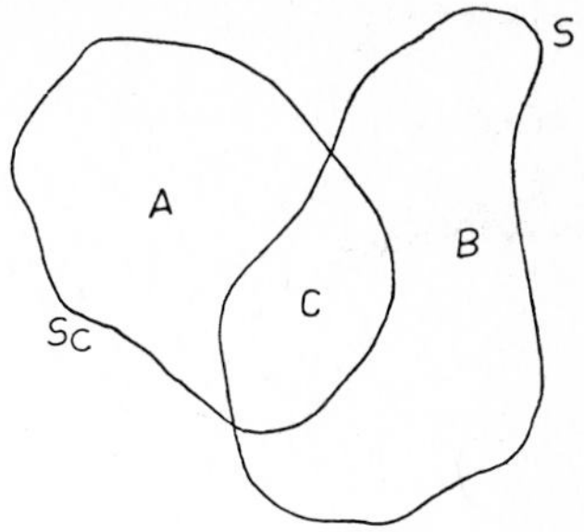

Fig. 8-5. The regions S and S_c.

Using the notation of fig. 8-5 we can write

$$\int_A f(X \mid H_0)\,dX = \int_{S_c} f(X \mid H_0)\,dX - \int_C f(X \mid H_0)\,dX$$

$$= \int_S f(X \mid H_0)\,dX - \int_C f(X \mid H_0)\,dX$$

$$= \int_B f(X \mid H_0)\,dX.$$

Since A is contained in S_c we can use (4.3)

$$\int_A f(X \mid H_0)\,dX \leqslant c \int_A f(X \mid H_1)\,dX.$$

Correspondingly, since B is outside S_c,

$$\int_B f(X \mid H_0)\,dX \geqslant c \int_B f(X \mid H_1)\,dX.$$

We can now express the power function (3.7) in terms of these integrals

$$M(S_c, \lambda_1) = \int_{S_c} f(X \mid H_1)\,dX = \int_A f(X \mid H_1)\,dX + \int_C f(X \mid H_1)\,dX$$

$$\geqslant \frac{1}{c} \int_A f(X \mid H_0)\,dX + \int_C f(X \mid H_1)\,dX$$

$$\geqslant \int_B f(X \mid H_1)\,dX + \int_C f(X \mid H_1)\,dX$$

$$\geqslant \int_S f(X \mid H_1)\,\mathrm{d}X = M(S, \lambda_1),$$

or directly

$$M(S_c, \lambda_1) \geqslant M(S, \lambda_1). \tag{4.4}$$

This is nothing else but the requirement (3.12) for a most powerful test. Since we have made no special assumptions about the alternative hypothesis $H_1(\lambda = \lambda_1)$ or the region S, we have in fact proved that the requirement (4.3) assures a *uniformly most powerful test* if it holds for all alternative hypotheses.

Example 8-4: *Most powerful test for the problem of example* 8-3

We continue the discussion of example 8-3, i.e. we discuss tests on a sample of size N drawn from a normal population with variance σ^2 and unknown mean λ. The conditional probability density of a point $X = (x^{(1)}, x^{(2)}, \ldots, x^{(N)})$ in E space is the joint probability density of the $x^{(j)}$ for given values of λ, i.e.

$$f(X \mid H_0) = \left(\frac{1}{\sqrt{(2\pi)}\,\sigma}\right)^N \exp\left[-\frac{N}{2\sigma^2}\sum_{j=1}^{N}(x^{(j)} - \lambda_0)^2\right], \tag{4.5}$$

and

$$f(X \mid H_1) = \left(\frac{1}{\sqrt{(2\pi)}\,\sigma}\right)^N \exp\left[-\frac{N}{2\sigma^2}\sum_{j=1}^{N}(x^{(j)} - \lambda_1)^2\right], \tag{4.6}$$

for the null hypothesis and an alternative hypothesis respectively. The quotient (4.3) becomes

$$Q = \frac{f(X \mid H_0)}{f(X \mid H_1)} = \exp\left[-\frac{1}{2\sigma^2}\left\{\sum_{j=1}^{N}(x^{(j)} - \lambda_0)^2 - \sum_{j=1}^{N}(x^{(j)} - \lambda_1)^2\right\}\right]$$

$$= \exp\left[-\frac{1}{2\sigma^2}\left\{N(\lambda_0^2 - \lambda_1^2) - 2(\lambda_0 - \lambda_1)\sum_{j=1}^{N}x^{(j)}\right\}\right].$$

The expression

$$\exp\left[-\frac{N}{2\sigma^2}(\lambda_0^2 - \lambda_1^2)\right] = k \geqslant 0,$$

is a non-negative constant. Therefore the condition (4.3) takes the form

$$k \exp\left[\frac{\lambda_0 - \lambda_1}{\sigma^2}\sum_{j=1}^{N}x^{(j)}\right] \begin{cases} \leqslant c \text{ for } X \in S_c, \\ \geqslant c \text{ for } X \,\overline{\in}\, S_c. \end{cases}$$

This is equivalent to

$$(\lambda_0 - \lambda_1)\bar{x} \begin{cases} \leqslant c' \text{ for } X \in S_c, \\ \geqslant c' \text{ for } X \bar{\in} S_c, \end{cases} \tag{4.7}$$

where c' is a constant different from c. Eq. (4.7) does not only place a condition on S_c but also tells us to use $\bar{x}$ as the test statistic. For any given λ_1, i.e. for any simple alternative hypothesis $H_1(\lambda = \lambda_1)$, relation (4.7) gives a clear definition of S_c or for U, the critical region of the statistic $\bar{x}$.

For $\lambda_1 < \lambda_0$ relation (4.7) becomes

$$\bar{x} \begin{cases} \leqslant c'' & \text{for } X \in S_c, \\ \geqslant c'' & \text{for } X \bar{\in} S_c. \end{cases}$$

This corresponds to the situation in fig. 8-4 (b3) with $c'' = \lambda^{\mathrm{IV}}$. Correspondingly for any alternative hypothesis with $\lambda_1 > \lambda_0$ the critical region of the most powerful test is given by

$$\bar{x} \geqslant c'''$$

(cf. fig. 8-4 (b2) with $c''' = \lambda'''$). There exists however no uniformly most powerful test, since the factor $(\lambda_0 - \lambda_1)$ in relation (4.7) changes sign at $\lambda_1 = \lambda_0$.

8-5. The likelihood ratio method

The Neyman–Pearson theorem gave a condition for a most powerful test. However such a test does not exist if the alternative hypothesis comprises values of the parameter both larger and smaller than the one of the null hypothesis. We found this for example 8-4. It can also be shown that this is true in general. The question is now: what is the best test to use if no uniformly most powerful test exists? Of course it is not easy to define the meaning of "best", and we do not attempt to do it. We describe instead a procedure that yields tests with desirable properties, and which has the advantage that it is easy to use.

We shall discuss the general case with p parameters $\boldsymbol{\lambda} = (\lambda_1, \lambda_2, \ldots, \lambda_p)$. The result of a sampling process, i.e. the point $X = (x^{(1)}, x^{(2)}, \ldots, x^{(N)})$ in E-space, is used to test a certain hypothesis. The (composite) null hypothesis is defined by a certain range of values for each parameter. We can span a p-dimensional space using $\lambda_1, \lambda_2, \ldots, \lambda_p$ as coordinates and consider the totality of all ranges allowed by the null hypothesis as a region in parameter space. We denote this region by ω. The region Ω in this space may describe all possible values of the parameters. The most general alternative hypothesis is then described by that part of Ω which does not contain ω. We write it for short as $\Omega - \omega$. We recall that in ch. 7 we introduced the maximum likelihood estimate $\tilde{\lambda}$ of a parameter λ. It was that value of λ for which the likelihood function took its maximum. The range of λ in which we looked for this maximum was tacitly assumed to include all possible values of λ. If only the maximum in a restricted range, say ω, is meant we write $\tilde{\lambda}^{(\omega)}$.

The *likelihood ratio test* now defines a statistic

$$T = \frac{f(x^{(1)}, x^{(2)}, \ldots, x^{(N)}; \tilde{\lambda}_1^{(\Omega)}, \tilde{\lambda}_2^{(\Omega)}, \ldots, \tilde{\lambda}_p^{(\Omega)})}{f(x^{(1)}, x^{(2)}, \ldots, x^{(N)}; \tilde{\lambda}_1^{(\omega)}, \tilde{\lambda}_2^{(\omega)}, \ldots, \tilde{\lambda}_p^{(\omega)})}. \tag{5.1}$$

Here $f(x^{(1)}, x^{(2)}, \ldots, x^{(N)}; \lambda_1, \lambda_2, \ldots, \lambda_p)$ is the joint density function of the $x^{(j)}$ $(j = 1, 2, \ldots, N)$ i.e. the likelihood function (7-1.4). The procedure of the likelihood ratio test specifies that we reject the null hypothesis if

$$T > T_{1-\alpha}, \tag{5.2}$$

where $T_{1-\alpha}$ is determined by

$$P(T > T_{1-\alpha} | H_0) = \int_{T_{1-\alpha}}^{\infty} g(T | H_0) \, \mathrm{d}T, \tag{5.3}$$

$g(T | H_0)$ being the conditional probability density of the statistic T. A theorem by WILKS [1938] gives the distribution function of T, or rather $-2 \ln T$, in the limit of large statistics:

> If a population is described by a probability density $f(x; \lambda_1, \lambda_2, \ldots, \lambda_p)$ which satisfies reasonable requirements of continuity and if the null hypothesis specifies r out of the p parameters
>
> $$H_0(\lambda_1 = \lambda_{10}, \lambda_2 = \lambda_{20}, \ldots, \lambda_r = \lambda_{r0}), \qquad r \leqslant p,$$
>
> then the statistic $-2 \ln T$ determined by a sample of size N follows the χ^2-distribution with $p - r$ degrees of freedom if $N \to \infty$. For a simple null hypothesis, i.e. $r = p$, the number of degrees of freedom is one.

We now apply this method to the problem of testing a normal population with known variance σ and unknown mean λ, which was already discussed in examples 8-3 and 8-4.

Example 8-5: *Power function for the test of example* 8-3

Since for the simple hypothesis $H_0(\lambda = \lambda_0)$ the region ω in parameter space degenerates into the point $\lambda = \lambda_0$, we have

$$\tilde{\lambda}^{(\omega)} = \lambda_0. \tag{5.4}$$

Considering the most general alternative hypothesis $H_1(\lambda = \lambda_1 \neq \lambda_0)$, we get as the maximum likelihood estimate of λ the sample mean $\bar{x}$. The likelihood ratio (5.1) therefore becomes

$$T = \frac{f(x^{(1)}, x^{(2)}, \ldots, x^{(N)}; \bar{x})}{f(x^{(1)}, x^{(2)}, \ldots, x^{(N)}; \lambda_0)}. \tag{5.5}$$

The joint probability density is given by eq. (7-2.6):

$$f(x^{(1)}, x^{(2)}, \ldots, x^{(N)}) = \left(\frac{1}{\sqrt{(2\pi)}\,\sigma}\right)^N \exp\left[-\frac{N}{2\sigma^2} \sum_{j=1}^{N} (x^{(j)} - \lambda)^2\right]. \tag{5.6}$$

Therefore

$$T = \exp\left[\frac{1}{2\sigma^2}\left\{-\sum_{j=1}^{N} (x^{(j)} - \bar{x})^2 + \sum_{j=1}^{N} (x^{(j)} - \lambda_0)^2\right\}\right]$$

$$= \exp\left[\frac{1}{2\sigma^2} \sum_{j=1}^{N} (\bar{x} - \lambda_0)^2\right]$$

$$= \exp\left[\frac{N}{2\sigma^2} (\bar{x} - \lambda_0)^2\right].$$

We have now to calculate $T_{1-\alpha}$ and reject the hypothesis H_0 if the inequality (5.2) holds. Since the logarithm of T is a monotonic function of T we can use

$$T' = 2 \ln T = \frac{N}{\sigma^2} (\bar{x} - \lambda_0)^2$$

as the test statistic and reject H_0 if

$$T' > T'_{1-\alpha}$$

with

$$\int_{T'_{1-\alpha}}^{\infty} h(T' \mid H_0)\,\mathrm{d}T' = \alpha.$$

To calculate the probability density $h(T' \mid H_0)$ of T' we start with the density $f(\bar{x})$ of the sample mean subiect to the condition $\lambda = \lambda_0$, which is

$$f(\bar{x}\,|\,H_0) = \left[\frac{N}{2\pi\sigma^2}\right]^{\frac{1}{2}} \exp\left(-\frac{N}{2\sigma^2}(\bar{x}-\lambda_0)^2\right).$$

To perform the transformation of variables given by eq. (4-4.1) we still need the derivative

$$\left|\frac{\mathrm{d}\bar{x}}{\mathrm{d}T'}\right| = \frac{1}{2}\left(\frac{\sigma^2}{N}\right)^{\frac{1}{2}} T'^{-\frac{1}{2}},$$

which is readily obtained from the definition of T'. Then

$$h(T'\,|\,H_0) = \left|\frac{\mathrm{d}\bar{x}}{\mathrm{d}T'}\right| f(\bar{x}\,|\,H_0) = \frac{1}{\sqrt{(2\pi)}}\, T'^{-\frac{1}{2}} \mathrm{e}^{-\frac{1}{2}T'}. \tag{5.7}$$

This is in fact a χ^2-distribution with one degree of freedom. In our example Wilks' theorem holds even for finite N. We also see that the likelihood ratio test yields the unbiased test of fig. 8-4 (b1). The test

$$T' = -\frac{N}{\sigma_2}(\bar{x}-\lambda_0)^2 > T'_{1-\alpha}$$

is equivalent to

$$\left(\frac{N}{\sigma^2}\right)^{\frac{1}{2}} |\bar{x}-\lambda_0| < \lambda', \qquad \left(\frac{N}{\sigma^2}\right)^{\frac{1}{2}} |\bar{x}-\lambda_0| > \lambda'', \tag{5.8}$$

with

$$-\lambda' = \lambda'' = (T'_{1-\alpha})^{\frac{1}{2}} = (\chi^2_{1-\alpha})^{\frac{1}{2}} = \chi_{1-\alpha}.$$

Using this result we can now calculate the power function of our test explicitly. For any value of the population mean the density of the sample mean is

$$f(\bar{x};\lambda) = \left(\frac{N}{2\pi\sigma^2}\right)^{\frac{1}{2}} \exp\left[-\frac{N(\bar{x}-\lambda)}{2\sigma^2}\right] = \phi_0\left(\frac{\bar{x}-\lambda}{\sigma/\sqrt{N}}\right).$$

Using (3.9) with (5.8) we have

$$M(S_c;\lambda) = \int_{-\infty}^{A} f(\bar{x};\lambda)\,\mathrm{d}\bar{x} + \int_{B}^{\infty} f(\bar{x};\lambda)\,\mathrm{d}\bar{x}$$

$$= \psi_0\left(\chi_{1-\alpha} - \frac{\lambda-\lambda_0}{\sigma/\sqrt{N}}\right) + \psi_0\left(\chi_{1-\alpha} + \frac{\lambda-\lambda_0}{\sigma/\sqrt{N}}\right), \tag{5.9}$$

where

$$A = -\chi_{1-\alpha}\,\sigma/\sqrt{N} - \lambda_0, \quad B = \chi_{1-\alpha}\,\sigma/\sqrt{N} - \lambda_0.$$

Here ϕ_0 and ψ_0 denote the probability density and distribution function of the standardized normal distribution. The power function (5.9) is presented for $\alpha = 0.05$ and different values of N/σ^2 in fig. 8-6.

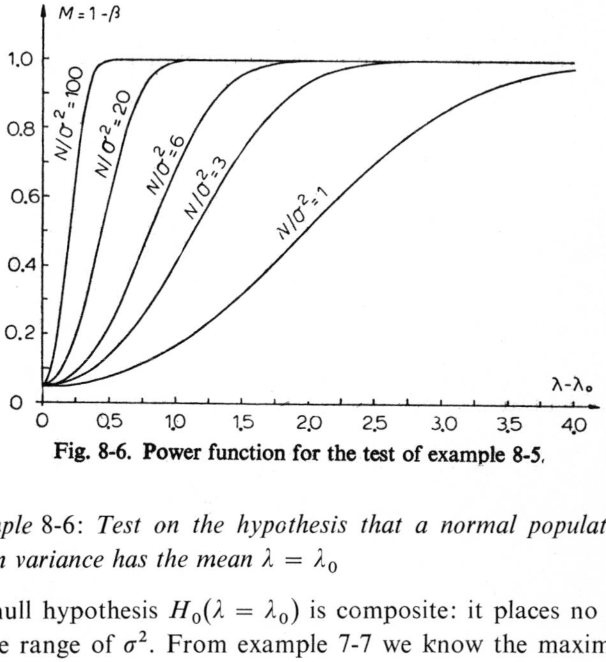

Fig. 8-6. Power function for the test of example 8-5.

Example 8-6: *Test on the hypothesis that a normal population of unknown variance has the mean* $\lambda = \lambda_0$

The null hypothesis $H_0(\lambda = \lambda_0)$ is composite: it places no restriction on the range of σ^2. From example 7-7 we know the maximum likelihood estimates in total parameter space:

$$\tilde{\lambda}^{(\Omega)} = \bar{x},$$

$$\sigma^{2(\Omega)} = \frac{1}{N}\sum_{j=1}^{N}(x^{(j)} - \bar{x})^2 = s'^2.$$

In the parameter space describing the null hypothesis we have

$$\tilde{\lambda}^{(\omega)} = \lambda_0,$$

$$\sigma^{2(\omega)} = \sigma^{2(\omega)} = \frac{1}{N}\sum_{j=1}^{N}(x^{(j)} - \lambda_0)^2.$$

The likelihood ratio (5.1) becomes

$$T = \left(\frac{\sum(x^{(j)} - \lambda_0)^2}{\sum(x^{(j)} - \bar{x})^2}\right)^{\frac{1}{2}N} \exp\left(-\frac{\sum(x^{(j)} - \bar{x})^2}{\sum(x^{(j)} - \bar{x})^2} + \frac{\sum(x^{(j)} - \lambda_0)^2}{\sum(x^{(j)} - \lambda_0)^2}\right)$$

$$= \left(\frac{\sum(x^{(j)} - \lambda_0)^2}{\sum(x^{(j)} - \bar{x})^2}\right)^{\frac{1}{2}N}.$$

Again we change to another statistic, T', which is a monotonic function of T defined by

$$T' = T^{2/N} = \frac{\sum(x^{(j)} - \lambda_0)^2}{\sum(x^{(j)} - \bar{x})^2} = \frac{\sum(x^{(j)} - \bar{x})^2 + N(\bar{x} - \lambda_0)^2}{\sum(x^{(j)} - \bar{x})^2},$$

$$T' = 1 + \frac{t^2}{N - 1}. \tag{5.10}$$

Here

$$t = N^{\frac{1}{2}} \frac{\bar{x} - \lambda_0}{\left(\frac{\sum(x^{(j)} - \bar{x})^2}{N - 1}\right)^{\frac{1}{2}}} = N^{\frac{1}{2}} \frac{\bar{x} - \lambda_0}{s_x}$$

is Student's statistic discussed in §2.

From eq. (5.10) we can therefore calculate the value of t for a particular sample and reject the null hypothesis if

$$|t| > t_{1-\frac{1}{2}\alpha}.$$

The very general likelihood ratio method has led us to Student's test which was carefully constructed to test the hypothesis that the sample originates from a normal distribution with known mean and unknown variance.

8-6. The χ^2-test on goodness of fit

So far we have performed tests on hypotheses which specified the value of one or several parameters of a population. We can call them *parameter tests*. The procedure was to construct a test statistic and to accept or reject a hypothesis on the basis of this statistic, i.e. on the basis of one single number determined from the sample. Another class of tests compares the distribution function of the sample directly with the distribution of the population – the latter could be based on some hypothesis. Such tests are *tests of fit*. We discuss only the most important test of this class – the so-called χ^2-*test*.

We denote the distribution function and probability density of the population by $F(x)$ and $f(x)$ respectively. The total range of the random variable x can be split into r intervals

$$\xi_1, \xi_2, \ldots, \xi_k, \ldots, \xi_r.$$

This is sketched in fig. 8-7. By integration of $f(x)$ over the individual intervals we obtain the probability of observing x in ξ_k

$$p_k = P(x \in \xi_k) = \int_{\xi_k} f(x)\,\mathrm{d}x; \qquad \sum_{k=1}^{r} p_k = 1. \tag{6.1}$$

We now draw a sample of size n and call n_k the number of elements of the sample which fall inside the interval ξ_k. Of course we have

$$\sum_{k=1}^{r} n_k = n. \tag{6.2}$$

From the (hypothetical) probability density we would expect the value

$$n\,p_k$$

for n_k. As a measure of the deviation of the sample distribution from the hypothetical distribution it seems reasonable to use the statistic

$$X^2 = \sum_{k=1}^{r} \frac{(n_k - n\,p_k)^2}{n\,p_k} = \frac{1}{n} \sum_{k=1}^{r} \frac{n_k^2}{p_k} - n. \tag{6.3}$$

The statistic X^2 is asymptotically (i.e. for $n \to \infty$) distributed as χ^2 with $r - 1$ degrees of freedom.

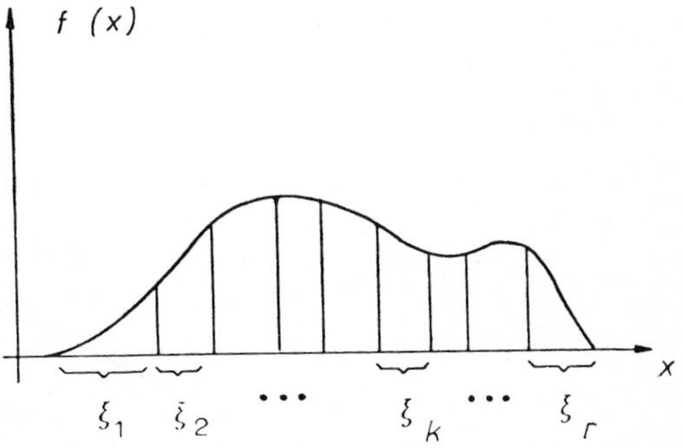

Fig. 8-7. Partition of the variable x into intervals ξ_k.

There are several quite different ways of demonstrating this important property of X^2. We want to sketch here a rather elegant proof due to Fisher. We observe that the eqs. (6.1), (6.2) and fig. 8-7 correspond to a multinomial distribution (ch. 5, §1). According to (5-1.10) the joint probability of observing n_1 elements of the sample in the interval ξ_1, n_2 elements in ξ_2, etc. is

$$P = \frac{n!}{\prod_{k=1}^{r} n_k!} \prod_{k=1}^{r} p_k^{n_k}. \tag{6.4}$$

We now consider r independent Poisson distributions with parameters $np_1, np_2, \ldots, np_r$, i.e. we assume that the population number n_k of the interval ξ_k is described by a Poisson distribution with parameter np_k. The probability of observing n_1 elements in ξ_1, n_2 in ξ_2 etc. is then

$$P(n_1, n_2, \ldots, n_r) = \prod_{k=1}^{r} \frac{(n\, p_k)^{n_k}}{n_k!} \mathrm{e}^{-np_k} = \mathrm{e}^{-n} n^n \prod_{k=1}^{r} \frac{p_k^{n_k}}{n_k!}. \tag{6.5}$$

Of course the condition (6.2) has to be fulfilled. According to example 5-5 the sum of r independent Poisson distributions is again a Poisson distribution, whose parameter is given by the sum of the parameters of the individual distributions

$$P(n) = \frac{n^n}{n!} \mathrm{e}^{-n}. \tag{6.6}$$

The probability of observing n_1 in ξ_1, n_2 in ξ_2, etc., subject to the condition that we observe a total of n elements, is now simply the quotient of (6.5) and (6.6) (cf. eq. (2-3.1))

$$P(n_1, n_2, \ldots, n_r \,|\, n) = \frac{n!}{\prod_{k=1}^{r} n_k!} \prod_{k=1}^{r} (n\, p_k)^{n_k}. \tag{6.7}$$

It is identical to (6.4). Our case is therefore described by the multinomial distribution or – equivalently – by r independent variables following Poisson distributions with mean and variance np_k. Instead of the n_k we can introduce reduced variables

$$u_k = \frac{n_k - n\, p_k}{\sqrt{(n\, p_k)}} \tag{6.8}$$

with zero mean and unit variance.

By virtue of the central limit theorem in the asymptotic limit the variables u_k follow a normalized Gaussian distribution. In this limit the expression (6.3)

$$X^2 = \sum_{k=1}^{r} \frac{(n_k - n\,p_k)^2}{np_k} = \sum_{k=1}^{r} u_k^2 \tag{6.9}$$

is a sum of squares from a normalized Gaussian distribution with one additional condition (eqs. (6.2) and (6.8))

$$\sum_{k=1}^{r} \sqrt{(np_k)}\, u_k = \sum_{k=1}^{r} n_k - \sum_{k=1}^{r} n\,p_k = 0. \tag{6.10}$$

We found in ch. 6, §§5 and 6 that such a sum of squares follows a χ^2-distribution with $r - 1$ degrees of freedom.

The χ^2-test can now be performed as follows. A total of n observations $x_1, x_2, \ldots, x_n$ are made of the random variable x. A partition of the range of x into r intervals $\xi_1, \xi_2, \ldots, \xi_r$ is chosen such that the number n_k of observations in each interval is sufficiently large to ensure that the X^2 can be regarded as distributed in accordance with a χ^2-distribution. On the other hand there should be enough intervals because otherwise the step function of observations is no longer very characteristic of the probability density $f(x)$. As a "rule of thumb" it is usually assumed that the n_k should be at least 4. Next a significance level α is chosen, the quantity X^2 is calculated and compared with the fractile $\chi^2_{1-\alpha}$ for $r - 1$ degrees of freedom. If

$$X^2 > \chi^2_{1-\alpha}$$

the hypothesis based on the density $f(x)$ is rejected.

So far we have assumed that this hypothesis is simple, i.e. that $f(x)$ is completely known. Sometimes a *composite* hypothesis has to be tested. In such cases the general structure of $f(x)$ may be known; but the values of one or several parameters of $f(x)$ have still to be obtained from the observations. We sketch the general case of p unknown parameters $\boldsymbol{\lambda} = (\lambda_1, \lambda_2, \ldots, \lambda_p)$ with $p < r$. Since the probabilities p_k are functions of $f(x; \boldsymbol{\lambda})$ (eq. (6.1)), we have

$$p_k = p_k(\boldsymbol{\lambda}).$$

The likelihood function is directly given by the joint probability (6.4)

$$L(\lambda) = \frac{n!}{\prod_{k=1}^{r} n_k!} \prod_{k=1}^{r} p_k(\boldsymbol{\lambda})^{n_k}.$$

We obtain a system of likelihood equations

$$\frac{\partial l}{\partial \lambda_i} = \sum_{k=1}^{r} \frac{n_k}{p_k} \frac{\partial p_k}{\partial \lambda_i} = 0; \qquad i = 1, 2, \ldots, p. \tag{6.11}$$

These relations are further restrictions on the variables (6.8), just as eq. (6.10) is. The number of degrees of freedom of the sum of squares (6.9) is therefore reduced by p (cf. ch. 6, §6). We have thus obtained the following rule for a χ^2-test on a composite hypothesis.

If the parameters $\boldsymbol{\lambda} = (\lambda_1, \lambda_2, \ldots, \lambda_p)$ are estimated from the observations by the method of maximum likelihood, then the quantity

$$X^2 = \sum_{k=1}^{r} \frac{(n_k - n\, p_k(\boldsymbol{\lambda}))^2}{n\, p_k(\boldsymbol{\lambda})} \tag{6.12}$$

(asymptotically) follows a χ^2-distribution with $f = r - p - 1$ degrees of freedom.

Example 8-7: *χ^2-test on the fit between an empirical frequency distribution and the Poisson distribution*

In an experimental study of photon–proton interactions, a hydrogen bubble chamber is exposed to a beam of photons (γ-quanta) of high energy. The photons also give rise to electron–positron pair productions, which are of no direct interest in this experiment but their

TABLE 8-3
Data for χ^2-test of example 8-7

Number of electron pairs on photographs k	Number of photographs with k electron pairs n_k	Prediction of Poisson distribution np_k	$\frac{(n_k - np_k)^2}{np_k}$
0	47	34.4	4.61
1	69	80.2	1.56
2	84	93.7	1.00
3	76	72.8	0.14
4	49	42.6	0.96
5	16	19.9	0.76
6	11	7.8	1.31
7	3	2.5	0.10
8	–	(0.7)	–
	$n = \sum n_k = 355$		$X^2 = 10.44$

average number per bubble chamber picture can serve as a measure of the intensity of the photon beam. The frequency of pictures with 0, 1, 2, ... has to follow a Poisson distribution (cf. examples in ch. 5, §4). From significant deviations from this distribution one would conclude the existence of observation losses and therefore of systematic errors in the experiment.

In column 2 of table 8-3 and in fig. 8-8 the results of the observation of $n = 355$ pictures are given. From example 7-4 we know that the maximum likelihood estimate of the parameter of the Poisson distribution is given by

$$\tilde{\lambda} = \sum_k k\, n_k / \sum_k n_k.$$

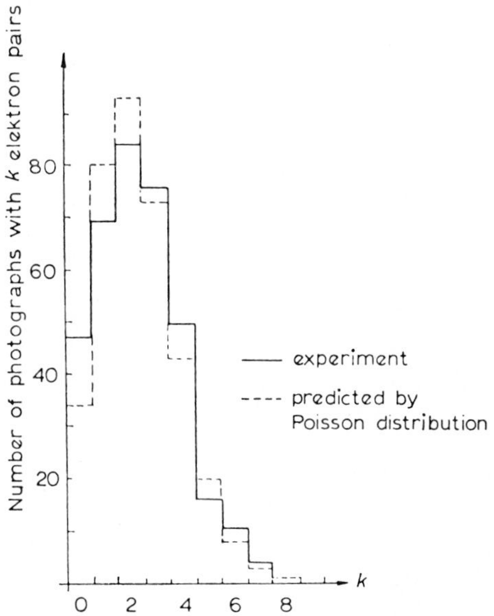

Fig. 8-8. Experimental distribution of example 8-7 compared with the Poisson distribution.

We find $\tilde{\lambda} = 2.33$. From table F-1 we can obtain the values p_k of the Poisson distribution with this parameter which, multiplied by n, are given in column 3. The value of $X^2 = 10.44$ is obtained by summing up the squares in column 4. There are 6 degrees of freedom since $r = 8$, $p = 1$. We choose $\alpha = 1\%$ and find $\chi^2_{0.99} = 16.81$ from table F-5. There is, therefore, no reason to reject the hypothesis of a Poisson distribution.

Example 8-8: *Contingency table*

A total of n experiments may have been performed whose results are characterized by the values of two random variables x and y. We assume that the variables are discrete and that they can take on the values $x_1, x_2, \ldots, x_k$; y_1 $y_2, \ldots, y_l$. (Continuous variables can be approximated by discrete ones by dividing their range into intervals as in fig. 8-7.) The number of experiments yielding $x = x_i$ and $y = y_j$ is n_{ij}. These numbers can be grouped into a matrix, called a *contingency table* (table 8-4).

TABLE 8-4
Contingency table

	y_1	y_2	...	y_l
x_1	n_{11}	n_{12}	...	n_{1l}
x_2	n_{21}	n_{22}	...	n_{2l}
⋮	⋮	⋮		⋮
x_k	n_{k1}	n_{k2}	...	n_{kl}

We denote the probability of obtaining $x = x_i$ by p_i and similarly $q_j = P(y = y_j)$. If the variables are independent, the probability of finding simultaneously $x = x_i$ and $y = y_j$ is the product $p_i q_j$. The maximum likelihood estimates of the p and q are

$$\tilde{p}_i = \frac{1}{n} \sum_{j=1}^{l} n_{ij}, \qquad \tilde{q}_j = \frac{1}{n} \sum_{i=1}^{k} n_{ij}. \tag{6.13}$$

Since

$$\sum_{j=1}^{l} \tilde{q}_j = \sum_{i=1}^{k} \tilde{p}_i = \frac{1}{n} \sum_{j=1}^{l} \sum_{i=1}^{k} n_{ij} = 1, \tag{6.14}$$

there are $k + l - 2$ independent estimates $\tilde{p}_i$, $\tilde{q}_j$.

We can now group the elements of the contingency table in a single string

$$n_{11}, n_{12}, \ldots, n_{1l}, n_{21}, n_{22}, \ldots, n_{2l}, \ldots, n_{kl},$$

and perform a χ^2-test. We have to compute the quantity

$$X^2 = \sum_{i=1}^{k} \sum_{j=1}^{l} \frac{(n_{ij} - n\,\tilde{p}_i\,\tilde{q}_j)^2}{n\,\tilde{p}_i\,\tilde{q}_j}, \tag{6.15}$$

and to compare it with the fractile $\chi^2_{1-\alpha}$ of the χ^2-distribution corresponding to a given level of significance, α. The number of degrees of freedom is obtained by reducing the number of elements of the contingency table by one and subtracting the number of estimated parameters

$$f = kl - 1 - (k + l - 2) = (k - 1)(l - 1). \tag{6.16}$$

If the variables are not independent, then n_{ij} will in general be quite different from $n\tilde{p}_i\tilde{q}_j$. Therefore we find that

$$X^2 > \chi^2_{1-\alpha},$$

and reject the hypothesis of independence.

Exercises

8-1: *F-test*
Given two samples
(1) 21, 19, 14, 27, 25, 23, 22, 18, 21 ($N_1 = 9$),
(2) 16, 24, 22, 21, 25, 21, 18 ($N_2 = 7$),
does sample (2) have smaller variance than sample (1) at the significance level $\alpha = 5\%$?

8-2: *Student's test*
Test the hypothesis that the 30 measurements of example 6-6 stem from a population with mean 25.5 at a significance level of $\alpha = 10\%$. Assume that the population is normal. (Use of a desk calculator is recommended!)

8-3: *χ^2-test on variance*
Use the likelihood ratio method to construct a test on the hypothesis H_0 ($\sigma^2 = \sigma_0^2$) that a sample originates from a normal distribution of unknown mean a and variance σ_0^2. The parameters are $\lambda = (a, \sigma)$. In ω one has $\tilde{\lambda}^{(\omega)} = (\bar{x}', \sigma_0)$, in Ω: $\tilde{\lambda}^{(\Omega)} = (\bar{x}, s)$.
(a) Form the likelihood ratio T.
(b) Show that rather than T the test statistic $T' = (N-1)s^2/\sigma_0^2$ can be used.
(c) Show that T' follows a χ^2-distribution with $(N-1)$ degrees of freedom, so that the test can be performed as indicated in table E-6.

8-4: *χ^2-test on goodness of fit*
(a) Determine mean $\tilde{a}$ and variance $\tilde{\sigma}^2$ from the histogram of fig. 6-6b, i.e.

x_k	193	195	197	199	201	203	205	207	209	211
n_k	1	2	9	12	23	25	11	9	6	2

Use the result of example 7-7 to construct the estimators. Write them down explicitly in terms of n_k and x_k.
(b) Perform a χ^2-test for the fit between the histogram and a normal distribution with mean $\tilde{a}$ and variance $\tilde{\sigma}^2$ for $\alpha = 10\%$. Use only those bins of the histogram for which $np_k \geqslant 4$. To find

p_k form the difference of two readings from table F-2a. Give an explicit formula for p_k in terms of x_k, Δx, $\tilde{a}$, $\tilde{\sigma}^2$ and ψ_0. To compute χ^2 build up a table containing columns for x_k, n_k, np_k and $(n_k - np_k)^2/np_k$.

8-5: *Contingency table*

(a) In an experiment in immunology (taken from the book of SOKAL and ROHLF) the effect of an antiserum on a certain type of bacteria was to be tested. 57 mice received a standard dose of bacteria as well as antiserum, while 54 mice received only bacteria. After some time the number of mice which died or survived were counted in both groups and entered into the following contingency table

	Dead	Alive	Sum
Bacteria and antiserum	13	44	57
Bacteria only	25	29	54
Sum	38	73	Total 111

Test the hypothesis that the antiserum had no influence on survival at $\alpha = 10\%$.

(b) While computing χ^2 in (a) you will have realized that the numerators in eq. (6.15) were all equal. Show that in general for 2×2 contingency tables

$$n_{ij} - n\tilde{p}_i\tilde{q}_j = \frac{1}{n}(n_{11}n_{22} - n_{12}n_{21}).$$

CHAPTER 9

THE METHOD OF LEAST SQUARES

The method of least squares goes back to works of Legendre and Gauss in 1805 and 1809, respectively. It is based on a prescription which in the simplest case takes the following form:

> The repeated measurements y_j may be regarded as the sum of the (unknown) quantity x and measurement errors ε_j
>
> $$y_j = x + \varepsilon_j.$$
>
> The ε_j are now to be determined in such a way that the sum of squares of the errors ε_j takes its minimum
>
> $$\sum_j \varepsilon_j^2 = \sum_j (x - y_j)^2 = \text{min}.$$

We shall find out that in many cases this prescription can be obtained from the maximum likelihood theory which was developed at a much later stage. But also in other cases it yields the most satisfactory results compared with other possible prescriptions. The method of least squares, which of all statistical methods is the one most widely used in practical problems, can also be used if the measured quantities y_j are not directly related to the unknown quantity x but indirectly, e.g. if they have to be regarded as a function of several unknowns. Rather than begin with the most general case we intend to discuss the different special cases first. Since all of them are of great practical interest we will be rather explicit and also work through some examples in detail.

9-1. Direct measurements with equal or different accuracy

The simplest case has already been sketched. A total of n measurements may have been performed on an unknown quantity x. The measured values y_j contain errors of measurements ε_j which we assume to be *normally* distributed about zero

$$y_j = x + \varepsilon_j, \quad E(\varepsilon_j) = 0, \quad E(\varepsilon_j^2) = \sigma^2. \tag{1.1}$$

This assumption seems to be justified in many cases by the central limit theorem.

The probability of measuring the value y_j (within a small interval dy) is

$$f_j \mathrm{d}y = \frac{1}{\sqrt{(2\pi)}\,\sigma} \exp\left(- \frac{(y_j - x)^2}{2\sigma^2} \right) \mathrm{d}y.$$

The logarithmic likelihood function for all n measurements becomes (cf. example 7-2)

$$l = - \frac{1}{2\sigma^2} \sum_{j=1}^{n} (y_j - x)^2 + \text{const.} \tag{1.2}$$

The maximum likelihood condition

$$l = \max$$

is therefore equivalent to

$$M = \sum_{j=1}^{n} (y_j - x)^2 = \sum_{j=1}^{n} \varepsilon_j^2 = \min. \tag{1.3}$$

This is just the least squares condition. As we have already shown in examples 7-2 and 7-4 it leads to the result that the best estimator for x is the arithmetic mean of the y_j

$$\tilde{x} = \bar{y} = \frac{1}{n} \sum_{j=1}^{n} y_j. \tag{1.4}$$

The variance of this result is

$$\sigma^2(\bar{y}) = \sigma^2/n, \tag{1.5}$$

or – identifying errors with standard deviations –

$$\Delta\tilde{x} = \Delta y/\sqrt{n}. \tag{1.6}$$

Also the more general case of *direct measurements with different accuracy* has already been considered in example 7-4. Again we assume that the errors of measurement are distributed normally about zero, i.e.

$$y_j = x + \varepsilon_j, \quad E(\varepsilon_j) = 0, \quad E(\varepsilon_j^2) = \sigma_j^2 = 1/g_j. \tag{1.7}$$

Comparison with eq. (7-2.7) shows that the maximum likelihood method requires

$$M = \sum_{j=1}^{n} \frac{(y_j - x)^2}{\sigma_j^2} = \sum_{j=1}^{n} g_j(y_j - x)^2 = \sum_{j=1}^{n} g_j \varepsilon_j^2 = \min. \tag{1.8}$$

The terms in the sum of squares are now *weighted* by the reciprocals of the variances. The best estimate for x is then (cf. eq. (7-2.8))

$$\tilde{x} = \frac{\sum_{j=1}^{n} g_j y_j}{\sum_{j=1}^{n} g_j}, \tag{1.9}$$

i.e. the weighted mean of the individual measurements. We can see that the larger the variance of a measurement, the smaller its contribution to the result. From eq. (7-3.20) we also know the variance of this estimate; it is

$$\sigma^2(\tilde{x}) = \left(\sum_{j=1}^{n} \frac{1}{\sigma_j^2}\right)^{-1} = \left(\sum_{j=1}^{n} g_j\right)^{-1}. \tag{1.10}$$

We can use the result (1.9) to calculate best estimates $\tilde{\varepsilon}_j$ for the original measurement errors ε_j defined in eq. (1.1). We get

$$\tilde{\varepsilon}_j = y_j - \tilde{x}.$$

We expect that these quantities are distributed normally about zero with variance σ_j^2, i.e. that the quantities $\tilde{\varepsilon}_j/\sigma_j$ follow the normalized Gaussian distribution. Then according to ch. 6, §5 the sum

$$M = \sum_{j=1}^{n} \left(\frac{\tilde{\varepsilon}_j}{\sigma_j}\right)^2 = \sum_{j=1}^{n} \frac{(y_j - \tilde{x})^2}{\sigma_j^2} = \sum_{j=1}^{n} g_j(y_j - \tilde{x})^2 \tag{1.11}$$

follows a χ^2-distribution with $n - 1$ degrees of freedom.

This property of the quantity M can now be used to perform a χ^2-test on the validity of our assumption (1.7). If, for a given level of significance α, the quantity M exceeds $\chi^2_{1-\alpha}$ we will have to reconsider the assumptions (1.7). Usually there should be no doubt that the quantities y_j are in fact measurements of the unknown x. The errors ε_j might not however be normally distributed. In particular the measurements might be biased, i.e. the expectation value of the error ε_j does not vanish.

The presence of such systematic errors – if they are different from one measurement to the next – can usually be expected if the χ^2-test fails.

Example 9-1: *Weighted average of measurements with different accuracies*

Best values for constants of fundamental importance are usually obtained by forming the weighted average of measurements performed by different experimental groups. For the properties of elementary particles such averages are established regularly by Rosenfeld et al. [cf. e.g. ROSENFELD et al., 1967]. We discuss the average value of the mass of the neutral K-meson (K^0), taken from this compilation. The results of four experiments, performed using different techniques, were used in the averaging. The calculations can be performed using a scheme indicated in table 9-1. The quantity M takes the value 7.2. We choose $\alpha = 5\%$. From table F-5 we find that for 3 degrees of freedom $\chi^2_{0.95} = 7.82$. We are therefore satisfied that the result $m_{K^0} = (497.9 \pm 0.2)$ MeV constitutes the best value of the K^0 mass as long as no new experiments are performed.

TABLE 9-1
Weighted average of 4 measurements of the mass of the neutral K-meson

	y_j (K^0-mass in MeV)	σ_j	$1/\sigma_j^2 = g_j$	$y_j g_j$	$y_j - \tilde{x}$	$(y_j - \tilde{x})^2 g_j$
1	498.1	0.4	6.3	3038.0	0.2	0.3
2	497.44	0.33	10	4974.4	−0.46	2.1
3	498.9	0.5	4	1995.6	1.0	4.0
4	497.44	0.5	4	1989.8	−0.46	0.8
			$\Sigma g_j = 24.3$	$\Sigma y_j g_j = 11997.8$		$M = \Sigma(y_j - \tilde{x})^2 g_j = 7.2$

$\tilde{x} = \Sigma y_j g_j / \Sigma g_j = 497.9$

$\Delta\tilde{x} = (\Sigma g_j)^{-\frac{1}{2}} = 0.20$

Let us consider a case in which the χ^2-test fails. As discussed above, we would usually assume that at least one of the measurements is biased. By considering the individual measurements one can sometimes observe that one or two measurements deviate very far from all the others. Such a case is sketched in fig. 9-1a in which a number of different measurements are plotted with their errors. (The measured quantity is given by the ordinate; the abscissa simply distinguishes the different measurements.) While a χ^2-test would fail, if all measurements of fig. 9-1a are used, this is no longer the case, if the measurements 4 and 6 are excluded from the averaging.

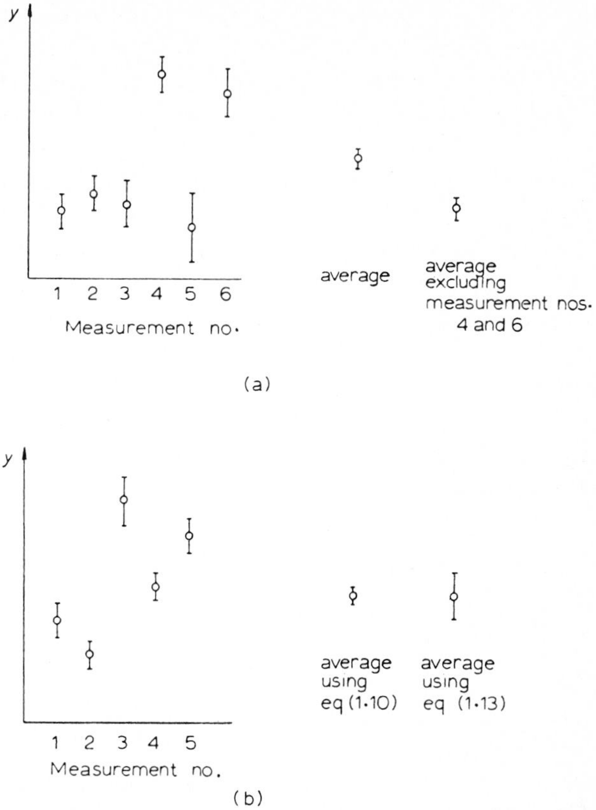

Fig. 9-1. Cases of weighted averaging with failing χ^2-test. (a) Abnormal deviation of a few measurements. (b) Standard deviations of individual measurements apparently too small.

Unfortunately, the situation is often not so clear-cut. In fig. 9-1b an example is given in which a χ^2-test would also fail. No single measurement can, however, be held responsible for this. The mathematically correct procedure would be not to attempt averaging and to make no statement on the best value of the unknown quantity until further measurements have been made.

In practice this is of course unsatisfactory. ROSENFELD et al. [1967] have proposed a procedure for increasing the individual errors by the *scale factor* $\surd[M/(n-1)]$, i.e. replacing σ_j by

$$\sigma_j' = \sigma_j \sqrt{\left(\frac{M}{n-1}\right)}. \tag{1.12}$$

The weighted average $\tilde{x}$ obtained with the use of these errors does not differ from expression (1.9). Its variance is, however, different from (1.10):

$$\sigma'^2(\tilde{x}) = \frac{M}{n-1} \left(\sum_{j=1}^{n} \frac{1}{\sigma_j} \right)^{-1}. \tag{1.13}$$

We now evaluate the expression corresponding to (1.11)

$$M' = \frac{n-1}{M} \sum_{j=1}^{n} \frac{(y_j - \tilde{x})^2}{\sigma_j^2} = \frac{n-1}{M} M = n - 1. \tag{1.14}$$

This is equal to the expectation value of χ^2 for $n - 1$ degrees of freedom. Eq. (1.14) gave rise to the relation (1.12). We remind the reader that this relation is not based on a rigorous mathematical foundation. It has to be used with care since it suppresses the effect of systematic differences between measurements. On the other hand in cases like the one of fig. 9-1b it yields reasonable errors of the averaged values, whereas the direct application of eq. (1.10) yields errors which are far too small to reflect the spread of the individual measurements. Both solutions are plotted in fig. 9-1b.

9-2. Indirect measurements

9-2.1. The linear case

We discuss now the more general case of several unknown quantities x_i ($i = 1, 2, \ldots, r$). Frequently the quantities of interest are not directly measurable. Instead a number of linear functions of the x_i can be measured

$$\eta_j = p_{j0} + p_{j1} x_1 + p_{j2} x_2 + \ldots + p_{jr} x_r. \tag{2.1}$$

We prefer to write these equations in a slightly different form

$$\eta_j + a_{j0} + a_{j1} x_1 + a_{j2} x_2 + \ldots + a_{jr} x_r = 0. \tag{2.2}$$

We can define a column vector

$$\boldsymbol{a}_j = \begin{pmatrix} a_{j1} \\ a_{j2} \\ \vdots \\ a_{jr} \end{pmatrix}, \tag{2.3}$$

and write eq. (2.2) in the concise form

$$\eta_j + a_{j0} + \boldsymbol{a}_j^{\mathrm{T}} \boldsymbol{x} = 0, \qquad j = 1, 2, \ldots, n. \tag{2.4}$$

Defining further

$$\boldsymbol{\eta} = \begin{pmatrix} \eta_1 \\ \eta_2 \\ \vdots \\ \eta_n \end{pmatrix}, \quad \boldsymbol{a}_0 = \begin{pmatrix} a_{10} \\ a_{20} \\ \vdots \\ a_{n0} \end{pmatrix}, \quad A = \begin{pmatrix} a_{11} & a_{12} & \dots & a_{1r} \\ a_{21} & a_{22} & \dots & a_{2r} \\ \vdots & & & \\ a_{n1} & a_{n2} & \dots & a_{nr} \end{pmatrix}, \tag{2.5}$$

the system (2.4) can be written as

$$\boldsymbol{\eta} + \boldsymbol{a}_0 + A\boldsymbol{x} = 0. \tag{2.6}$$

Again we assume that each measurement contains an error ε_j, which is normally distributed. The actual measurements are again called* y_j

$$\begin{aligned} y_j &= \eta_j + \varepsilon_j, \\ E(\varepsilon_j) &= 0, \\ E(\varepsilon_j^2) &= \sigma_j^2 = 1/g_j. \end{aligned} \tag{2.7}$$

Since the y_j are *independent* measurements we can group the variances σ_j^2 into a *diagonal* covariance matrix of the y_j or ε_j

$$C_y = C_\varepsilon = \begin{pmatrix} \sigma_1^2 & & & 0 \\ & \sigma_2^2 & & \\ & & \ddots & \\ 0 & & & \sigma_n^2 \end{pmatrix}. \tag{2.8}$$

By analogy with eq. (1.7) we call the inverse matrix the weight matrix

$$G_y = G_\varepsilon = C_y^{-1} = C_\varepsilon^{-1} = \begin{pmatrix} g_1 & & & 0 \\ & g_2 & & \\ & & \ddots & \\ 0 & & & g_n \end{pmatrix}. \tag{2.9}$$

Constructing also n-vectors of measurements and errors we have from eq. (2.7)

$$\boldsymbol{y} = \boldsymbol{\eta} + \boldsymbol{\varepsilon}, \tag{2.10}$$

then, using eq. (2.6), we obtain

$$\boldsymbol{y} - \boldsymbol{\varepsilon} + \boldsymbol{a}_0 + A\boldsymbol{x} = 0. \tag{2.11}$$

We attempt to solve this set of equations for the unknowns $\boldsymbol{x}$ using the maximum likelihood method. According to our assumption (2.7) the measurements y_j are normally distributed with the probability density

* For simplicity we no longer distinguish random variables by special type letters. It will always be clear from the context which variables are random ones.

$$f(y_j) = \frac{1}{\sqrt{(2\pi)}\,\sigma_j} \exp\left(-\frac{(y_j - \eta_j)^2}{2\sigma_j^2}\right) = \frac{1}{\sqrt{(2\pi)}\,\sigma_j} \exp\left(-\frac{\varepsilon_j^2}{2\sigma_j^2}\right). \quad (2.12)$$

For all n measurements we therefore obtain the likelihood functions

$$L = \prod_{j=1}^{n} f(y_j) = (2\pi)^{-\frac{1}{2}n} \left(\prod_{j=1}^{n} \sigma_j^{-1}\right) \exp\left(-\frac{1}{2}\sum_{j=1}^{n} \frac{\varepsilon_j^2}{\sigma_j^2}\right), \quad (2.13)$$

$$l = \ln L = -\tfrac{1}{2}n \ln 2\pi + \ln\left(\prod_{j=1}^{n} \sigma_j^{-1}\right) - \frac{1}{2}\sum_{j=1}^{n} \frac{\varepsilon_j^2}{\sigma_j^2}. \quad (2.14)$$

The latter expression clearly takes its maximum value when

$$M = \sum_{j=1}^{n} \frac{\varepsilon_j^2}{\sigma_j^2} = \sum_{j=1}^{n} \frac{(y_j + \boldsymbol{a}_j^{\mathrm{T}} \boldsymbol{x} + a_{j0})^2}{\sigma_j^2} = \min. \quad (2.15)$$

Using eqs. (2.9) and (2.11) we can rewrite this expression as

$$M = \varepsilon^{\mathrm{T}} G_y \varepsilon = \min, \quad (2.16)$$

or

$$M = (\boldsymbol{y} + \boldsymbol{a}_0 + A\boldsymbol{x})^{\mathrm{T}} G_y (\boldsymbol{y} + \boldsymbol{a}_0 + A\boldsymbol{x}) = \min. \quad (2.17)$$

The function M is minimized if the partial derivatives with respect to the x_i vanish simultaneously, i.e. if

$$\frac{\partial M}{\partial x_i} = 0, \qquad i = 1, 2, \ldots, r. \quad (2.18)$$

This requirement leads to

$$2A^{\mathrm{T}} G_y(\boldsymbol{y} + \boldsymbol{a}_0 + A\boldsymbol{x}) = 0. \quad (2.19)$$

If $r \leqslant n$ this system can be solved (cf. appendix B §3). The solution is

$$\tilde{\boldsymbol{x}} = -(A^{\mathrm{T}} G_y A)^{-1} A^{\mathrm{T}} G_y (\boldsymbol{y} + \boldsymbol{a}_0). \quad (2.20)$$

We have expressed the best estimates $\tilde{\boldsymbol{x}}$ of the unknowns $\boldsymbol{x}$ in terms of the known coefficients A and $\boldsymbol{a}_0$, the measured quantities $\boldsymbol{y}$ and their covariance matrix G_y^{-1}. To conform with the notation of the following paragraphs we introduce the abbreviation

$$\boldsymbol{c} = \boldsymbol{y} + \boldsymbol{a}_0. \quad (2.21)$$

Eq. (2.11) can then be written as

$$Ax + c - \varepsilon = 0, \tag{2.22}$$

and the solution (2.20) becomes

$$\tilde{x} = -(A^{T}G_{y}A)^{-1}A^{T}G_{y}\,c. \tag{2.23}$$

The solution of course contains the special cases of §1. For the case of direct measurements of different accuracy x has only one element, a_0 vanishes and A reduces to an n-component column vector with all elements equal to -1. We can easily verify that in fact

$$\tilde{x} = \left[(1, 1, \ldots 1)\begin{pmatrix} g_1 & & 0 \\ & g_2 & \\ & & \ddots \\ 0 & & g_n \end{pmatrix}\begin{pmatrix} 1 \\ 1 \\ \vdots \\ 1 \end{pmatrix}\right]^{-1}\begin{pmatrix} g_1 & & 0 \\ & g_2 & \\ & & \ddots \\ 0 & & g_n \end{pmatrix}\begin{pmatrix} y_1 \\ y_2 \\ \vdots \\ y_n \end{pmatrix}$$

is equivalent to eq. (1.9). If moreover all diagonal elements of the matrix G_y are the same, the result reduces to (1.4).

We shall now study the influence of the measurement errors on the unknowns x. Since eq. (2.20) assures a linear relation between the $\tilde{x}$ and the measurements y we can make use of the law of propagation of errors found in ch. 4, §5. With eqs. (4-5.2) and (4-5.4) we immediately get that

$$G_{\tilde{x}}^{-1} = [(A^{T}G_{y}A)^{-1}A^{T}G_{y}]G_{y}^{-1}[(A^{T}G_{y}A)^{-1}A^{T}G_{y}]^{T}.$$

The matrices G_y, G_y^{-1} and $(A^{T}G_{y}A)$ are symmetric, i.e. each is identical to its transpose. Using the rule (B-2.14) we can therefore simplify the above expression

$$\begin{aligned} G_{\tilde{x}}^{-1} &= (A^{T}G_{y}A)^{-1}A^{T}G_{y}\,G_{y}^{-1}\,G_{y}A(A^{T}G_{y}A)^{-1} \\ &= (A^{T}G_{y}A)^{-1}(A^{T}G_{y}A)(A^{T}G_{y}A)^{-1}, \end{aligned}$$

$$G_{\tilde{x}}^{-1} = (A^{T}G_{y}A)^{-1}. \tag{2.24}$$

We have obtained a simple form for the covariance matrix of the best estimates $\tilde{x}$ of the unknowns x. The square roots of the diagonal elements can be considered the "errors of measurement" of the $\tilde{x}$ although these quantities are not directly measured.

Using the result (2.23) we can now also improve the original measurements y. Inserting (2.23) into (2.22) we can determine a vector of best estimates of the measurement errors ε:

$$\tilde{\varepsilon} = A\tilde{x} + c = -A(A^{T}G_{y}A)^{-1}A^{T}G_{y}\,c + c. \tag{2.25}$$

They can be used to correct the original measurements $\boldsymbol{y}$ and to calculate a vector of improved measurements

$$\tilde{\boldsymbol{\eta}} = \boldsymbol{y} - \tilde{\boldsymbol{\varepsilon}} = \boldsymbol{y} + A(A^{\mathrm{T}}G_yA)^{-1}A^{\mathrm{T}}\,G_y\,\boldsymbol{c} - \boldsymbol{c},$$

$$\tilde{\boldsymbol{\eta}} = A(A^{\mathrm{T}}G_yA)^{-1}A^{\mathrm{T}}\,G_y\,\boldsymbol{c} - \boldsymbol{a}_0. \tag{2.26}$$

This is again linear in $\boldsymbol{y}$. We can therefore use error propagation to determine the covariance matrix of the improved measurements

$$G_{\tilde{\eta}}^{-1} = [A(A^{\mathrm{T}}G_yA)^{-1}A^{\mathrm{T}}G_y]G_y^{-1}[A(A^{\mathrm{T}}G_yA)^{-1}A^{\mathrm{T}}G_y]^{\mathrm{T}},$$

$$G_{\tilde{\eta}}^{-1} = A(A^{\mathrm{T}}G_yA)^{-1}A^{\mathrm{T}} = AG_{\tilde{x}}^{-1}A^{\mathrm{T}}. \tag{2.27}$$

The improved measurements $\tilde{\eta}$ fulfil the eqs. (2.1) exactly if the unknowns are replaced by the least squares solutions $\tilde{\boldsymbol{x}}$.

As a first example we discuss the determination of the "best" straight line through a few measured points. Such *curve fitting* problems are typical in data analysis.

Example 9-2: *Fit of a straight line*

We consider measured quantities which are a function of a variable t. The values t_j of the independent variable are known accurately, i.e. with negligible errors. We assume a linear dependence

$$\eta_j = y_j - \varepsilon_j = x_1 + x_2\,t_j, \quad \text{or} \quad \boldsymbol{\eta} - x_1 - x_2\,\boldsymbol{t} = 0, \tag{2.28}$$

and want to determine the unknowns

$$\boldsymbol{x} = \begin{pmatrix} x_1 \\ x_2 \end{pmatrix}$$

from the 4 measurements of table 9-2. The accuracy in y is 0.5, i.e. we have

$$C_y = \begin{pmatrix} 0.5^2 & & & 0 \\ & 0.5^2 & & \\ & & 0.5^2 & \\ 0 & & & 0.5^2 \end{pmatrix} = 0.25\,I, \qquad G_y = C_y^{-1} = 4\,I.$$

We have still to construct the matrix A and the vector $\boldsymbol{a}_0$ of coefficients for our problem. From eq. (2.28) we have

$$\eta_j - x_1 - x_2\,t_j = 0.$$

Comparison with (2.2) yields

TABLE 9-2
Data of example 9-2

j	1	2	3	4
t_j	0.0	1.0	2.0	3.0
y_j	1.4	1.5	3.7	4.1
σ_j	0.5	0.5	0.5	0.5

$$\boldsymbol{\eta} + A\boldsymbol{x} = 0,$$

i.e. $\boldsymbol{a}_0 = 0$ and

$$A = -\begin{pmatrix} 1 & 0 \\ 1 & 1 \\ 1 & 2 \\ 1 & 3 \end{pmatrix}.$$

Since

$$\boldsymbol{y} = \begin{pmatrix} 1.4 \\ 1.5 \\ 3.7 \\ 4.1 \end{pmatrix},$$

eq. (2.16) leads to the solution

$$\tilde{\boldsymbol{x}} = \left[\begin{pmatrix} 1 & 1 & 1 & 1 \\ 0 & 1 & 2 & 3 \end{pmatrix}\begin{pmatrix} 1 & 0 \\ 1 & 1 \\ 1 & 2 \\ 1 & 3 \end{pmatrix}\right]^{-1}\begin{pmatrix} 1 & 1 & 1 & 1 \\ 0 & 1 & 2 & 3 \end{pmatrix}\begin{pmatrix} 1.4 \\ 1.5 \\ 3.7 \\ 4.1 \end{pmatrix}$$

$$= \begin{pmatrix} 4 & 6 \\ 6 & 14 \end{pmatrix}^{-1}\begin{pmatrix} 10.7 \\ 21.2 \end{pmatrix}.$$

Using eq. (B-3.12) to invert the square matrix of order 2, we get

$$\tilde{\boldsymbol{x}} = \frac{1}{20}\begin{pmatrix} 14 & -6 \\ -6 & 4 \end{pmatrix}\begin{pmatrix} 10.7 \\ 21.2 \end{pmatrix} = \begin{pmatrix} 1.13 \\ 1.03 \end{pmatrix}.$$

Or explicitly

$$\tilde{x}_1 = 1.13, \qquad \tilde{x}_2 = 1.03.$$

We now apply (2.24) to obtain the covariance matrix of $\tilde{\boldsymbol{x}}$

$$G_{\tilde{x}}^{-1} = C_{\tilde{x}} = 0.25\left[\begin{pmatrix}1 & 1 & 1 & 1\\ 0 & 1 & 2 & 3\end{pmatrix}\begin{pmatrix}1 & 0\\ 1 & 1\\ 1 & 2\\ 1 & 3\end{pmatrix}\right]^{-1}$$

$$= \begin{pmatrix}0.175 & -0.075\\ -0.075 & 0.175\end{pmatrix}.$$

The improved measurements can be simply calculated using eq. (2.28)

$$\tilde{\eta} = -A\tilde{x} = \begin{pmatrix}1 & 0\\ 1 & 1\\ 1 & 2\\ 1 & 3\end{pmatrix}\begin{pmatrix}1.13\\ 1.03\end{pmatrix} = \begin{pmatrix}1.13\\ 2.16\\ 3.19\\ 4.22\end{pmatrix}.$$

They lie on the straight line described by $\tilde{\eta} = -A\tilde{x}$, which will of course in general be different from the "true" solution. The *residual errors* of the $\tilde{\eta}$ obtained from eq. (2.19) are

$$G_{\tilde{\eta}}^{-1} = C_{\tilde{\eta}} = \begin{pmatrix}1 & 0\\ 1 & 1\\ 1 & 2\\ 1 & 3\end{pmatrix}\begin{pmatrix}0.175 & -0.075\\ -0.075 & 0.05\end{pmatrix}\begin{pmatrix}1 & 1 & 1 & 1\\ 0 & 1 & 2 & 3\end{pmatrix}$$

$$= \begin{pmatrix}0.175 & -0.075\\ 0.1 & -0.025\\ 0.025 & -0.025\\ -0.5 & -0.175\end{pmatrix}\begin{pmatrix}1 & 1 & 1 & 1\\ 0 & 1 & 2 & 3\end{pmatrix}$$

$$= \begin{pmatrix}0.175 & 0.1 & 0.025 & -0.05\\ 0.1 & 0.075 & 0.05 & 0.025\\ 0.025 & 0.05 & 0.075 & 0.1\\ -0.05 & 0.025 & 0.1 & 0.175\end{pmatrix}.$$

The square roots of the diagonal elements are

$$\Delta\tilde{\eta}_1 = 0.42, \quad \Delta\tilde{\eta}_2 = 0.27, \quad \Delta\tilde{\eta}_3 = 0.27, \quad \Delta\tilde{\eta}_4 = 0.42.$$

The fitting procedure which made use of more measurements (4) than were necessary for the determination of the unknowns (2) has significantly decreased the individual errors of measurement which were $\Delta y_j = 0.5$ for all points. The results of the fitting procedure are presented in fig. 9-2. The measurements y_j are drawn as a function of the variable t. The vertical bars indicate the standard deviations. They extend over the region $y_j \pm \sigma(y_j)$. The straight line corresponds to the results $\tilde{x}_1$, $\tilde{x}_2$. The improved measurements lie on this line. They are plotted in fig. 9-2b together with the residual errors $\Delta\tilde{\eta}_j$. To illustrate the error in

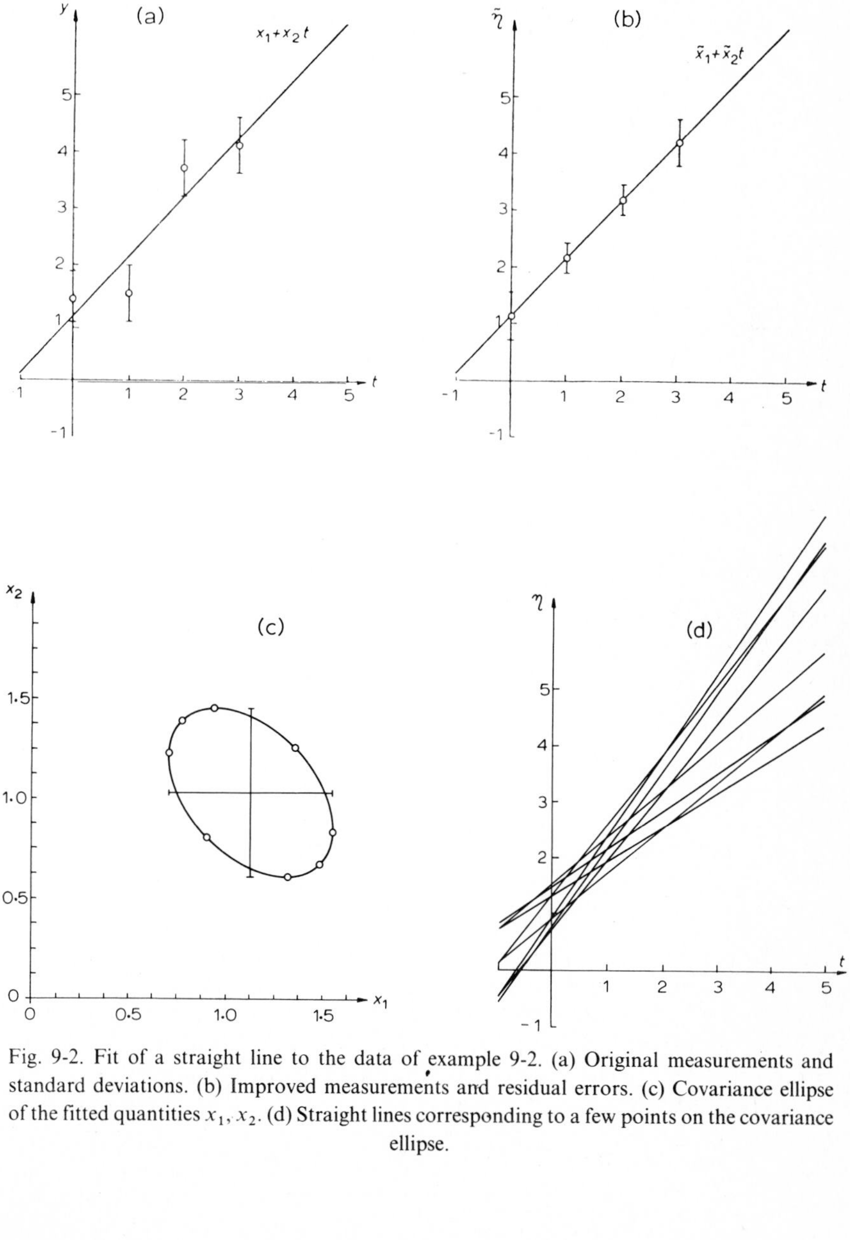

Fig. 9-2. Fit of a straight line to the data of example 9-2. (a) Original measurements and standard deviations. (b) Improved measurements and residual errors. (c) Covariance ellipse of the fitted quantities x_1, x_2. (d) Straight lines corresponding to a few points on the covariance ellipse.

the $\tilde{x}_1$, $\tilde{x}_2$ we consider the covariance matrix $C_{\tilde{x}}$. It determines a covariance ellipse (cf. ch. 5, §10) in a plane spanned by the variables $\tilde{x}_1$, $\tilde{x}_2$. The ellipse is drawn in fig. 9-2c. Points on the ellipse correspond to equal probability. Each of them determines a particular line in the t–y-plane. A few points are marked by small circles in fig. 9-2c, the corresponding lines are drawn in fig. 9-2d. The points on the covariance matrix thus determine a bundle of straight lines. The line corresponding to the true value of the unknowns is inside this bundle with probability $1 - e^{-\frac{1}{2}}$ (cf. eq. (5-10.17)).

9-2.2. The non-linear case

So far we have assumed that the relation between measurable quantities and unknown parameters is of the form (2.2), i.e. linear. Frequently, however, this relation is not linear. In general we have

$$f_j(\boldsymbol{x}, \boldsymbol{\eta}) = \eta_j - g_j(x_1, x_2, \ldots, x_r) = 0, \qquad j = 1, 2, \ldots, n, \tag{2.29}$$

or

$$\boldsymbol{f}(\boldsymbol{x}, \boldsymbol{\eta}) = 0. \tag{2.30}$$

We can now reduce the general case to the linear one by performing a Taylor series expansion of the f_j and considering only the linear terms of this expansion. The expansion is performed at a "point" $\boldsymbol{x}_0 = (x_{10}, x_{20}, \ldots, x_{r0})$ which describes a first approximation of the unknowns that has been obtained in some way.

$$f_j(\boldsymbol{x}, \boldsymbol{\eta}) = f_j(\boldsymbol{x}_0, \boldsymbol{\eta}_0) + \left(\frac{\partial f_j}{\partial x_1}\right)_{\boldsymbol{x}_0}(x_1 - x_{10}) + \ldots + \\ + \left(\frac{\partial f_j}{\partial x_r}\right)_{\boldsymbol{x}_0}(x_r - x_{r0}) + \ldots. \tag{2.31}$$

Introducing the definitions

$$\boldsymbol{\xi} = \boldsymbol{x} - \boldsymbol{x}_0 = \begin{pmatrix} x_1 - x_{10} \\ x_2 - x_{20} \\ \vdots \\ x_r - x_{r0} \end{pmatrix}, \tag{2.32}$$

$$a_{jl} = \left(\frac{\partial f_j}{\partial x_l}\right)_{\boldsymbol{x}_0} = -\left(\frac{\partial g_j}{\partial x_l}\right)_{\boldsymbol{x}_0}, \qquad A = \begin{pmatrix} a_{11} & a_{12} & \ldots & a_{1r} \\ a_{21} & a_{22} & \ldots & a_{2r} \\ \vdots & & & \\ a_{n1} & a_{n2} & \ldots & a_{nr} \end{pmatrix}, \tag{2.33}$$

$$c_j = f_j(\boldsymbol{x}_0, \boldsymbol{y}) = y_j - g_j(\boldsymbol{x}_0), \qquad \boldsymbol{c} = \begin{pmatrix} c_1 \\ c_2 \\ \vdots \\ c_n \end{pmatrix}, \tag{2.34}$$

and making use of the relation (2.10) we obtain

$$f_j(\boldsymbol{x}_0, \boldsymbol{\eta}) = f_j(\boldsymbol{x}_0, \boldsymbol{y} - \boldsymbol{\varepsilon}) = f_j(\boldsymbol{x}_0, \boldsymbol{y}) - \boldsymbol{\varepsilon}. \tag{2.35}$$

These relationships then enable us to restate the system of eqs. (2.31) in the form:

$$\boldsymbol{f} = A\boldsymbol{\xi} + \boldsymbol{c} - \boldsymbol{\varepsilon}. \tag{2.36}$$

This matrix equation is completely analogous to eq. (2.22). The least squares solution for the corrections $\tilde{\boldsymbol{\xi}}$ is therefore (cf. (2.23))

$$\tilde{\boldsymbol{\xi}} = -(A^{\mathrm{T}}G_y A)^{-1}A^{\mathrm{T}}G_y\,\boldsymbol{c}. \tag{2.37}$$

The corresponding covariance matrix is (cf. eq. (2.27))

$$G_{\tilde{\xi}}^{-1} = C_{\tilde{\xi}} = (A^{\mathrm{T}}G_y A)^{-1}. \tag{2.38}$$

As in the linear case we can obtain a vector $\tilde{\boldsymbol{\varepsilon}}$ of measurement errors, a vector $\tilde{\boldsymbol{\eta}}$ of improved measurements and their covariance matrix $G_{\tilde{\eta}}^{-1}$. The notation was chosen in such a way that the results are already given by eqs. (2.25)–(2.27).

Since the $\boldsymbol{x}_0$ were fixed, the covariance matrix of the $\boldsymbol{x}_1 = \boldsymbol{x}_0 + \tilde{\boldsymbol{\xi}}$ is identical to $G_{\tilde{\xi}}^{-1}$

$$G_{x_1}^{-1} = G_{\tilde{\xi}}^{-1} = (A^{\mathrm{T}}G_y A)^{-1}. \tag{2.39}$$

We can now replace $\boldsymbol{x}_0$ by $\boldsymbol{x}_1$ and repeat the procedure to obtain a better approximation to the unknowns. This iteration process can be repeated until the corrections $\tilde{\boldsymbol{\xi}}$ become so small that the improvement resulting from an additional step would be negligible. Suppose we have used s steps in the iteration then we consider the final results to be

$$\begin{aligned} \tilde{\boldsymbol{x}} &= \boldsymbol{x}_s, \\ G_{\tilde{x}}^{-1} &= G_{\tilde{\xi}}^{-1} = (A^{\mathrm{T}}G_y A)^{-1}, \\ \tilde{\eta} &= \boldsymbol{y} + A(A^{\mathrm{T}}G_y A)^{-1}A^{\mathrm{T}}G_y\,\boldsymbol{c}, \end{aligned}$$

the elements of A and $\boldsymbol{c}$ having been calculated by using the intermediate result $\boldsymbol{x}_{s-1}$.

We have not shown that the procedure yields some unique solution $\boldsymbol{x}$. This indeed cannot be proved in general. The convergence has to be tested in each case by forming the differences of the solutions provided by consecutive steps. It is an obvious condition that the functions f_j must not deviate much from linearity in a region around $\boldsymbol{x}_0$, which also contains the solution $\tilde{\boldsymbol{x}}$. In other words: if the f_j differ only by small amounts from linear functions, $\boldsymbol{x}_0$ (i.e. the starting values) may be quite different from the final solution; if the f_j differ strongly from linear functions the procedure will converge only if $\boldsymbol{x}_0$ is already a very good approximation. In fact the art of using the least squares method with non-linear problems is to provide sufficiently good first approximations. No general rule for doing this can be given since it clearly depends entirely on the nature of the problem.

Example 9-3: *Fit of a sine curve*

The quantity η is measured as a function of the variable t, which is assumed to be free of error as in example 9-2. The expected dependence of η on t is of the form

$$\eta = a \sin(2\pi \nu t),$$

i.e. we expect an oscillation with amplitude a and frequency ν. We write this equation in the form (2.29)

$$f_j = \eta_j - x_1 \sin(2\pi\, x_2\, t_j), \qquad j = 1, 2, \ldots, n.$$

According to (2.33) and (2.34) we get

$$A = \begin{pmatrix} -\sin(2\pi\, x_{20}\, t_1) & -2\pi\, t_1\, x_{10} \cos(2\pi\, x_{20}\, t_1) \\ -\sin(2\pi\, x_{20}\, t_2) & -2\pi\, t_2\, x_{10} \cos(2\pi\, x_{20}\, t_2) \\ \vdots & \\ -\sin(2\pi\, x_{20}\, t_n) & -2\pi\, t_n\, x_{10} \cos(2\pi\, x_{20}\, t_n) \end{pmatrix}$$

and

$$\boldsymbol{c} = \begin{pmatrix} y_1 - x_{10} \sin(2\pi\, x_{20}\, t_1) \\ y_2 - x_{10} \sin(2\pi\, x_{20}\, t_2) \\ \vdots \\ y_n - x_{10} \sin(2\pi\, x_{20}\, t_n) \end{pmatrix}.$$

The five measurements of table 9-3 may be given. The σ_j were obtained by assuming a relative measurement error of 10%, i.e. $\sigma_j = y_j/10$. The measurements are plotted in fig. 9-3a. As usual the vertical bars indicate the measurement errors, i.e. they mark the region $y_j \pm \sigma_j$. Drawing a rough sine curve by hand through the measured points we can read off as a first approximation

$$a_0 = x_{10} = 2, \qquad \nu_0 = x_{20} = 1.$$

TABLE 9-3
Data of example 9-3

j	1	2	3	4	5
t_j	0.1	0.2	0.3	0.4	0.5
y_j	1.2	1.9	2.0	2.0	1.4
σ_j	0.12	0.19	0.20	0.20	0.14

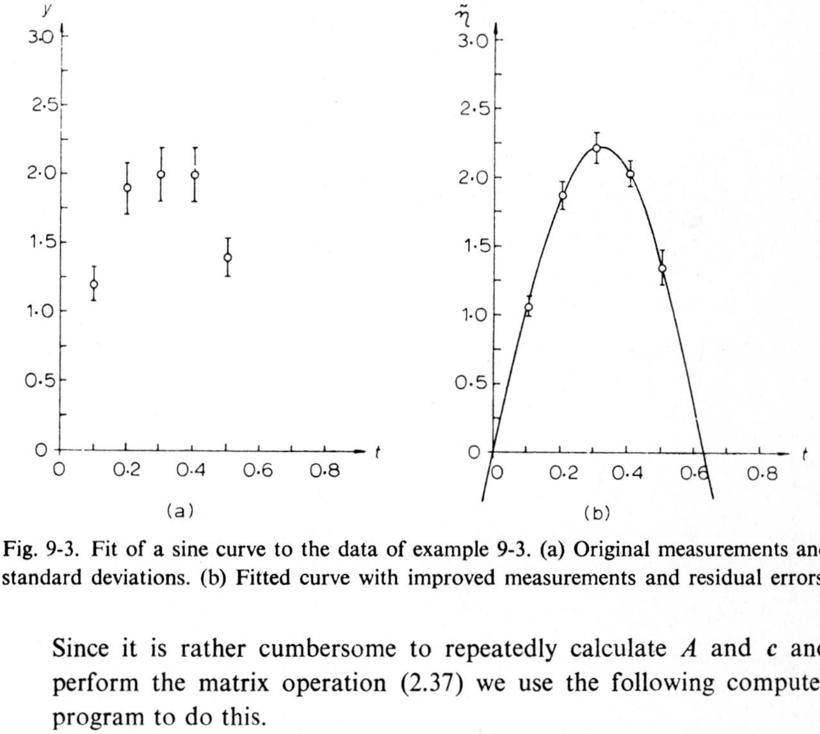

Fig. 9-3. Fit of a sine curve to the data of example 9-3. (a) Original measurements and standard deviations. (b) Fitted curve with improved measurements and residual errors.

Since it is rather cumbersome to repeatedly calculate A and c and perform the matrix operation (2.37) we use the following computer program to do this.

Program 9-1: *Main program* SINFIT *fitting a sine curve to experimental data*

```
C
C     SINE FIT USING    LEAST SQUARES
C
      DIMENSION T(5),Y(5),GY(25),CY(25),X(2),XI(2),CX(4)
      DIMENSION EPSLN(5),ETA(5),CETA(25),A(10),C(5)
      DIMENSION S1(25),S2(25)
C
C     DEFINE NUMBER OF MEASUREMENTS (N) AND UNKNOWNS (NR)
C
      N=5
      NR=2
C
C     INTRODUCE DATA
```

```
C
      DATA T,Y /0.1,0.2,0.3,0.4,0.5,1.2,1.9,2.0,2.0,1.4/
      NN=N**2
      DO 10 I=1,NN
   10 CY(I)=0.
      DO 20 I=1,N
      II=(I-1)*N+I
   20 CY(II)=(Y(I)/10.)**2
      CALL MTXINV (CY,GY,N)
C
C     FIRST APPROXIMATIONS OF UNKNOWNS
C
      X(1)=2.
      X(2)=1.
C
C     PERFORM 3 STEPS OF ITERATION
C
      DO  40 NSTEP=1,3
C
C     CALCULATE MATRIX OF DERIVATIVES A AND VECTOR C
C
      DO  30 I=1,N
      IK=(I-1)*NR+1
      A(IK)=-SIN(6.28318*X(2)*T(I))
      A(IK+1)=-6.28318*T(I)*X(1)*COS(6.28318*X(2)*T(I))
   30 C(I)=Y(I)-X(1)*SIN(6.28318*X(2)*T(I))
      WRITE(6,1000) NSTEP
      CALL MTXWRT(A,N,NR)
      WRITE(6,1100)
      CALL MTXWRT(C,1,N)
C
C     CALCULATE  VECTOR XI OF CORRECTIONS  (EQS. (2.37) AND (2.24))
C
      CALL MTXMLT (GY,A,S1,N,N,NR)
      CALL MTXMAT (A,S1,S2,NR,N,NR)
      CALL MTXINV (S2,CX,NR)
      CALL MTXMBT (CX,A,S2,NR,NR,N)
      CALL MTXMLT (S2,GY,S1,NR,N,N)
      CALL MTXMLT (S1,C,XI,NR,N,1)
      CALL MTXMSC (XI,XI,-1.,NR,1)
C
C     CALCULATE NEW VALUE FOR X
C
      CALL MTXADD (X,XI,X,NR,1)
C
C     CALCULATE MINIMUM FUNCTION XM
C
      CALL MTXMLT (A,XI,S1,N,NR,1)
      CALL MTXADD (S1,C,EPSLN,N,1)
      CALL MTXMAT (EPSLN,GY,S1,1,N,N)
      CALL MTXMLT (S1,EPSLN,XM,1,N,1)
C
C     WRITE RESULTS OF THIS STEP
C
   40 WRITE (6,1200) XM,X
C
C     CALCULATE  ETA  AND COVARIANCE MATRIX CETA (EQS. (2.26) AND (2.27))
C
      CALL MTXSUB (Y,EPSLN,ETA,N,1)
      CALL MTXMLT (A,CX,S1,N,NR,NR)
      CALL MTXMBT (S1,A,CETA,N,NR,N)
C
C     WRITE FINAL RESULTS
C
      WRITE (6,1300) ETA
      WRITE (6,1400)
      CALL MTXWRT (CX,NR,NR)
      WRITE (6,1500)
      CALL MTXWRT (CETA,N,N)
      STOP
C
C     FORMAT STATEMENTS
```

```
1000 FORMAT(/15H ITERATION STEP,I2,10H, MATRIX A/)
1100 FORMAT(/9H VECTOR C/)
1200 FORMAT(/19H MINIMUM FUNCTION =F8.5,14H, UNKNOWNS X =,2F10.5)
1300 FORMAT(/29H IMPROVED MEASUREMENTS ETA = 5F10.5)
1400 FORMAT(/30H COVARIANCE MATRIX OF UNKNOWNS/)
1500 FORMAT(/43H COVARIANCE MATRIX OF IMPROVED MEASUREMENTS/)
     END
```

The program is composed of three parts. Extensive use is made of the matrix handling subroutines of Appendix B, §4. In the initial stage the number of measurements and unknowns are set and the values t_j and y_j are defined. The covariance matrix $C_y = G_y^{-1}$ (denoted by CY in the program) is set up. Finally the first approximations of the unknowns are set. In the second part an iteration process of three steps is performed. In each step the matrix A of derivatives and the vector $\boldsymbol{c}$ are computed using the current approximations of $\boldsymbol{x}$. Using the subroutine MTXWRT these quantities are printed for every step. Next a vector $\boldsymbol{\xi}$ of corrections is computed which, added to the current approximation of $\boldsymbol{x}$, provides a new approximation of the unknowns. The value of the minimum function M (denoted by XM in the program) is calculated and printed together with the values of the unknowns for every step. Intermediate matrices are stored using auxiliary arrays denoted by S1 and S2. In the third part of the program the improved measurements $\tilde{\boldsymbol{\eta}}$ and their covariance matrix $C_{\tilde{\eta}} = G_{\tilde{\eta}}^{-1}$ (denoted by ETA and CETA, respectively) are computed and printed together with the covariance matrix $C_{\tilde{x}} = G_{\tilde{x}}^{-1}$ of the unknowns. The print out of the program is reproduced in table 9-4.

We observe that a satisfactory result is reached after two steps, since in the third step the minimum function decreases only by a very small amount. This can also be seen from the behaviour of the elements of $\boldsymbol{c}$, which does not change much in the third step. By definition (eq. (2.34)) these elements indicate the deviation from the exact fulfilment of eqs. (2.29). Once these elements cannot be reduced further, there is no sense in continuing the iteration process. In a more general program we might therefore repeat the iteration until the values of M in two consecutive steps differ only by an amount less than some specified small number. Such a program and further convergence criteria are discussed in §5.

Taking the square root of the diagonal elements of C_x we have the results $\tilde{a} = \tilde{x}_1 = 2.23 \pm 0.10$ and $\tilde{\nu} = \tilde{x}_2 = 0.80 \pm 0.02$. A curve with these parameters is drawn in fig. 9-3b. The improved measurements $\tilde{\eta}$ lie on this curve. Their errors – the square roots of the diagonal elements of the last matrix in table 9-4 – are also drawn.

TABLE 9-4

Computer print-out for example 9-3

```
ITERATION STEP 1, MATRIX A

 -0.58778  -1.01664
 -0.95106  -0.77665
 -0.95106   1.16495
 -0.58779   4.06655
 -0.00000   6.28318

VECTOR C

  0.02443  -0.00211   0.09789   0.82443   1.39999

MINIMUM FUNCTION = 4.27797, UNKNOWNS X =   2.12551   0.79193

ITERATION STEP 2, MATRIX A

 -0.47730  -1.17355
 -0.83885  -1.45398
 -0.99696  -0.31236
 -0.91328   2.17602
 -0.60810   5.30098

VECTOR C

  0.18548   0.11701  -0.11904   0.05882   0.10747

MINIMUM FUNCTION = 2.67794, UNKNOWNS X =   2.23210   0.79245

ITERATION STEP 3, MATRIX A

 -0.47759  -1.23219
 -0.83920  -1.52537
 -0.99703  -0.32394
 -0.91275   2.29179
 -0.60681   5.57373

VECTOR C

  0.13397   0.02681  -0.22548  -0.03734   0.04553

MINIMUM FUNCTION = 2.67793, UNKNOWNS X =   2.23211   0.79243

IMPROVED MEASUREMENTS ETA =   1.06601   1.87317   2.22548   2.03740   1.35458

COVARIANCE MATRIX OF UNKNOWNS

  0.01119   0.00083
  0.00083   0.00059

COVARIANCE MATRIX OF IMPROVED MEASUREMENTS

  0.00443   0.00706   0.00672   0.00323  -0.00241
  0.00706   0.01139   0.01115   0.00606  -0.00245
  0.00672   0.01115   0.01172   0.00809   0.00124
  0.00323   0.00606   0.00809   0.00895   0.00836
 -0.00241  -0.00245   0.00124   0.00836   0.01686
```

9-2.3. Properties of the least squares solution; χ^2-test

So far the method of least squares has been merely an application of the maximum likelihood method to linear or linearized problems. The prescription of least squares (eq. (2.15)) was obtained directly by minimizing the likelihood function (2.14). In order to construct the likelihood function the probability density of the measurements was assumed to be known. We have assumed that the latter is a normal distribution. If the normal distribution of the measurements around the true value is not assured, we can of course still try to use the relation (2.14), and with it the other formulae of this section. This would appear to lack a theoretical justification. The Gauss–Markov theorem states, however, that even in this case the method of least squares yields results with good properties. Before discussing these we briefly review the properties of a maximum likelihood solution.

(a) The solution $\tilde{\boldsymbol{x}}$ is unbiased, i.e.

$$E(\tilde{x}_i) = x_i, \qquad i = 1, 2, \ldots, r.$$

(b) It has minimum variance, i.e.

$$\sigma^2(\tilde{x}_i) = E\{(\tilde{x}_i - x_i)^2\} = \text{min}.$$

(c) The quantity (2.16)

$$M = \boldsymbol{\varepsilon}^{\mathrm{T}} G_y \boldsymbol{\varepsilon}$$

follows a χ^2-distribution with $n - r$ degrees of freedom.

The properties (a) and (b) are familiar from ch. 7. We demonstrate the validity of (c) for the simple case of direct measurement ($r = 1$), in which G_y is a diagonal matrix:

$$G_y = \begin{pmatrix} 1/\sigma_1^2 & & & 0 \\ & 1/\sigma_2^2 & & \\ & & \ddots & \\ 0 & & & 1/\sigma_n^2 \end{pmatrix}.$$

The quantity M then becomes simply a sum of squares

$$M = \sum_{j=1}^{n} \varepsilon_j^2/\sigma_j^2. \tag{2.40}$$

Since each ε_j originates from a normal distribution with zero mean and variance σ_j^2 the expressions ε_j/σ_j follow the standardized Gaussian distribution. In this case the sum of squares (2.40) follows a χ^2-distribution with $n - 1$ degrees of freedom (ch. 6, §6).

If the distribution of the errors ε_j is not known the least squares solution has the following properties.

(a) The solution is unbiased.

(b) Of all the solutions $\boldsymbol{x}^*$ that are unbiased estimates of the $\boldsymbol{x}$, and linear combinations of the measurements $\boldsymbol{y}$, the least squares solution has the

smallest variance. (This is the Gauss–Markov theorem.)
(c) The expectation value of

$$M = \boldsymbol{\varepsilon}^{\mathrm{T}} G_y \boldsymbol{\varepsilon}$$

is equal to

$$E(M) = n - r.$$

(This is the equal to the expectation value of χ^2 in the case of $n - r$ degrees of freedom.)

The quantity M is frequently called simply χ^2, although it does not necessarily follow the χ^2-distribution. Together with the matrices $C_{\tilde{x}}$ and $C_{\tilde{\eta}}$ it provides a convenient measure of the quality of the least squares fit. If the value obtained for M is much larger than $n - r$ we shall have to check carefully the assumptions on which the calculations were based.

The number $f = n - r$ is called the *number of degrees of freedom of the fit* or the *number of constraints* of the fit. The latter expression can be understood from the special case discussed in §3. It is clear (cf. Appendix B, §3) that a least squares problem can only be solved for $f \geqslant 0$. Only for $f > 0$ is the quantity M meaningful and can be used to test the quality of the fit.

In those cases in which a normal distribution of the errors can be assumed, the least squares fit can be combined with a χ^2-test. We reject the result of the fit if

$$M = \boldsymbol{\varepsilon}^{\mathrm{T}} G_y \boldsymbol{\varepsilon} > \chi^2_{1-\alpha}(n - r), \tag{2.41}$$

i.e. if M exceeds the critical value of χ^2 belonging to a level of significance α and $n - r$ degrees of freedom. The rejection might have been caused by one of the following reasons (apart from the possibility of an error of the first kind):
(a) The assumed functional dependence $\boldsymbol{f}(\boldsymbol{x}, \boldsymbol{\eta}) = 0$ between measurable quantities $\boldsymbol{\eta}$ and the unknown parameters $\boldsymbol{x}$ does not apply. Either the function $\boldsymbol{f}(\boldsymbol{x}, \boldsymbol{\eta})$ is totally wrong or some parameters thought to be known are incorrect.
(b) The function $\boldsymbol{f}(\boldsymbol{x}, \boldsymbol{\eta})$ is correct, but the Taylor expansion with only one term does not represent it adequately over the required range.
(c) The first approximation $\boldsymbol{x}_0$ is too far from the true solution $\boldsymbol{x}$. Better values for $\boldsymbol{x}_0$ could lead to an acceptable value of M. Clearly this point is closely related to (b).
(d) The covariance matrix G_y^{-1} of the measured variables which is frequently based only on a rough estimation, or even on guesswork, is incorrect.

These four points have to be carefully considered if the least squares method is to be successful. Often a least squares calculation is repeated many times with different sets of measured data. One can then study the distribution of M and compare it to the χ^2-distribution with the appropriate number of degrees of freedom. This comparison can be helpful for the estimation of G_y^{-1}. In general if a new experiment is started, the measuring instrument, which yields the values y_j, is tested by performing measurements of known quantities. In this way the $\boldsymbol{x}$ will be known for several sets of data. The covariance matrix can now be adjusted such that the distribution of M for the measurements with known results coincides with the χ^2-distribution.

9-3. Constrained measurements

We now return to the case of §1 in which the interesting quantities were directly measurable. However we now no longer assume that all n measurements are completely independent. They may, for instance, be related by q constraint equations. As an example we can consider the three angles of a triangle which may be directly measured. The constraint equation states that their sum must equal 180°. Again we want to determine the best values of the quantities $\tilde{\eta}_j$.

The measurements yield the quantities $\boldsymbol{y}$

$$y_j = \eta_j + \varepsilon_j, \qquad j = 1, 2, \ldots, n. \tag{3.1}$$

As before, we assume a normal distribution for the errors ε_j centred around zero:

$$E(\varepsilon_j) = 0, \qquad E(\varepsilon_j^2) = \sigma_j^2.$$

The q constraint equations have the form

$$f_k(\boldsymbol{\eta}) = 0, \qquad k = 1, 2, \ldots, q. \tag{3.2}$$

We first consider the simple case of linear constraint equations. The eqs. (3.2) then take the form

$$\begin{aligned} b_{10} + b_{11}\eta_1 + b_{12}\eta_2 + \ldots + b_{1n}\eta_n &= 0, \\ b_{20} + b_{21}\eta_1 + b_{22}\eta_2 + \ldots + b_{2n}\eta_n &= 0, \\ &\vdots \\ b_{q0} + b_{q1}\eta_1 + b_{q2}\eta_2 + \ldots + b_{qn}\eta_n &= 0, \end{aligned} \tag{3.3}$$

or, in matrix notation

$$B\boldsymbol{\eta} + \boldsymbol{b}_0 = 0. \tag{3.4}$$

9-3.1. The method of elements

We can use these q equations to eliminate q of the n quantities $\boldsymbol{\eta}$. The remaining $n - q$ quantities α_i ($i = 1, 2, \ldots, n - q$) are called *elements*. They can be arbitrarily selected from the original $\boldsymbol{\eta}$ or they can be l ıear combinations of them. We can then again express the complete vect $\boldsymbol{\eta}$ as a set of linear combinations of the elements

$$\eta_j = f_{j0} + f_{j1}\,\alpha_1 + f_{j2}\,\alpha_2 + \ldots + f_{j,n-q}\,\alpha_{n-q}, \qquad j \,\colon \quad 2, \ldots, n, \tag{3.5}$$

or

$$\boldsymbol{\eta} = F\boldsymbol{\alpha} + \boldsymbol{f}_0. \tag{3.6}$$

Eq. (3.6) is of the same type as eq. (2.2) the solution must therefore be of the form (2.20), i.e.

$$\tilde{\boldsymbol{\alpha}} = (F^{\mathrm{T}}G_yF)^{-1}F^{\mathrm{T}}G_y(\boldsymbol{y} - \boldsymbol{f}_0). \tag{3.7}$$

Eq. (3.7) describes the least squares estimate of the elements $\boldsymbol{\alpha}$. The corresponding covariance matrix is

$$G_{\tilde{\boldsymbol{\alpha}}}^{-1} = (F^{\mathrm{T}}G_yF)^{-1} \tag{3.8}$$

(cf. eq. (2.24)). The improved measurements are obtained by substituting (3.7) into (3.6):

$$\tilde{\boldsymbol{\eta}} = F\tilde{\boldsymbol{\alpha}} + \boldsymbol{f}_0 = F(F^{\mathrm{T}}G_yF)^{-1}F^{\mathrm{T}}G_y(\boldsymbol{y} - \boldsymbol{f}_0) + \boldsymbol{f}_0. \tag{3.9}$$

By error propagation the covariance matrix is

$$G_{\tilde{\boldsymbol{\eta}}}^{-1} = F(F^{\mathrm{T}}G_yF)^{-1}F^{\mathrm{T}} = FG_{\tilde{\boldsymbol{\alpha}}}^{-1}F^{\mathrm{T}}. \tag{3.10}$$

Example 9-4: Angles of a triangle constrained by the sum of angles

Measurements of the angles of a triangle may have given the figures $y_1 = 89°$, $y_2 = 31°$, $y_3 = 62°$, i.e.

$$\boldsymbol{y} = \begin{pmatrix} 89 \\ 31 \\ 62 \end{pmatrix}.$$

The linear constraint equation is

$$\eta_1 + \eta_2 + \eta_3 = 180.$$

This can be written as

$$B\boldsymbol{\eta} + \boldsymbol{b}_0 = 0,$$

with

$$B = (1, 1, 1), \qquad \boldsymbol{b}_0 = b_0 = -180.$$

As elements we choose η_1 and η_2. The system (3.5) becomes

$$\eta_1 = \alpha_1, \quad \eta_2 = \alpha_2, \quad \eta_3 = 180 - \alpha_1 - \alpha_2,$$

or

$$\eta = \begin{pmatrix} 1 & 0 \\ 0 & 1 \\ -1 & -1 \end{pmatrix} \boldsymbol{\alpha} + \begin{pmatrix} 0 \\ 0 \\ 180 \end{pmatrix},$$

i.e.

$$F = \begin{pmatrix} 1 & 0 \\ 0 & 1 \\ -1 & -1 \end{pmatrix}, \qquad \boldsymbol{f}_0 = \begin{pmatrix} 0 \\ 0 \\ 180 \end{pmatrix}.$$

We assume an error of 1° in angular measurement, i.e.

$$C_y = \begin{pmatrix} 1 & 0 & 0 \\ 0 & 1 & 0 \\ 0 & 0 & 1 \end{pmatrix} = I, \qquad G_y = C_y^{-1} = I.$$

Application of (3.7) yields

$$\tilde{\boldsymbol{\alpha}} = \left[\begin{pmatrix} 1 & 0 & -1 \\ 0 & 1 & -1 \end{pmatrix} I \begin{pmatrix} 1 & 0 \\ 0 & 1 \\ -1 & -1 \end{pmatrix} \right]^{-1} \begin{pmatrix} 1 & 0 & -1 \\ 0 & 1 & -1 \end{pmatrix} I \begin{pmatrix} 89 \\ 31 \\ -118 \end{pmatrix}$$

$$= \begin{pmatrix} 2 & 1 \\ 1 & 2 \end{pmatrix}^{-1} \begin{pmatrix} 207 \\ 149 \end{pmatrix} = \frac{1}{3} \begin{pmatrix} 2 & -1 \\ -1 & 2 \end{pmatrix} \begin{pmatrix} 207 \\ 149 \end{pmatrix}$$

$$= \begin{pmatrix} 88\frac{1}{3} \\ 30\frac{1}{3} \end{pmatrix}.$$

Using eq. (3.9) we have

$$\tilde{\boldsymbol{\eta}} = F\tilde{\boldsymbol{\alpha}} + \boldsymbol{f}_0 = \begin{pmatrix} 1 & 0 \\ 0 & 1 \\ -1 & -1 \end{pmatrix} \begin{pmatrix} 88\frac{1}{3} \\ 30\frac{1}{3} \end{pmatrix} + \begin{pmatrix} 0 \\ 0 \\ 180 \end{pmatrix} = \begin{pmatrix} 88\frac{1}{3} \\ 30\frac{1}{3} \\ 61\frac{1}{3} \end{pmatrix}.$$

Of course this result was to be expected. The "excess" of the measured angles of 2° was subtracted in equal parts from each of the measurements. If, however, different variances had been assumed for the angles, the result would have been different. The reader is encouraged to perform the calculations for this case.

By applying of eq. (3.10) we can determine the residual errors of the improved measurements

$$G_{\tilde{\eta}}^{-1} = \begin{pmatrix} 1 & 0 \\ 0 & 1 \\ -1 & -1 \end{pmatrix} \left[\begin{pmatrix} 1 & 0 & 1 \\ 0 & 1 & -1 \end{pmatrix} I \begin{pmatrix} 1 & 0 \\ 0 & 1 \\ -1 & -1 \end{pmatrix} \right]^{-1} \begin{pmatrix} 1 & 0 & 1 \\ 0 & 1 & -1 \end{pmatrix}$$

$$= \frac{1}{3} \begin{pmatrix} 1 & 0 \\ 0 & 1 \\ -1 & -1 \end{pmatrix} \begin{pmatrix} 2 & -1 \\ -1 & 2 \end{pmatrix} \begin{pmatrix} 1 & 0 & 1 \\ 0 & 1 & -1 \end{pmatrix}$$

$$= \frac{1}{3} \begin{pmatrix} 2 & -1 & -1 \\ -1 & 2 & -1 \\ -1 & -1 & 2 \end{pmatrix}.$$

The residual error of each angle is equal to $\sqrt{\frac{2}{3}} \approx 0.82$.

At this point we want to make a general statement on measurements which are related by certain constraint equations. Although till now statistical methods have not helped us in correcting against, or even in detecting *systematic errors*, constraint equations provide us with a means of doing so. If, for instance, in a large number of measurements the sum of angles exceeds 180° much more often than it falls below this limit, we will be justified in suspecting a systematic defect in the measuring instrument.

9-3.2. The method of Lagrangian multipliers

Instead of performing the calculation with the help of elements we can use the method of Lagrangian multipliers. While of course both methods yield identical results the latter has the advantage of treating all unknowns on the same footing, thus relieving the user from the choice of the elements. The method of Lagrangian multipliers is familiar from differential calculus.

We begin again with the system of linear constraint equations (3.4)

$$B\boldsymbol{\eta} + \boldsymbol{b}_0 = 0,$$

and remember that the measured values are the sum of the true values $\boldsymbol{\eta}$ and the errors $\boldsymbol{\varepsilon}$

$$\boldsymbol{y} = \boldsymbol{\eta} + \boldsymbol{\varepsilon}.$$

Therefore

$$By - B\varepsilon + b_0 = 0. \tag{3.11}$$

Since the y are known from measurements and B and b_0 are also composed of known coefficients, we can construct the column vector c of q components

$$c = By + b_0, \tag{3.12}$$

which contains no unknowns. Eq. (3.11) can then be written

$$c - B\varepsilon = 0. \tag{3.13}$$

We now introduce another q-component column vector whose elements are the as yet unknown *Lagrangian multipliers*

$$\mu = \begin{pmatrix} \mu_1 \\ \mu_2 \\ \vdots \\ \mu_q \end{pmatrix}, \tag{3.14}$$

and extend the original minimum function (2.12)

$$M = \varepsilon^{\mathrm{T}} G_y \varepsilon$$

to

$$L = \varepsilon^{\mathrm{T}} G_y \varepsilon + 2\mu^{\mathrm{T}}(c - B\varepsilon). \tag{3.15}$$

The function L is called a *Lagrangian* (function). The requirement

$$M = \min$$

with the constraints

$$c - B\varepsilon = 0$$

is fulfilled if the total differential of the Lagrangian vanishes, i.e. if

$$\mathrm{d}L = 2\varepsilon^{\mathrm{T}} G_y \,\mathrm{d}\varepsilon - 2\mu^{\mathrm{T}} B \,\mathrm{d}\varepsilon = 0.$$

This is equivalent to

$$\varepsilon^{\mathrm{T}} G_y - \mu^{\mathrm{T}} B = 0. \tag{3.16}$$

The system (3.16) is composed of n equations which contain altogether $n + q$ unknowns $\varepsilon_1, \varepsilon_2, \ldots, \varepsilon_n$ and $\mu_1, \mu_2, \ldots, \mu_q$. In addition we still have the q constraint equations (3.13). Transposing (3.16) we get

$$G_y \varepsilon = B^{\mathrm{T}} \mu, \quad \text{or} \quad \varepsilon = G_y^{-1} B^{\mathrm{T}} \mu. \tag{3.17}$$

Substituting this in (3.13) we obtain

$$c - BG_y^{-1}B^T\mu = 0,$$

which can be easily solved for μ:

$$\tilde{\mu} = (BG_y^{-1}B^T)^{-1}c. \tag{3.18}$$

Using (3.17) we have least squares estimates of the deviations ε

$$\tilde{\varepsilon} = G_y^{-1}B^T(BG_y^{-1}B^T)^{-1}c. \tag{3.19}$$

The best estimates for the unknowns η are given by (3.1)

$$\tilde{\eta} = y - \tilde{\varepsilon} = y - G_y^{-1}B^T(BG_y^{-1}B^T)^{-1}c.$$

Introducing the abbreviation

$$G_B = (BG_y^{-1}B^T)^{-1},$$

this becomes

$$\tilde{\eta} = y - G_y^{-1}B^TG_B\, c.$$

The covariance matrices of the μ and η are readily obtained by applying the propagation of errors to the linear systems (3.18) and (3.19).

$$G_{\tilde{\mu}}^{-1} = (BG_y^{-1}B^T)^{-1} = G_B, \tag{3.20}$$

$$G_{\tilde{\eta}}^{-1} = G_y^{-1} - G_y^{-1}B^TG_BBG_y^{-1}. \tag{3.21}$$

Example 9-5: *Lagrangian multipliers applied to example* 9-4

We apply the method of Lagrangian multipliers to the problem in example 9-4. We have

$$c = By + b_0 = (1, 1, 1)\begin{pmatrix}89\\31\\62\end{pmatrix} - 180 = 182 - 180 = 2.$$

Furthermore

$$G_B = (BG_yB^T)^{-1} = \left[(1, 1, 1)I\begin{pmatrix}1\\1\\1\end{pmatrix}\right]^{-1} = 3^{-1} = \tfrac{1}{3},$$

and

$$G_yB^T = I\begin{pmatrix}1\\1\\1\end{pmatrix} = \begin{pmatrix}1\\1\\1\end{pmatrix}.$$

We can now calculate (3.20)

$$\tilde{\boldsymbol{\eta}} = \begin{pmatrix} 89 \\ 31 \\ 62 \end{pmatrix} - \begin{pmatrix} 1 \\ 1 \\ 1 \end{pmatrix} \frac{2}{3} = \begin{pmatrix} 88\frac{1}{3} \\ 30\frac{1}{3} \\ 61\frac{1}{3} \end{pmatrix}.$$

The covariance matrices become

$$G_{\tilde{\mu}}^{-1} = \tfrac{1}{3},$$

$$G_{\tilde{\eta}}^{-1} = I - I \begin{pmatrix} 1 \\ 1 \\ 1 \end{pmatrix} \frac{1}{3} (1, 1, 1) I$$

$$= I - \frac{1}{3} \begin{pmatrix} 1 & 1 & 1 \\ 1 & 1 & 1 \\ 1 & 1 & 1 \end{pmatrix} = \frac{1}{3} \begin{pmatrix} 2 & -1 & -1 \\ -1 & 2 & -1 \\ -1 & -1 & 2 \end{pmatrix}.$$

We now generalize the method of Lagrangian multipliers to non-linear constraint equations of the general form (3.2), i.e.

$$f_k(\boldsymbol{\eta}) = 0, \qquad k = 1, 2, \ldots, q.$$

We may develop these equations in a Taylor series in a neighbourhood of $\boldsymbol{\eta}_0$, which denotes a first approximation of the true values $\boldsymbol{\eta}$:

$$f_k(\boldsymbol{\eta}) = f_k(\boldsymbol{\eta}_0) + \left(\frac{\partial f_k}{\partial \eta_1}\right)_{\boldsymbol{\eta}_0} (\eta_1 - \eta_{10}) + \ldots + \left(\frac{\partial f_k}{\partial \eta_n}\right)_{\boldsymbol{\eta}_0} (\eta_n - \eta_{n0}). \quad (3.22)$$

Introducing

$$b_{kl} = \left(\frac{\partial f_k}{\partial \eta_l}\right)_{\boldsymbol{\eta}_0}, \qquad B = \begin{pmatrix} b_{11} & b_{12} & \ldots & b_{1n} \\ b_{21} & b_{22} & \ldots & b_{2n} \\ \vdots & & & \\ b_{q1} & b_{q2} & \ldots & b_{qn} \end{pmatrix},$$

$$c_k = f_k(\boldsymbol{\eta}_0), \qquad \mathbf{c} = \begin{pmatrix} c_1 \\ c_2 \\ \vdots \\ c_q \end{pmatrix},$$

$$\delta_k = (\eta_k - \eta_{k0}), \qquad \boldsymbol{\delta} = \begin{pmatrix} \delta_1 \\ \delta_2 \\ \vdots \\ \delta_n \end{pmatrix},$$

we can write (3.22)

$$B\boldsymbol{\delta} + \boldsymbol{c} = 0. \tag{3.23}$$

It is not difficult to find a first approximation $\boldsymbol{\eta}_0$, since the y_j are measurements of the η_j. We can therefore use them directly as a first approximation and set

$$\boldsymbol{\eta}_0 = \boldsymbol{y}. \tag{3.24}$$

The vector $\boldsymbol{\delta}$ then becomes

$$\boldsymbol{\delta} = \boldsymbol{\eta} - \boldsymbol{\eta}_0 = \boldsymbol{\eta} - \boldsymbol{y} = -\boldsymbol{\varepsilon}. \tag{3.25}$$

Eq. (3.23) therefore takes the form

$$\boldsymbol{c} - B\boldsymbol{\varepsilon} = 0,$$

which is identical to (3.13). The solution $\tilde{\boldsymbol{\varepsilon}}$ is given by (3.19). Usually one will, however, not be satisfied with one step in the non-linear case, but will perform an iteration. From the second step on, the simple relation $\boldsymbol{\delta} = -\boldsymbol{\varepsilon}$ no longer holds. We solve (3.23) by the same reasoning as in the linear case by requiring that the expression

$$M' = \boldsymbol{\delta}^{\mathrm{T}} G_y \boldsymbol{\delta} \tag{3.26}$$

takes a minimum under the constraint equations (3.23). The requirement is equivalent to saying that the weighted sum of squares of the differences between two approximations for the η should be minimum. The solution is then given, by analogy with (3.19), by

$$\tilde{\boldsymbol{\delta}} = G_y^{-1} B^{\mathrm{T}} (B G_y^{-1} B^{\mathrm{T}})^{-1} \boldsymbol{c}. \tag{3.27}$$

The result of each step is used to get a better approximation of $\boldsymbol{\eta}$. If the iteration is broken off after s steps we have

$$\tilde{\boldsymbol{\eta}} = \boldsymbol{\eta}_s = \boldsymbol{\eta}_{s-1} + \tilde{\boldsymbol{\delta}}. \tag{3.28}$$

The corresponding covariance matrix is still given by (3.21), i.e.

$$G_{\tilde{\eta}}^{-1} = G_y^{-1} - G_y B^{\mathrm{T}} G_B B G_y^{-1}, \tag{3.29}$$

if for the elements of B the values computed in the last step are substituted. The least squares estimates for the measurement errors can now be calculated from

$$\tilde{\boldsymbol{\varepsilon}} = \boldsymbol{y} - \tilde{\boldsymbol{\eta}}. \tag{3.30}$$

From these the original minimum function

$$\tilde{M} = \tilde{\varepsilon}^{\mathrm{T}} G_y \tilde{\varepsilon}$$

can finally be obtained. This quantity can again be used for a χ^2-test with q degrees of freedom.

9-4. The general case of least squares fitting

After the foregoing discussion we can generalize the method quite simply to the case of indirect measurements with constraint equations.

We first repeat our notation. There are r unknown parameters grouped in the vector $\boldsymbol{x}$. The measurable quantities are grouped in an n-vector $\boldsymbol{\eta}$. The actually measured quantities $\boldsymbol{y}$ deviate from the $\boldsymbol{\eta}$ by the errors $\boldsymbol{\varepsilon}$. A normal distribution of the individual ε_j $(j = 1, 2, \ldots, n)$ is assumed. The $\boldsymbol{x}$ and $\boldsymbol{\eta}$ are related by m functions

$$f_k(\boldsymbol{x}, \boldsymbol{\eta}) = f_k(\boldsymbol{x}, \boldsymbol{y} + \boldsymbol{\varepsilon}) = 0, \qquad k = 1, 2, \ldots, m. \tag{4.1}$$

We also assume that we already have a first approximation $\boldsymbol{x}_0$ of the unknowns. As a first approximation of the $\boldsymbol{\eta}$ we take $\boldsymbol{\eta}_0 = \boldsymbol{y}$ as in §3. Finally, we require that the functions f_k can be reasonably well approximated by linear functions in a region around $(\boldsymbol{x}_0, \boldsymbol{\eta}_0)$ determined by the differences $\boldsymbol{x} - \boldsymbol{x}_0$ and $\boldsymbol{\eta} - \boldsymbol{\eta}_0$. We can then write

$$\begin{aligned} f_k(\boldsymbol{x}, \boldsymbol{\eta}) &= f_k(\boldsymbol{x}_0, \boldsymbol{\eta}_0) \\ &\quad + \left(\frac{\partial f_k}{\partial x_1}\right)_{\boldsymbol{x}_0, \boldsymbol{\eta}_0} (x_1 - x_{10}) + \ldots + \left(\frac{\partial f_k}{\partial x_r}\right)_{\boldsymbol{x}_0, \boldsymbol{\eta}_0} (x_r - x_{r0}) \\ &\quad + \left(\frac{\partial f_k}{\partial \eta_1}\right)_{\boldsymbol{x}_0, \boldsymbol{\eta}_0} (\eta_1 - \eta_{10}) + \ldots + \left(\frac{\partial f_k}{\partial \eta_n}\right)_{\boldsymbol{x}_0, \boldsymbol{\eta}_0} (\eta_n - \eta_{n0}). \end{aligned} \tag{4.2}$$

Introducing the abbreviations

$$a_{kl} = \left(\frac{\partial f_k}{\partial x_l}\right)_{\boldsymbol{x}_0, \boldsymbol{\eta}_0}, \qquad A = \begin{pmatrix} a_{11} & a_{12} & \cdots & a_{1r} \\ a_{21} & a_{22} & \cdots & a_{2r} \\ \vdots & & & \\ a_{m1} & a_{m2} & \cdots & a_{mr} \end{pmatrix}, \tag{4.3}$$

$$b_{kl} = \left(\frac{\partial f_k}{\partial \eta_l}\right)_{\boldsymbol{x}_0, \boldsymbol{\eta}_0}, \qquad B = \begin{pmatrix} b_{11} & b_{12} & \cdots & b_{1n} \\ b_{21} & b_{22} & \cdots & b_{2n} \\ \vdots & & & \\ b_{m1} & b_{m2} & \cdots & b_{mn} \end{pmatrix}, \tag{4.4}$$

$$c_k = f_k(\boldsymbol{x}_0, \boldsymbol{\eta}_0), \qquad \boldsymbol{c} = \begin{pmatrix} c_1 \\ c_2 \\ \vdots \\ c_m \end{pmatrix}, \tag{4.5}$$

$$\boldsymbol{\xi} = \boldsymbol{x} - \boldsymbol{x}_0, \qquad \boldsymbol{\delta} = \boldsymbol{\eta} - \boldsymbol{\eta}_0, \tag{4.6}$$

the system (4.2) can be written as

$$A\boldsymbol{\xi} + B\boldsymbol{\delta} + \boldsymbol{c} = 0. \tag{4.7}$$

The Lagrangian is

$$L = \boldsymbol{\delta}^{\mathrm{T}} G_y \boldsymbol{\delta} + 2\boldsymbol{\mu}^{\mathrm{T}}(A\boldsymbol{\xi} + B\boldsymbol{\delta} + \boldsymbol{c}), \tag{4.8}$$

$\boldsymbol{\mu}$ being an m-vector of Lagrangian multipliers. Setting the total differential of (4.8) with respect to $\boldsymbol{\delta}$ equal to zero is equivalent to

$$G_y \boldsymbol{\delta} + B^{\mathrm{T}} \boldsymbol{\mu} = 0,$$

or

$$\boldsymbol{\delta} = - G_y^{-1} B^{\mathrm{T}} \boldsymbol{\mu}. \tag{4.9}$$

Substitution into (4.7) yields

$$A\boldsymbol{\xi} - BG_y^{-1}B^{\mathrm{T}}\boldsymbol{\mu} + \boldsymbol{c} = 0, \tag{4.10}$$

or

$$\boldsymbol{\mu} = G_B(A\boldsymbol{\xi} + \boldsymbol{c}), \tag{4.11}$$

where we have defined

$$G_B = (BG_y^{-1}B^{\mathrm{T}})^{-1}. \tag{4.12}$$

With (4.9) we can now write

$$\boldsymbol{\delta} = - G_y^{-1} B^{\mathrm{T}} G_B (A\boldsymbol{\xi} + \boldsymbol{c}). \tag{4.13}$$

Since the Lagrangian L takes its minimum also with respect to the $\boldsymbol{\xi}$, we have

$$\frac{\partial L}{\partial \boldsymbol{\xi}} = 2\boldsymbol{\mu}^{\mathrm{T}} A = 0.$$

Transposing this and substituting (4.11) we get

$$2A^{\mathrm{T}} G_B (A\boldsymbol{\xi} + \boldsymbol{c}) = 0,$$

or

$$\tilde{\boldsymbol{\xi}} = - (A^{\mathrm{T}} G_B A)^{-1} A^{\mathrm{T}} G_B \, \boldsymbol{c}. \tag{4.14}$$

Substitution of (4.14) into (4.13) and (4.11) gives the least squares estimates for the deviations $\boldsymbol{\delta}$ and the Lagrangian multipliers $\boldsymbol{\mu}$

$$\tilde{\boldsymbol{\delta}} = - G_y^{-1} B^{\mathrm{T}} G_B(\boldsymbol{c} - A(A^{\mathrm{T}} G_B A)^{-1} A^{\mathrm{T}} G_B \boldsymbol{c}), \tag{4.15}$$

$$\tilde{\boldsymbol{\mu}} = G_B A(\boldsymbol{c} - A(A^{\mathrm{T}} G_B A)^{-1} A^{\mathrm{T}} G_B\, \boldsymbol{c}). \tag{4.16}$$

The estimates for the parameters $\boldsymbol{x}$ and for the improved measurements $\boldsymbol{\eta}$ become

$$\tilde{\boldsymbol{x}} = \boldsymbol{x}_0 + \tilde{\boldsymbol{\xi}}, \tag{4.17}$$

$$\tilde{\boldsymbol{\eta}} = \boldsymbol{\eta}_0 + \tilde{\boldsymbol{\delta}}. \tag{4.18}$$

From (4.14), (4.4) and (4.5) we get

$$\frac{\partial \tilde{\boldsymbol{\xi}}}{\partial \boldsymbol{y}} = - (A^{\mathrm{T}} G_B A)^{-1} A^{\mathrm{T}} G_B \frac{\partial \boldsymbol{c}}{\partial \boldsymbol{y}} = - (A^{\mathrm{T}} G_B A)^{-1} A^{\mathrm{T}} G_B B.$$

By propagation of errors we obtain the covariance matrix

$$G_{\tilde{x}}^{-1} = G_{\tilde{\xi}}^{-1} = (A^{\mathrm{T}} G_B A)^{-1}. \tag{4.19}$$

Correspondingly we find that

$$G_{\tilde{\eta}}^{-1} = G_y^{-1} - G_y^{-1} B^{\mathrm{T}} G_B B G_y^{-1} + G_y^{-1} B^{\mathrm{T}} G_B A (A^{\mathrm{T}} G_B A)^{-1} A^{\mathrm{T}} G_B B G_y^{-1}. \tag{4.20}$$

It can be shown that, under the conditions that were always assumed to hold, i.e. sufficient linearity of (4.1) and normal distribution of errors, the minimum function M, which can also be written

$$M = (B\tilde{\boldsymbol{\varepsilon}})^{\mathrm{T}}\, G_B (B\tilde{\boldsymbol{\varepsilon}}), \tag{4.21}$$

follows a χ^2-distribution with $m - r$ degrees of freedom.

If the eqs. (4.1) are already linear, the final solution is given by (4.17)–(4.20). In non-linear cases we usually consider them as better approximations; we therefore replace $\boldsymbol{x}_0$, $\boldsymbol{\eta}_0$ by $\tilde{\boldsymbol{x}}$, $\tilde{\boldsymbol{\eta}}$, evaluate the matrices (4.3)–(4.5) on this basis and get still better approximations by introducing them into (4.17)–(4.20). This iteration process can be repeated until a satisfactory solution is obtained. In general it is not easy to determine when a solution is satisfactory. The simplest method is to investigate the value of M in each step and break off the iteration process if M either falls below a given value or decreases by less than a given fraction from one step to the next. Of course it is also possible that M does not decrease at all, i.e. that the iteration process diverges. In this case the points (a)–(d) mentioned at the end of §2 have to be considered. More elaborate criteria for the acceptance of a solution can only be set up if the particular nature of the problem at hand is known.

9-5. A FORTRAN program for general least squares fitting; examples

In this section we discuss a rather simple FORTRAN program to solve the general least squares problem of §4 using the iteration technique. The program can be regarded as an extension of the sine fitting program of example 9-3 to the general case. It is, however, more flexible in the following respects.

The measurements $\boldsymbol{y}$, their covariance matrix $C_y = G_y^{-1}$, and the function $\boldsymbol{f}(\boldsymbol{\eta}, \boldsymbol{x})$ are not specified in the program. To solve a particular problem, two special subroutines have to be written by the user of the program. In the first, called INVAL, the measurements and their covariance matrix are placed into the arrays Y and CY, respectively. Also the number of measurements, N, is specified. The subroutine DERIV is used in every step of the iteration process to compute the matrices A and B of the derivatives, and the vector C. In the starting step the routine DERIV also provides the first approximation of the unknowns X. There are problems where it is useful to try fits to the data of different functions $\boldsymbol{f}(\boldsymbol{\eta}, \boldsymbol{x})$, i.e. to try different hypotheses. The integer variable NHYP, which is also an argument of the subroutine DERIV, is used to count these hypotheses. Each time DERIV is called, it performs the operations described above for the hypothesis numbered NHYP. If all hypotheses have been tried and DERIV is called again, NHYP is set negative. In this case the main program comes to STOP.

Program 9-2: *Main program* LSQFIT *solving the general case of least squares fitting*

```
C
C      PRCGRAM FOR GENERAL LEAST SQUARES FITTING
C
       DIMENSION X(10),Y(100), CY (1000), A(1000), B(1000), C(50),
      *ETA(100),XI(10),DELTA(100),EPSLN(100),CX(100),CETA(1000),GB(1000),
      *S1(1000),S2(1000),S3(1000),S4(1000),S5(1000)
C
C      SUBROUTINE INVAL PREPARES INITIAL VALUES.
C      IT STORES THE MEASUREMENTS IN Y , THEIR COVARIANCE MATRIX IN CY
C      AND THE NUMBER OF MEASUREMENTS IN N.
C
       CALL INVAL(Y,CY,N)
C
C      WRITE INITIAL VALUES
C
       WRITE(6,1400)
       CALL MTXWRT (Y,1,N)
       WRITE(6,1500)
       CALL MTXWRT (CY,N,N)
       NHYP=0
     5 NHYP=NHYP+1
C
C      SET FIRST APPROXIMATION OF ETA EQUAL TO Y.
C
       CALL MTXTRA(Y,ETA,N,1)
       NSTEP=0
    10 NSTEP=NSTEP+1
```

```
C     SUBROUTINE DERIV SETS UP ONE HYPOTHESIS AT A TIME. THE NUMBER
C     NHYP INDICATES THE CURRENT HYPOTHESIS. IT IS SET NEGATIVE, IF
C     NO MORE HYPOTHESES ARE TC BE TRIED. DERIV DETERMINES THE NUMBER
C     NR OF UNKNOWNS AND THE NUMBER M OF EQUATIONS. IT PLACES FIRST
C     APPROXIMATIONS FOR THE UNKNOWNS IN X AND ALSO FIXES THE
C     MAXIMUM NUMBER MAXSTP OF ITERATION STEPS TO BE PERFORMED.
C     DERIV CALCULATES THE MATRICES A, B OF DERIVATIVES AND THE VECTOR
C     C FOR THE CURRENT VALUES OF X AND ETA.
C
      CALL DERIV (X,ETA,A,B,C,NR,N,M,NHYP,NSTEP,MAXSTP)
      NF=M-NR
C
C     STOP PROGRAM IF 'NHYP' NEGATIVE.
C
      IF (NHYP) 500,11,11
  500 STOP
C
C     WRITE FIRST APPROXIMATION OF UNKNOWNS.
C
   11 IF(NSTEP-1) 15,12,15
   12 IF(NR) 13,15,13
   13 WRITE(6,1300) NHYP,M,NR,NF
      CALL MTXWRT (X,1,NR)
C
C     CALCUTATE  GB  (EQ. 4.12).
C
   15 CALL MTXMBT (CY,B,S1,N,N,M)
      CALL MTXMLT (B,S1,S2,M,N,M)
      CALL MTXINV (S2,GB,M)
      IF(NR) 17,16,17
   16 CALL MTXTRA (C,S2,M,1)
      GO TO 18
C
C     CALCULATE MATRIX CX (EQ. 4.19).
C
   17 CALL MTXMLT (GB,A,S1,M,M,NR)
      CALL MTXMAT (A,S1,S2,NR,M,NR)
      CALL MTXINV (S2,CX,NR)
C
C     CALCULATE VECTOR OF CORRECTIONS XI (EQ. 4.14).
C
      CALL MTXMBT (CX,A,S2,NR,NR,M)
      CALL MTXMLT (S2,GB,S1,NR,M,M)
      CALL MTXMLT (S1,C,XI,NR,M,1)
      CALL MTXMSC (XI.XI,-1.,NR,1)
C
C     CALCULATE NEW VALUES FOR X (EQ. 4.17).
C
      CALL MTXADD (X,XI,X,NR,1)
C
C     CALCULATE VECTCR OF CORRECTIONS DELTA (EQ. 4.15).
C
      CALL MTXMLT (A,XI,S1,M,NR,1)
      CALL MTXADD (S1,C,S2,M,1)
   18 CALL MTXMLT (GB,S2,S1,M,M,1)
      CALL MTXMAT (B,S1,S2,N,M,1)
      CALL MTXMLT (CY,S2,DELTA,N,N,1)
C
C     CALCULATE NEW VALUES FOR ETA (EQ. 4.18).
C
      CALL MTXSUB (ETA,DELTA,ETA,N,1)
C
C     CALCULATE VECTCR EPSLN
C
      CALL MTXSUB (ETA,Y,EPSLN,N,1)
C
C     CALCULATE MINIMUM FUNCTICN XM (EQ. 4.21).
C
      CALL MTXMLT (B,EPSLN,S1,M,N,1)
      CALL MTXMLT (GB,S1,S2,M,M,1)
      CALL MTXMAT (S1,S2,XM,1,M,1)
```

```
C     WRITE RESULTS OF THIS STEP
C
      WRITE(6,1700) NHYP,NSTEP,XM
      IF(NR) 19,20,19
   19 WRITE(6,1900)
      CALL MTXWRT (X,1,NR)
   20 WRITE(6,1800)
      CALL MTXWRT (ETA,1,N)
C
C     BREAK OFF ITERATION, IF MAXIMUM NUMBER OF STEPS PERFORMED.
C
      IF (NSTEP-MAXSTP)25,80,80
C
C     IF THIS WAS FIRST STEP,STORE VALUE OF XM IN XMO AND REPEAT ITERATION.
C
   25 IF (NSTEP-1) 30,30,40
   30 XMO=XM
      GO TO 10
C
C     IF MINIMUM FUNCTION DIFFERS FROM PREVIOUS VALUE BY
C     LESS THAN 0.1 PER CENT,BREAK OFF ITERATION.
C
   40 TEST= (XM-XMO)/XMO
   50 IF (ABS(TEST)-0.001) 90,60,60
   60 XMO=XM
C
C     AFTER FIRST 3 STEPS COMPARE MINIMUM FUNCTION WITH RESULT OF PREVIOUS STEP.
C     GIVE UP THIS HYPOTHESIS, IF LARGER BY MORE THAN 1 PER CENT.
C     OTHERWISE PERFORM NEXT STEP.
C
      IF(NSTEP-3) 10,70,70
   70 IF(TEST-0.01)10,10,75
   75 WRITE(6,1000) NHYP
      GO TO 5
   80 WRITE (6,1100)
C
C     CALCULATE COVARIANCE MATRIX OF ETA (EQ. 4.20).
C
   90 CALL MTXMLT (B,CY,S1,M,N,N)
      CALL MTXMLT (GB,S1,S2,M,M,N)
      IF (NR) 92,95,92
   92 CALL MTXMAT (A,S2,S3,NR,M,N)
      CALL MTXMLT (CX,S3,S5,NR,NR,N)
      CALL MTXMAT (S3,S5,S4,N,NR,N)
   95 CALL MTXMAT (S1,S2,S5,N,M,N)
      CALL MTXSUB (CY,S5,S1,N,N)
      IF(NR) 97,96,97
   96 CALL MTXTRA(S1,CETA,N,N)
      GO TO 100
   97 CALL MTXADD (S1,S4,CETA,N,N)
C
C     WRITE RESULTS OF CURRENT HYPOTHESIS.
C
      WRITE (6,1200)
      CALL MTXWRT (CX,NR,NR)
  100 WRITE (6,1600)
      CALL MTXWRT (CETA,N,N)
C
C     TRY NEXT HYPOTHESIS.
C
      GO TO 5
C
C     FORMAT STATEMENTS
```

```
1000 FORMAT(//40H PROCESS DIVERGES FOR HYPOTHESIS NUMBER ,I2)
1100 FORMAT (//69H ITERATION WAS BROKEN OFF BECAUSE MAXIMUM NUMBER OF S
    *TEPS WAS REACHED )
1200 FORMAT (//31H COVARIANCE MATRIX OF UNKNOWNS /)
1300 FORMAT(//18H HYPOTHESIS NUMBER,I2/
    *32H  NUMBER OF EQUATIONS =          I3/
    *32H  NUMBER OF UNKNOWNS =           I3/
    *32H  NUMBER OF DEGREES OF FREEDOM = I3//
    *34H  FIRST APPROXIMATION OF UNKNOWNS /)
1400 FORMAT(15H1INITIAL VALUES///15H    MEASUREMENTS/)
1500 FORMAT(//36H    COVARIANCE MATRIX OF MEASUREMENTS/)
1600 FORMAT(//43H COVARIANCE MATRIX OF IMPROVED MEASUREMENTS /)
1700 FORMAT(//18H HYPOTHESIS NUMBER,I2,13H, STEP NUMBER,I2,20H, MINIMUM
    * FUNCTION =,F10.5 )
1800 FORMAT(//24H    IMPROVED MEASUREMENTS/)
1900 FORMAT(//11H    UNKNOWNS/)
     END
```

In this way all calculations relevant to the special problem are kept out of the main program, which can thus be used for a large variety of problems. The program begins with a call to the subroutine INVAL, described above, and writes out the measurements and their covariance matrix. The rest of the program consists of a nest of 2 loops, the outer one (starting at statement number 7) is performed once for each hypothesis and begins by transferring the measurements into the array ETA. The inner loop (starting at statement number 10) runs over all steps of the iteration process, numbered by the integer variable NSTEP. The inner loop begins by calling the subroutine DERIV. As described above DERIV provides a first approximation of the unknowns in the array X and computes A, B and C. It also provides the number of equations (M) and unknowns (NR). In the case of constrained measurements without unknowns, as discussed in §3, the variable NR is put equal to zero. This is used in IF-statements further on which control the program in such a way as to neglect all terms containing the matrix A, which has no meaning in that case. The following section of the program performs the matrix calculations necessary to derive corrections (XI, DELTA) and improved values for unknowns and measurements, which are stored again in the array X and ETA. It computes a vector of deviations between original and improved measurements (EPSLN) and uses it to determine the minimum function for this step (XM).

The next section (up to statement number 80) determines whether the iteration process needs to be repeated. The following simple criteria are used. A sufficient degree of accuracy has been reached if the minimum function has not appreciably decreased (i.e. by less than 1%) in the current step. In this case the iteration is called *convergent*. The iteration process is broken off if a certain maximum number of steps (defined in DERIV and stored in the variable MAXSTP) has been performed. In both cases the flow of the program is transferred to statement number 90 and onwards, where the covariance matrix of the improved measurements (CETA) is calculated and the results for the current hypothesis are written. If, on the other hand, the minimum function has increased (by more than 1 %) in this step the process is thought to *diverge* and no results for this hypothesis are written. An increase of the minimum function is tolerated during the first 3 steps since the iteration process sometimes shows a certain oscillation of M about its final value.

Example 9-6: *Fit of polynomials*

Our general program can of course also be applied to linear problems. In this case only one step is necessary and the choice of the first approximation of the unknowns is unimportant. We can put them equal to zero in the subroutine DERIV and fix MAXSTP at one. (We could also choose MAXSTP larger to perform several steps and check if the results still change after the first step. If this is the case the subroutine DERIV contains mistakes in the programmed arithmetic.) In example 9-2 the fitting of n measured points $y_j(t_j)$ to a straight line was discussed, where t_j was a variable free of errors. We now study the general problem of fitting a polynominal of order l to these measurements:

$$f_j(\boldsymbol{\eta}, \boldsymbol{x}) = \eta_j - x_1 - x_2 t_j - \dots - x_{l+1} t_j^l, \qquad j = 1, 2, \dots, n.$$

This equation is linear in the components of $\boldsymbol{\eta}$ and $\boldsymbol{x}$. The matrix B is simply a unit matrix of order n. The matrix A is an $(n \times (l + 1))$ matrix and takes the form

$$A = -\begin{pmatrix} 1 & t_1 & t_1^2 & \dots & t_1^l \\ 1 & t_2 & t_2^2 & \dots & t_2^l \\ \vdots & & & & \\ 1 & t_n & t_n^2 & \dots & t_n^l \end{pmatrix}.$$

Finally, since $\boldsymbol{\eta}_0 = \boldsymbol{y}$ and $\boldsymbol{x}_0 = 0$, the vector $\boldsymbol{c}$ is identical to $\boldsymbol{y}$.

Often it is interesting to try fits of the measured data to polynomials of different order, as long as the number of degrees of freedom of the fit is at least one, i.e. $n > l + 1$.

For a numerical example we use measurements taken from an experiment in particle physics. The elastic scattering of negative K-mesons on protons is studied for a fixed value of the K-meson energy. The distribution of the cosine of the scattering angle θ in the centre-of-mass system of the collision is characteristic of the angular momentum of possible intermediate states in the collision. In particular, if the distribution is regarded as a polynomial in cos θ, the order of the polynomial can serve to determine the spin quantum number of such intermediate states.

The measurements y_j ($j = 1, 2, \ldots, 10$) are simply the number of decays observed with $\cos\theta$ within a small interval near the values $t_j = \cos\theta_j$. The statistical errors, i.e. the square roots of the observed numbers, are taken as the measurement errors. The variances are therefore identical to the measurements. Since the individual measurements are assumed to be independent the covariance matrix contains only diagonal elements. The subroutine INVAL takes the simple form reproduced below, where the measurements are introduced by a DATA statement.

Program 9-3: *Subroutine* INVAL *defining initial values for the fit of polynomials*

```
      SUBROUTINE INVAL (Y,CY,N)
C
C     POLYNOMIAL FIT.
C
      DIMENSION Y(1),CY(1),A(10)
      DATA  A/81.,50.,35.,27.,26.,60.,106.,189.,318.,520./
      N=10
      CALL MTXTRA (A,Y,10,1)
      CALL MTXUNT  (CY,10)
      DO 10 J=1,N
      JJ=(J-1)*N+J
   10 CY(JJ)=Y(J)
      RETURN
      END
```

The subroutine DERIV is written in such a way that polynomial fits of zero (a constant) up to 5th order, i.e. a total of 6 hypotheses, are tried

Program 9-4: *Subroutine* DERIV *setting up the matrices of derivatives for the fit of polynomials*

```
      SUBROUTINE DERIV(X,Y,A,B,C,NR,N,M,NHYP,NSTEP,MAXSTP)
C
C     POLYNOMIAL FIT.
C
      DIMENSION X(1),Y(1),A(1),B(1),C(1),T(10)
      DATA T/-0.9,-0.7,-0.5,-0.3,-0.1,0.1,0.3,0.5,0.7,0.9/
      IF(NHYP-6) 5 , 5,80
    5 MAXSTP=1
      M=N
      CALL MTXUNT(B,M)
      NR=NHYP
      DO 20 J=1,NR
```

```
20 X(J)=0.
   DO 60 I=1,N
   IK=(I-1)*NR
   A(IK+1)=-1.
   C(I)=Y(I)
   IF(NR-1) 60,60,40
40 DO 50 K=2,NR
50 A(IK+K)=-T(I)**(K-1)
60 CONTINUE
70 RETURN
80 NHYP=-1
   GO TO 70
   END
```

The first part of the print-out from the program is reproduced in table 9-5. The results of all hypotheses are summarized in table 9-6 and fig. 9-4. We observe that the first two hypotheses (constant and straight line) give no agreement with the experimental data. This is seen from fig. 9-4 and is reflected in the values of the minimum function.

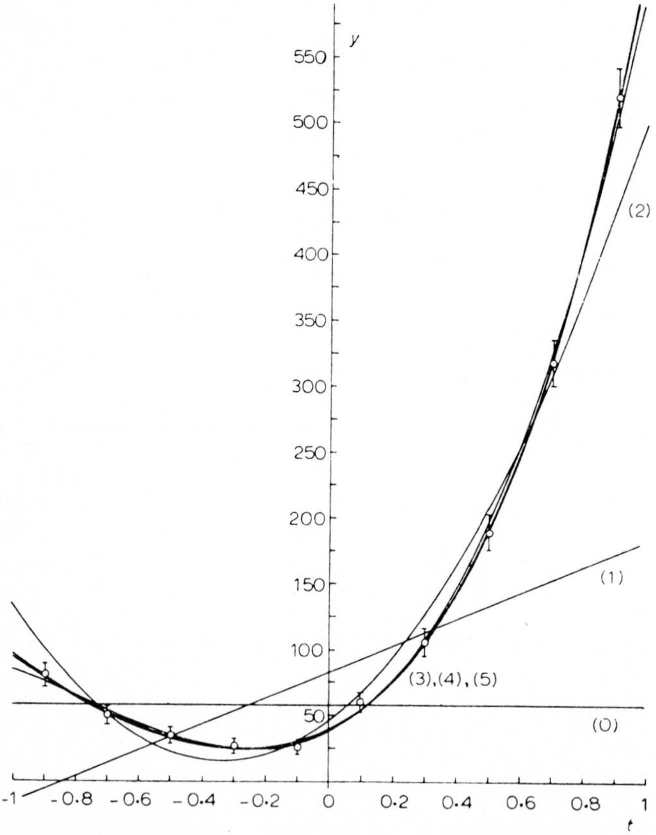

Fig. 9-4. Fit of polynomials of different order (0,1,...,5) to the data of example 9-6.

TABLE 9-5

Part of print-out for example 9-6

```
INITIAL VALUES

 MEASUREMENTS

81.00000  50.00000  35.00000  27.00000  26.00000  60.00000 106.00000 189.00000 318.00000 520.00000

 COVARIANCE MATRIX OF MEASUREMENTS

81.00000   0.0       0.0       0.0       0.0       0.0       0.0       0.0       0.0       0.0
 0.0      50.00000   0.0       0.0       0.0       0.0       0.0       0.0       0.0       0.0
 0.0       0.0      35.00000   0.0       0.0       0.0       0.0       0.0       0.0       0.0
 0.0       0.0       0.0      27.00000   0.0       0.0       0.0       0.0       0.0       0.0
 0.0       0.0       0.0       0.0      26.00000   0.0       0.0       0.0       0.0       0.0
 0.0       0.0       0.0       0.0       0.0      60.00000   0.0       0.0       0.0       0.0
 0.0       0.0       0.0       0.0       0.0       0.0     106.00000   0.0       0.0       0.0
 0.0       0.0       0.0       0.0       0.0       0.0       0.0     189.00000   0.0       0.0
 0.0       0.0       0.0       0.0       0.0       0.0       0.0       0.0     318.00000   0.0
 0.0       0.0       0.0       0.0       0.0       0.0       0.0       0.0       0.0     520.00000

HYPOTHESIS NUMBER 1
 NUMBER OF EQUATIONS =          10
 NUMBER OF UNKNOWNS =            1
 NUMBER OF DEGREES OF FREEDOM =  9

 FIRST APPROXIMATION OF UNKNOWNS

  0.0

HYPOTHESIS NUMBER 1, STEP NUMBER 1, MINIMUM FUNCTION = 833.54541

 UNKNOWNS

57.84526

 IMPROVED MEASUREMENTS

57.84528  57.84525  57.84525  57.84523  57.84525  57.84526  57.84529  57.84532  57.84570  57.84570
```

Table 9-5 (continued)

ITERATION WAS BROKEN OFF BECAUSE MAXIMUM NUMBER OF STEPS WAS REACHED

COVARIANCE MATRIX OF UNKNOWNS

5.78453

COVARIANCE MATRIX OF IMPROVED MEASUREMENTS

5.78456	5.78453	5.78453	5.78453	5.78453	5.78453	5.78453	5.78453	5.78453	5.78453
5.78453	5.78455	5.78453	5.78453	5.78453	5.78453	5.78453	5.78453	5.78453	5.78453
5.78453	5.78453	5.78455	5.78453	5.78453	5.78453	5.78453	5.78453	5.78453	5.78453
5.78453	5.78453	5.78453	5.78455	5.78453	5.78453	5.78453	5.78453	5.78453	5.78453
5.78453	5.78453	5.78453	5.78453	5.78455	5.78453	5.78453	5.78453	5.78453	5.78453
5.78453	5.78453	5.78453	5.78453	5.78453	5.78455	5.78453	5.78453	5.78453	5.78453
5.78453	5.78453	5.78453	5.78453	5.78453	5.78453	5.78457	5.78453	5.78453	5.78453
5.78453	5.78453	5.78453	5.78453	5.78453	5.78453	5.78453	5.78459	5.78453	5.78453
5.78453	5.78453	5.78453	5.78453	5.78453	5.78453	5.78453	5.78453	5.78477	5.78453
5.78453	5.78453	5.78453	5.78453	5.78453	5.78453	5.78453	5.78453	5.78453	5.78477

HYPOTHESIS NUMBER 2
NUMBER OF EQUATIONS = 10
NUMBER OF UNKNOWNS = 2
NUMBER OF DEGREES OF FREEDOM = 8

FIRST APPROXIMATION OF UNKNOWNS

0.0 0.0

HYPOTHESIS NUMBER 2, STEP NUMBER 1, MINIMUM FUNCTION = 585.44751

UNKNOWNS

82.65517 99.09996

IMPROVED MEASUREMENTS

-6.53476 13.28522 33.10519 52.92517 72.74515 92.56514 112.38513 132.20517 152.02516 171.84546

Table 9-5 (continued)

```
ITERATICN WAS BROKEN OFF BECAUSE MAXIMUM NUMBER OF STEPS WAS REACHED

COVARIANCE MATRIX OF UNKNOWNS

  8.26552   9.90999
  9.91000  39.58414

COVARIANCE MATRIX OF IMPROVED MEASUREMENTS

 22.49068  17.34750  12.20437   7.06124   1.91809  -3.22504  -8.36816 -13.51132 -18.65446 -23.79762
 17.34750  13.78776  10.22796   6.66819   3.10841  -0.45136  -4.01114  -7.57090 -11.13069 -14.69049
 12.20438  10.22797   8.25157   6.27514   4.29873   2.32232   0.34591  -1.63051  -3.60692  -5.58334
  7.06124   6.66819   6.27514   5.88211   5.48905   5.09600   4.70295   4.30990   3.91685   3.52379
  1.91810   3.10841   4.29873   5.48905   6.67938   7.86968   9.05999  10.25031  11.44063  12.63093
 -3.22504  -0.45136   2.32232   5.09600   7.86968  10.64337  13.41703  16.19070  18.96439  21.73805
 -8.36817  -4.01114   0.34591   4.70295   9.05999  13.41703  17.77411  22.13110  26.48816  30.84518
-13.51131  -7.57091  -1.63050   4.30990  10.25031  16.19070  22.13110  28.07158  34.01193  39.95233
-18.65443 -11.13069  -3.60691   3.91686  11.44063  18.96440  26.48816  34.01192  41.53596  49.05948
-23.79758 -14.69047  -5.58332   3.52381  12.63095  21.73808  30.84520  39.95233  49.05949  58.16685
```

TABLE 9-6
Summary of results of example 9-6

Hypothesis No.	Unknowns $\tilde{x}_1$	$\tilde{x}_2$	$\tilde{x}_3$	$\tilde{x}_4$	$\tilde{x}_5$	$\tilde{x}_6$	Degrees of freedom	Minimum function
1	57.85						9	833.55
2	82.66	99.10					8	585.45
3	47.27	185.96	273.61				7	36.41
4	37.94	126.55	312.02	137.59			6	2.85
5	39.62	119.10	276.49	151.91	52.60		5	1.68
6	39.88	121.39	273.19	136.58	56.90	16.72	4	1.66

Hypothesis 3, a polynomial of second order gives qualitative agreement. Most of the measured points, however, do not fall on the fitted parabola within their error bars, i.e. within one standard deviation. A χ^2-test, performed at $\alpha = 0.001$ with the computed value of the minimum function, fails. For the hypotheses 4, 5 and 6, on the other hand, the agreement is very good. The fitted curves pass through the error bars and are nearly identical. The χ^2-test would not lead to rejection even at $\alpha = 0.5$. We can therefore conclude that a third order polynomial is sufficient to describe the distribution. The coefficients $\tilde{x}_5$ and $\tilde{x}_6$ in hypotheses 5 and 6 do not contribute significantly. Inspection of the complete results shows that they are small compared to their errors, i.e. to the square roots of the elements (5,5) and (6,6) of the covariance matrix of the unknowns.

Example 9-7: *Fit of a circle*

As an example of a non-linear problem we consider a number of measurements of points s_i, t_i on a plane with axes s and t. The points lie on a circle with unknown centre s_0, t_0 and unknown radius r (fig. 9-5). Such a circle is described by the equation

$$(s - s_0)^2 + (t - t_0)^2 = r^2,$$

or

$$s^2 + t^2 - 2ss_0 - 2tt_0 + s_0^2 + t_0^2 - r^2 = 0.$$

We can express this equation in the notation of our general least squares procedure. We use the unknowns

$x_1 = s_0, \quad x_2 = t_0, \quad x_3 = r.$

The measured coordinate pairs s_i, t_i ($i = 1, 2, \ldots, m$) are called

$y_{2i-1} = s_i, \qquad y_{2i} = t_i,$

i.e.

$y_1 = s_1, \quad y_2 = t_1, \quad y_3 = s_2, \quad y_4 = t_2, \quad \ldots .$

The equations (4.1) therefore become

$$y_{2i-1}^2 + y_{2i}^2 - 2x_1 y_{2i-1} - 2x_2 y_{2i} + x_1^2 + x_2^2 - x_3^2 = 0, \qquad i = 1, 2, \ldots m.$$

The matrices (4.3) and (4.4) become

$$A = \begin{pmatrix} -2y_1 + 2x_{10}, & -2y_2 + 2x_{20}, & -2x_{30} \\ -2y_3 + 2x_{10}, & -2y_4 + 2x_{20}, & -2x_{30} \\ \vdots & & \\ -2y_{2m-1} + 2x_{10}, & -2y_{2m} + 2x_{20}, & -2x_{30} \end{pmatrix},$$

and

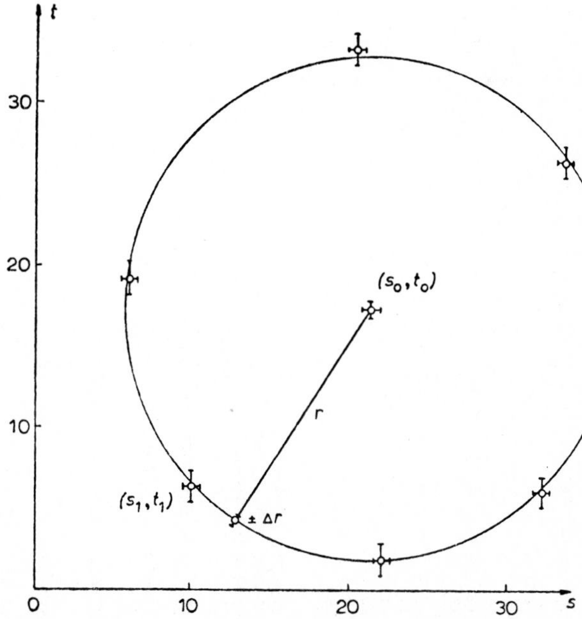

Fig. 9-5. Fit of a circle to the data of example 9-7.

TABLE 9-7
Print-out for example 9-7

```
INITIAL VALUES

 MEASUREMENTS

 10.20000   6.40000  22.09999   1.90000  32.20000   6.10000  33.50000  26.50000  20.20000  33.29999   6.00000  19.20000

 COVARIANCE MATRIX OF MEASUREMENTS

 0.25000   0.0       0.0       0.0       0.0       0.0       0.0       0.0       0.0       0.0       0.0       0.0
 0.0       1.00000   0.0       0.0       0.0       0.0       0.0       0.0       0.0       0.0       0.0       0.0
 0.0       0.0       0.25000   0.0       0.0       0.0       0.0       0.0       0.0       0.0       0.0       0.0
 0.0       0.0       0.0       1.00000   0.0       0.0       0.0       0.0       0.0       0.0       0.0       0.0
 0.0       0.0       0.0       0.0       0.25000   0.0       0.0       0.0       0.0       0.0       0.0       0.0
 0.0       0.0       0.0       0.0       0.0       1.00000   0.0       0.0       0.0       0.0       0.0       0.0
 0.0       0.0       0.0       0.0       0.0       0.0       0.25000   0.0       0.0       0.0       0.0       0.0
 0.0       0.0       0.0       0.0       0.0       0.0       0.0       1.00000   0.0       0.0       0.0       0.0
 0.0       0.0       0.0       0.0       0.0       0.0       0.0       0.0       0.25000   0.0       0.0       0.0
 0.0       0.0       0.0       0.0       0.0       0.0       0.0       0.0       0.0       1.00000   0.0       0.0
 0.0       0.0       0.0       0.0       0.0       0.0       0.0       0.0       0.0       0.0       0.25000   0.0
 0.0       0.0       0.0       0.0       0.0       0.0       0.0       0.0       0.0       0.0       0.0       1.00000

HYPOTHESIS NUMBER 1
 NUMBER OF EQUATIONS =          6
 NUMBER OF UNKNOWNS =           3
 NUMBER OF DEGREES OF FREEDOM = 3

 FIRST APPROXIMATION OF UNKNOWNS

 21.35919  17.92542  16.04251

HYPOTHESIS NUMBER 1, STEP NUMBER 1, MINIMUM FUNCTION =   0.46391

 UNKNOWNS

 21.29190  17.33159  15.49220

 IMPROVED MEASUREMENTS

 10.22099   6.48673  22.10045   1.86022  32.15938   6.27723  33.59576  26.77057  20.20985  32.77652   5.91352  19.22870
```

Table 9-7 (continued)

HYPOTHESIS NUMBER 1, STEP NUMBER 2, MINIMUM FUNCTION = 0.44326

UNKNOWNS

21.29671	17.33029	15.49808

IMPROVED MEASUREMENTS

10.22159	6.48908	22.10054	1.85306	32.15958	6.27639	33.59340	26.76332	20.20961	32.79022	5.91522	19.22786

HYPOTHESIS NUMBER 1, STEP NUMBER 3, MINIMUM FUNCTION = 0.44328

UNKNOWNS

21.29669	17.33029	15.49808

IMPROVED MEASUREMENTS

10.22159	6.48907	22.10054	1.85308	32.15958	6.27639	33.59340	26.76332	20.20961	32.79021	5.91522	19.22786

COVARIANCE MATRIX OF UNKNOWNS

0.16513	-0.00794	0.03522
-0.00794	0.23195	0.01188
0.03522	0.01188	0.09978

COVARIANCE MATRIX OF IMPROVED MEASUREMENTS

0.21694	-0.12944	-0.00078	0.05992	-0.00927	0.03773	0.00904	0.02773	-0.00037	0.02105	0.03444	-0.01699
-0.12944	0.49318	-0.00305	0.23462	-0.03629	0.14773	0.03538	0.10857	-0.00145	0.08244	0.13485	-0.06654
-0.00078	-0.00305	0.24988	0.00891	0.00102	-0.00415	-0.00002	-0.00005	0.00003	-0.00174	-0.00014	0.00007
0.05992	0.23462	0.00891	0.31340	-0.07858	0.31987	0.00121	0.00372	-0.00235	0.13377	0.01089	-0.00537
-0.00927	-0.03629	0.00102	-0.07858	0.22750	0.09158	0.01495	0.04589	0.00026	-0.01493	0.01553	-0.00766
0.03773	0.14773	-0.00415	0.31987	0.09158	0.62723	-0.06087	-0.18677	-0.00107	0.06075	-0.06322	0.03120
0.00904	0.03538	-0.00002	0.00121	0.01495	-0.06087	0.22522	-0.07605	-0.00179	0.10160	0.00259	-0.00128
0.02773	0.10857	-0.00005	0.00372	0.04589	-0.18677	-0.07605	0.76666	-0.00548	0.31175	0.00796	-0.00393
-0.00037	-0.00145	0.00003	-0.00235	0.00026	-0.00107	-0.00179	-0.00548	0.24980	0.01137	0.00206	-0.00102
0.02105	0.08244	-0.00174	0.13377	-0.01493	0.06075	0.10160	0.31175	0.01137	0.35337	-0.11736	0.05791
0.03444	0.13485	-0.00014	0.01089	0.01553	-0.06322	0.00259	0.00796	0.00206	-0.11736	0.19551	0.02689
-0.01699	-0.06654	0.00007	-0.00537	-0.00766	0.03120	-0.00128	-0.00393	-0.00102	0.05791	0.02689	0.98673

$$B=\begin{pmatrix} 2y_1-2x_{10}, & 2y_2-2x_{20}, & 0, & 0, & \ldots, & 0, & 0 \\ 0, & 0, & 2y_3-2x_{10}, & 2y_4-2x_{20}, & \ldots, & 0, & 0 \\ 0, & 0, & 0, & 0, & \ldots, & 2y_{2m-1}-2x_{10}, & 2y_{2m}-2x_{20} \end{pmatrix}$$

We now need a first approximation for the unknowns s_0, t_0, r. We use a simple geometrical procedure, consider the first 3 points and construct the perpendicular bisector on the two secants determined by neighbouring points. The intersection of these would be the centre of the circle if the three points were measured without error. The radius is given by the distance between the centre and any (we take the first) point on the circle. The necessary formulae are well known from elementary geometry. We do not give them explicitly; they can be read off from the program.

The subroutine DERIV takes the following form.

Program 9-5: *Subroutine* DERIV *setting up the matrices of derivatives for the fit of a circle*

```
      SUBROUTINE DERIV(X,Y,A,B,C,NR,N,M,NHYP,NSTEP,MAXSTP)
C
C     FIT OF CIRCLE TO MEASURED POINT COORDINATES
C
      DIMENSION X(1),Y(1),A(1),B(1),C(1)
      IF(NHYP-1) 60,5,60
    5 IF(NSTEP-1) 8,7,8
    7 NR=3
      M=N/2
      MAXSTP=10
C
C     DERIVE FIRST APPOXIMATION
      SA=(Y(1)+Y(3))/2.
      TA=(Y(2)+Y(4))/2.
      SB=(Y(3)+Y(5))/2.
      TB=(Y(4)+Y(6))/2.
      SLOPEA=-(Y(3)-Y(1))/(Y(4)-Y(2))
      SLOPEB=-(Y(5)-Y(3))/(Y(6)-Y(4))
      SO=(TA-TB-SLOPEA*SA+SLCPEB*SB)/(SLOPEB-SLOPEA)
      TO =TA+ SLOPEA*(SO-SA)
      RSQU=(SO-Y(1))**2+(TO-Y(2))**2
      X(1)=SO
      X(2)=TO
      X(3)=SQRT(RSQU)
C
C     CALCULATE MATRICES A,B,C
C
    8 DO 10 I=1,M
      IJ=(I-1)*3+1
      A(IJ)=-2.*Y(2*I-1)+2.*X(1)
      A(IJ+1)=-2.*Y(2*I)+2.*X(2)
   10 A(IJ+2)=-2.*X(3)
      DO 30 I=1,M
      DO 30 J=1,N
      IJ=(I-1)*N+J
      IF(2*I-1-J) 20,15,20
   15 B(IJ)=2.*Y(J)-2.*X(1)
      GO TO 30
   20 IF(2*I-J) 28,25,28
   25 B(IJ)=2.*Y(J)-2.*X(2)
      GO TO 30
```

```
28 B(IJ)=0.
30 CONTINUE
   DO 40 I=1,M
40 C(I)=Y(2*I-1)**2+Y(2*I)**2-2.*X(1)*Y(2*I-1)-2.*X(2)*Y(2*I)
  *+X(1)**2+X(2)**2 -X(3)**2
50 RETURN
60 NHYP=-1
   GO TO 50
   END
```

For our example 6 points on a circle with unknown centre and radius were measured with an estimated accuracy of 0.5 mm in s and 1 mm in t, i.e. $\sigma^2(s_i) = 0.25\,\text{mm}^2$, $\sigma^2(t_i) = 1\,\text{mm}^2$. The measurements were assumed to be uncorrelated. The print-out resulting from the program is reproduced in table 9-7. We observe that the convergence is rather fast. In fact the only appreciable correction to the first approximation is done in the first step. We also note that the differences between improved and original measurements (i.e. the $\tilde{\varepsilon}_j$) are in general larger for the t-coordinates. This follows from our starting assumption that $\sigma^2(t_i) > \sigma^2(s_i)$.

Exercises

9-1: *Direct measurements*
Different measurements of the same quantity x yield the results

$$
\begin{aligned}
y_1 &= 10.23 \pm 0.3\\
y_2 &= 11.14 \pm 0.4\\
y_3 &= 9.82 \pm 0.2\\
y_4 &= 9.71 \pm 0.2\\
y_5 &= 10.78 \pm 0.5
\end{aligned}
$$

Compute the weighted average $\tilde{x}$ and its error and perform a χ^2-test at $\alpha = 5\%$ on the hypothesis that all measurements stem from the same population (have no systematic errors). Make up a table similar to table 9-1. You will notice that the test fails. Therefore compute $\Delta\tilde{x}' = \Delta\tilde{x}\sqrt{[M/(n-1)]}$ according to eq. (1.13).

9-2: *Indirect measurements, linear case*
(a) Fit a straight line through the origin to the measurements of example 9-2.
(b) Compare the slope and its error of the solution to the values obtained in example 9-2.
(c) Compute the minimum function for this exercise and for example 9-2 and perform a χ^2-test on both fits at 10% level of significance.
(d) Try to explain why the χ^2-test fails although the error of the slope is very small.

9-3: *Indirect measurements, non-linear case*
Given the measurements

j	1	2	3	4	5
t_j	0.5	2.0	4.0	5.0	7.0
y_j	−2.0	3.0	6.0	11.0	22.0
σ_j	1.0	1.0	1.0	1.0	1.0

fit the exponential function

$$\eta_j = x_1 t_j^{x_2}, \qquad j = 1, 2, \ldots, 5$$

to the data. To obtain a first approximation to the unknowns a logarithmic plot of the data could be made corresponding to

$$\ln \eta_j = \ln (\eta_j + \varepsilon_j) = x_2 \ln t_j + x_1$$

from which x_{20} and x_{10} could be read off after fitting a straight line by hand. However, in order to have the same initial values as in the solution of this exercise use $x_{10} = 0.5$, $x_{20} = 2$. Compute only the next approximation $\tilde{x}_{11}$, $\tilde{x}_{21}$.

9-4: *FORTRAN programs*
Write the subroutines DERIV and INVAL required to solve exercise 9-3 performing an iteration of 5 steps.

9-5: *Constrained measurements*
An opinion poll before an election yields the following result:
votes for party A: $y_1 = (42 \pm 3)\%$
votes for party B: $y_2 = (28 \pm 4)\%$
votes for other parties: $y_3 = (\ 8 \pm 5)\%$
abstention: $y_4 = (15 \pm 3)\%$
Increase the accuracy of prediction and errors by making use of the fact that the four numbers must add up to 100%.

9-6: *Important constraint equation: sum of measurements is constant*
The constraint equation

$$\sum_{i=1}^{n} \eta_i - b_0 = 0$$

is particularly important. It occurs in the form of conservation laws (of energy etc).
(a) Show that the correction to the measurement y_k is obtained by weighting the difference between the sum of the measurements and the predicted sum by the variance of y_k, i.e.

$$\tilde{\delta}_k = \tilde{\eta}_k - y_k = -\frac{\sigma_k^2}{\sum \sigma_i^2} \left(\sum y_i - b_0\right).$$

(b) Derive the covariance matrix $G_{\tilde{\eta}}^{-1}$ of improved measurements.

(c) Give a qualitative argument why the covariance between any two improved measurements is negative?

9-7: *General case*

Mass m, velocity v and momentum p of an object are measured. Identify these quantities with the measurements y_1, y_2 and y_3, respectively and assume them to be uncorrelated, i.e.

$$G_y^{-1} = \begin{pmatrix} \sigma_1^2 & & 0 \\ & \sigma_2^2 & \\ 0 & & \sigma_3^2 \end{pmatrix}.$$

The kinetic energy $x = E = \frac{1}{2}mv^2$ is unknown. From classical mechanics one has the constraint equation $p = mv$.

(a) Set up the functions (4.1) and the matrices A, B, G_B and the vector c. Assume $x_0 = \frac{1}{2}y_1 y_2^2$ as first approximation for x.

(b) Perform one step of the iteration to find an improved value of x and its variance using $m = 1$ kg, $v = 1\ \mathrm{ms}^{-1}$, $p = 0.8\ \mathrm{kg\,ms}^{-1}$,

$$G_y^{-1} = \begin{pmatrix} 0.01 & 0 & 0 \\ 0 & 0.04 & 0 \\ 0 & 0 & 0.09 \end{pmatrix}.$$

CHAPTER 10

SOME REMARKS ON MINIMIZATION

10-1. Parameter estimation and minimization

In the previous chapter we have determined unknown quantities $\boldsymbol{x}$ from measurements $\boldsymbol{y}$ by looking for that vector $\boldsymbol{x}$ for which a certain scalar function of $\boldsymbol{x}$ and $\boldsymbol{y}$ was minimized:

$$M(\boldsymbol{x}, \boldsymbol{y}) = \min. \tag{1.1}$$

The function M was of different complexity for the various special cases of least squares as listed in table E-7.

The original maximum likelihood method of ch. 7 provided an extremum condition similar to (1.1) by requiring

$$L = \max, \tag{1.2}$$

or

$$l = \ln L = \max, \qquad L = \prod_{j=1}^{N} f(\boldsymbol{y}^{(j)}, \boldsymbol{x}),$$

being the likelihood function, i.e. the joint conditional probability density of N observations $\boldsymbol{y}^{(j)}$ for given values of the parameters* $\boldsymbol{x}$. Thus parameter estimation is closely linked to the problem of finding an extremum. By changing the sign of the functions (1.2) we can always turn the problem into one of minimization.

Such problems are not only of considerable interest in data analysis but find several other applications in science, engineering and economics. They all have in common the feature that a certain *objective function*, $M(\boldsymbol{x})$, is constructed which contains a number of variable parameters $\boldsymbol{x}$. The values of $\boldsymbol{x}$, which minimize M are required. The objective function might be the profit (with a negative sign) of an economic project, with the parameters $\boldsymbol{x}$ describing such variables as personnel costs, delivery time, financing etc. The minimization procedure has a special tradition in lens design, where it is usually referred to as the *optimization* of design parameters.

* As in ch. 9 but unlike ch. 7 we denote measurements by y and unknowns by x.

In principle the Gaussian method used in the last chapter could be applied to derive the minimum of any objective function. In many practical cases, however, this is not possible for a number of reasons. For instance, the Gaussian method may not converge for certain functions M if the first approximation of the unknowns is not good enough. Also M may be such that it can not be differentiated analytically, so that its partial derivatives have to be approximated by difference quotients. The most important reason is that the convergence can be quite slow in the case of many parameters and that other methods, designed in a more ad hoc way to find the point in parameter space, where $M(\boldsymbol{x})$ becomes smallest, can be much faster.

The computer literature of the last decade consists, to a considerable extent, of the presentation and discussion of various minimization procedures. It is, therefore, quite beyond the scope of this book to give an adequate review of all the aspects of the subject. We simply want to discuss the main ideas of some procedures.

A number of fairly general minimization programs have been written by various authors and are usually available from their institutions in the form of FORTRAN card decks. The choice of a particular minimization method or program will depend on the characteristic features of the particular function $M(\boldsymbol{x})$ to be minimized. No general recommendation can be given. The efficiency of some procedures has been compared for a number of model functions by FLETCHER [1965] and LEON [1966a]. An extensive bibliography on optimization is presented by LEON [1966b].

10-2. Different minimization procedures

To solve the general problem

$$M(\boldsymbol{x}) = \min, \qquad \boldsymbol{x} = (x_1, x_2, \ldots, x_r), \tag{2.1}$$

all methods use *successive linear minimization*, i.e. starting from some approximation $\boldsymbol{x}_0$ a step is performed in parameter space along some direction $\boldsymbol{p}_1$ to yield a new set of parameters $\boldsymbol{x}_1 = \boldsymbol{x}_0 + \alpha_1 \boldsymbol{p}_1$ where the step length α_1 is chosen such that $M(\boldsymbol{x}_1) < M(\boldsymbol{x}_0)$. Usually we try to determine α_1 such that $M(\boldsymbol{x}_1)$ is the absolute minimum of M along the direction $\boldsymbol{p}_1$. A second step is then performed along a direction $\boldsymbol{p}_2$. A general step of the iteration takes the form

$$\boldsymbol{x}_{j+1} = \boldsymbol{x}_j + \alpha_j \boldsymbol{p}_j, \qquad M(\boldsymbol{x}_{j+1}) < M(\boldsymbol{x}_j). \tag{2.2}$$

The iteration is continued until some convergence criterion is satisfied. We might, for example, accept $\boldsymbol{x}_{j+1}$ as the solution if $M(\boldsymbol{x}_j) - M(\boldsymbol{x}_{j+1}) < \varepsilon$, where ε is some predetermined number, or if no direction is found for which $M(\boldsymbol{x}_{j+1}) < M(\boldsymbol{x}_j)$.

The methods vary drastically in the procedures used to determine stepping direction and step length, i.e. in the *strategy* applied to reach the minimum.

Eq. (2.2) can be interpreted geometrically. The function M defines a hypersurface in $(r + 1)$-dimensional space. For two unknowns, for example, $M(\boldsymbol{x}) = M(x_1, x_2)$ defines a surface in 3-dimensional space, spanned by x_1, x_2 and M. Now (2.2) describes the stepwise probing of this surface for a minimum. It is clear that the problem is very difficult if the surface has several local minima and if $\boldsymbol{x}_0$ is not already near to the absolute minimum. In order to assure a good first approximation it is often necessary to perform a *search* in parameter space: The function $M(\boldsymbol{x})$ is evaluated systematically using a "lattice" of points over a region of parameter space. As a first approximation the point $\boldsymbol{x}_0$ is used which yields the smallest value of M, provided the neighbouring points indicate a continuous behaviour of M with only one minimum. If this is not the case the search is repeated around the point $\boldsymbol{x}_0$ with a much finer mesh.

The simplest strategy is that of *linear minimization along the coordinate*

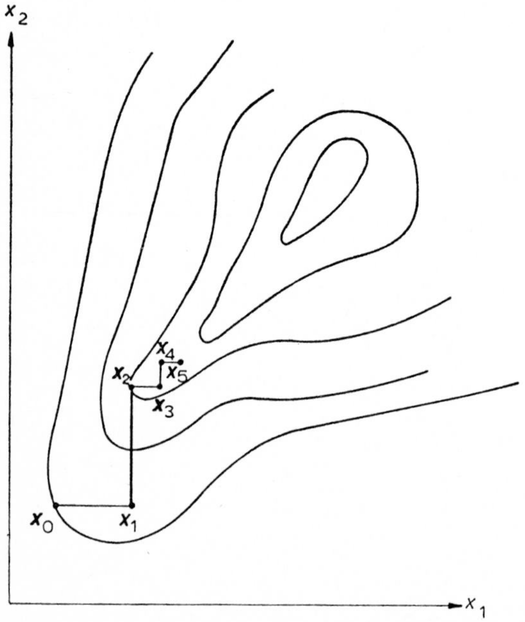

Fig. 10-1. Linear minimization along coordinate directions.

directions. It is sketched in fig. 10-1. Contour lines, i.e. lines for equal values of M, are shown for the case of two variables. Starting from the point $\boldsymbol{x}_0$ the minimum is first sought along the direction x_1 by solving

$$\left(\frac{\partial M}{\partial x_1}\right)_{x_2=x_{20}} = 0.$$

The next step is performed along the direction x_2, etc. The progress of the method is reasonably fast if there is no correlation between the variables. If, however, such a correlation exists, as in the case sketched in fig. 10-1, the size of each step is small and the convergence is slow.

The natural consequence of this observation is that we have to follow the local gradient of the function M, i.e. to use the method of *steepest descent*. As the direction $\boldsymbol{p}_j$ the local gradient in the point $\boldsymbol{x}_j$ is used, having the components

$$p_{ij} = \frac{\partial M}{\partial x_i}\left[\sum_{i=1}^{r}\left(\frac{\partial M}{\partial x_i}\right)^2\right]^{-\frac{1}{2}}, \qquad i = 1, 2, \ldots, r. \tag{2.3}$$

The minimum along the direction $\boldsymbol{p}_j$ is found at the point $\boldsymbol{x}_{j+1}$; here the new gradient $\boldsymbol{p}_{j+1}$ is calculated etc. Although the method of steepest descent

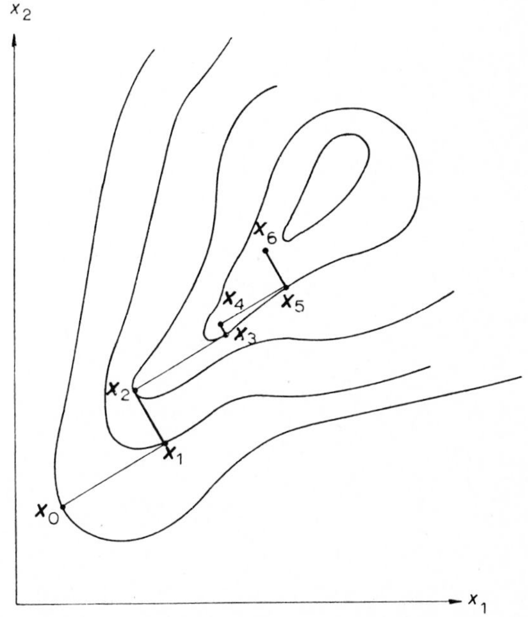

Fig. 10-2. Minimization using steepest descent.

seems to be superior to the minimization along the coordinate directions, it is in fact not much different, at least in the case of only two variables, as can be seen from fig. 10-2. By definition the gradient is perpendicular to the contour line at every point. Starting from $\boldsymbol{x}_0$ along the local gradient, the minimum of M is found at a point where the contour line is parallel to the search direction. The gradient at this point $\boldsymbol{x}_1$ is of course perpendicular to the contour line and therefore perpendicular to the original direction. For only two variables the situation is therefore completely analogous to minimization along the axes and the convergence is slow. This occurs because the path taken by the iteration procedure oscillates too much about the actual gradient line. Various modifications to the method have been proposed. It has been suggested for example [BOOTH, 1957], not to go all the way from $\boldsymbol{x}_0$ to the linear minimum $\boldsymbol{x}_1$ in fig. 10-2 but only a fraction, say, 9/10ths of that path, to determine the gradient there etc. In this way the angles between the directions can be kept smaller than 90° and the oscillations are diminished.

Another procedure, which achieves the same effect, is employed in the widely used program MINFUN written by HUMPHREY [1962] and described by ROSENFELD and HUMPHREY [1963]. The method which was probably first proposed by GELFAND [1961] is illustrated in fig. 10-3. Starting from

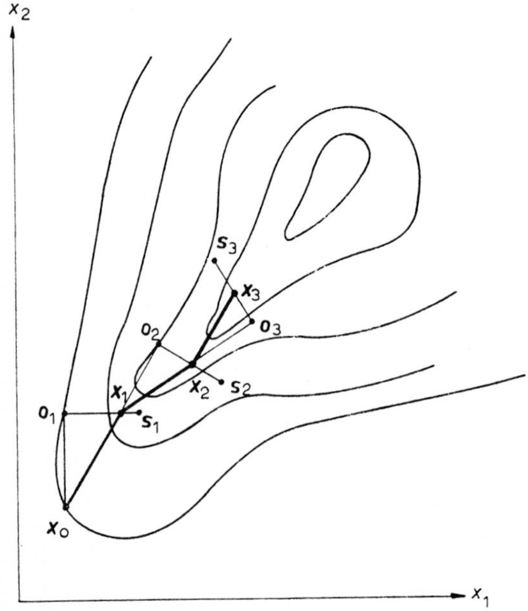

Fig. 10-3. Minimization using the ravine stepping procedure.

a point $\boldsymbol{x}_0$ a first step ("overstep") is taken along a coordinate direction to $\boldsymbol{o}_1$. At $\boldsymbol{o}_1$ a step ("sidestep") perpendicular to $(\boldsymbol{x}_0 - \boldsymbol{o}_1)$ is carried out up to $\boldsymbol{s}_1$. From the value of the function M at the points $\boldsymbol{o}_1$ and $\boldsymbol{s}_1$, and its derivative at $\boldsymbol{o}_1$, the minimum of M is predicted at the point $\boldsymbol{x}_1$ along the direction $(\boldsymbol{s}_1 - \boldsymbol{o}_1)$. A quadratic variation of M is assumed along this line. $M(\boldsymbol{x}_1)$ is evaluated to verify whether it is indeed smaller than $M(\boldsymbol{o}_1)$ and $M(\boldsymbol{s}_1)$. If this is not the case the size of the sidestep is decreased until a minimum is found. The next overstep is now taken starting from point $\boldsymbol{x}_1$ along the direction $(\boldsymbol{x}_1 - \boldsymbol{x}_0)$ up to $\boldsymbol{o}_2$, where a sidestep up to $\boldsymbol{s}_2$ is performed. Along $(\boldsymbol{s}_2 - \boldsymbol{o}_2)$ a new minimum $\boldsymbol{x}_2$ is found. The procedure follows a "*ravine*" in the function $M(\boldsymbol{x})$. It does not stop at the minimum of M but continues along the ravine, thus mapping M in the region near the minimum. Convergence to the minimum can be obtained by decreasing the step size and inverting the overstep direction every time $M(\boldsymbol{x})$ has increased.

A number of minimization programs (for example MINROS written by SHEPPEY [1966]) use a procedure suggested by ROSENBROCK [1964]. Successive linear minimization is performed in r *orthogonal directions.* Then a new set of orthogonal directions is derived, which seems particularly suited for further minimization. Let us denote the first set of orthogonal directions, which could for example be the coordinate directions in parameter space, by the unit vectors $\boldsymbol{p}_1, \boldsymbol{p}_2, \ldots, \boldsymbol{p}_r$. Starting from a point $\boldsymbol{x}_0$ a linear minimum is looked for along $\boldsymbol{p}_1$ and found at $\boldsymbol{x}_0 + \alpha_1 \boldsymbol{p}_1$. From there a search is carried out along $\boldsymbol{p}_2$ etc. The total of r searches in called a *stage* by Rosenbrock. Its intermediate steps can be characterized by the vectors

$$\begin{aligned} \boldsymbol{a}_1 &= \alpha_1 \boldsymbol{p}_1 + \alpha_2 \boldsymbol{p}_2 + \ldots + \alpha_r \boldsymbol{p}_r, \\ \boldsymbol{a}_2 &= \qquad\quad \alpha_2 \boldsymbol{p}_2 + \ldots + \alpha_r \boldsymbol{p}_r, \\ &\vdots \\ \boldsymbol{a}_r &= \qquad\qquad\qquad\qquad \alpha_r \boldsymbol{p}_r. \end{aligned} \tag{2.4}$$

Thus $\boldsymbol{a}_1$ is the vector summarizing all steps of the stage, $\boldsymbol{a}_2$ contains all but the first step, and $\boldsymbol{a}_r$ is just the last step of the stage. The result of the stage could have been reached in a single step along the direction $\boldsymbol{q}_1 = \boldsymbol{a}_1/|\boldsymbol{a}_1|$. By an orthogonal transformation a new set of directions $\boldsymbol{q}_1, \boldsymbol{q}_2, \ldots, \boldsymbol{q}_r$ is defined by

$$\boldsymbol{q}_1 = \frac{\boldsymbol{a}_1}{|\boldsymbol{a}_1|},$$

$$\begin{aligned} \boldsymbol{q}_2 &= \frac{\boldsymbol{a}_2 - (\boldsymbol{a}_1 \boldsymbol{q}_1)\boldsymbol{q}_1}{|\boldsymbol{a}_2 - (\boldsymbol{a}_1 \boldsymbol{q}_1)\boldsymbol{q}_1|}, \\ &\vdots \\ \boldsymbol{q}_r &= \frac{\boldsymbol{a}_r - \sum_{j=1}^{r-1} (\boldsymbol{a}_j \boldsymbol{q}_j)\boldsymbol{q}_j}{|\boldsymbol{a}_r - \sum_{j=1}^{r-1} (\boldsymbol{a}_j \boldsymbol{q}_j)\boldsymbol{q}_j|}. \end{aligned} \tag{2.5}$$

Thus, based on the results of the previous stage, $\boldsymbol{q}_1$ is the direction in which fastest advance is expected, $\boldsymbol{q}_2$ the best direction orthogonal to it and so on. For many types of functions $M(\boldsymbol{x})$ this method reduces the number of steps considerably.

Finally we want to discuss the method of *conjugate directions* [POWELL, 1964]. One of the computer programs based on this method is MINCON [SHEPPEY, 1966]. As in Rosenbrock's method the process is divided into stages of r linear minimization steps with the determination of a new set of directions after every stage. The choice of directions is based on the assumption that over the region explored in one stage $M(\boldsymbol{x})$ can be approximated by a quadratic function

$$M(\boldsymbol{x}) = \boldsymbol{x}A\boldsymbol{x} + \boldsymbol{b}\boldsymbol{x} + c. \tag{2.6}$$

Directions $\boldsymbol{p}$ and $\boldsymbol{q}$ are called *conjugate* if

$$\boldsymbol{p}A\boldsymbol{q} = 0, \tag{2.7}$$

i.e. they are orthogonal with respect to a metric defined by the matrix A (in Euclidean metric A is the unit matrix).

The first stage employing the directions $\boldsymbol{p}_1, \boldsymbol{p}_2, \ldots, \boldsymbol{p}_r$ proceeds as before yielding the result

$$\boldsymbol{x}_r = \boldsymbol{x}_0 + \boldsymbol{a}_1,$$

$\boldsymbol{a}_1$ being defined in (2.4). Another linear minimization is then performed in the direction $\boldsymbol{a}/|\boldsymbol{a}_1|$ giving the point $\boldsymbol{x}_{r+1}$. The next set of directions chosen is

$$\begin{aligned} \boldsymbol{q}_1 &= \boldsymbol{p}_2, \\ \boldsymbol{q}_2 &= \boldsymbol{p}_3, \\ &\vdots \\ \boldsymbol{q}_{r-1} &= \boldsymbol{p}_r, \\ \boldsymbol{q}_r &= \boldsymbol{a}_1/|\boldsymbol{a}_r|. \end{aligned} \tag{2.8}$$

It was proved by POWELL [1964] that after a total of r stages the directions (2.8) are mutually conjugate and that the minimum is found after r stages. It is required in the proof that $M(\boldsymbol{x})$ be of the form (2.6). Usually this is approximately true in the region between $\boldsymbol{x}_0$ and the minimum, which can then be reached after a relatively small number of steps. The method also has the advantage that the transformation (2.8) of directions involves less computational work than (2.5).

CHAPTER 11

ANALYSIS OF VARIANCE

The analysis of variance method, due to R. A. Fisher, is used to test the hypothesis of equal means of a number of samples. This problem arises, for example, when a series of measurements is performed under different conditions, or when samples for production control are taken from different machines. It is important to detect a possible influence of the variation of the *external variables* (experimental condition, number of machine) on the sample. For only 2 samples the problem can be tackled using Student's difference test (ch. 8, §2).

We speak of an analysis of variance with *one-way classification* if only one external variable is varied. As an example one might evaluate series of measurements of an object micrometer performed on several different microscopes. A *two-* (or more) *way classification* applies in the case of several variables being varied. If in the above example several observers carry out a number of measurements on each microscope, a two-way classification could reveal possible influences of both observer and instrument on the result.

11-1. One-way classification

We consider a sample of size n, which according to some criterion A can be divided into t groups or classes. The criterion, of course, is given by the sampling or measuring process. We say that the groups are formed by the *classification* A. The populations from which the t subsamples are drawn are assumed to be normally distributed with equal variance σ^2. We want to test the hypothesis that the means of these populations are also equal. If the hypothesis were true the samples would stem from the same distribution. We apply the results of ch. 6, §3 (sampling from partitioned distributions) to this situation. Using the same notation as there, we have t groups of size n_i with

$$n = \sum_{i=1}^{t} n_i,$$

and we denote the jth element of the ith group by x_{ij}. Then the sample mean of the ith group is

$$\bar{x}_i = \frac{1}{n_i} \sum_{j=1}^{n_i} x_{ij}, \tag{1.1}$$

and the mean of the total sample

$$\bar{x} = \frac{1}{n} \sum_{i=1}^{t} \sum_{j=1}^{n_i} x_{ij} = \frac{1}{n} \sum_{i=1}^{t} n_i \bar{x}_i. \tag{1.2}$$

We now construct the sum of squares

$$\begin{aligned} Q = \sum_{i=1}^{t} \sum_{j=1}^{n_i} (x_{ij} - \bar{x})^2 &= \sum_{i=1}^{t} \sum_{j=1}^{n_i} (x_{ij} - \bar{x}_i + \bar{x}_i - \bar{x})^2 \\ &= \sum_{i=1}^{t} \sum_{j=1}^{n_i} (x_{ij} - \bar{x}_i)^2 + \sum_{i=1}^{t} \sum_{j=1}^{n_i} (\bar{x}_i - \bar{x})^2 \\ &\quad + 2 \sum_{i=1}^{t} \sum_{j=1}^{n_i} (x_{ij} - \bar{x}_i)(\bar{x}_i - \bar{x}). \end{aligned}$$

The last term vanishes because of eqs. (1.1) and (1.2). Therefore

$$Q = \sum_{i=1}^{t} \sum_{j=1}^{n_i} (x_{ij} - \bar{x})^2 = \sum_{i=1}^{t} n_i(\bar{x}_i - \bar{x})^2 + \sum_{i=1}^{t} \sum_{j=1}^{n_i} (x_{ij} - \bar{x}_i)^2,$$

$$Q = Q_A + Q_W. \tag{1.3}$$

The first term is the *sum of squares between groups* formed by the classification A, the second is the sum over all *sums of squares within groups.* The sum of squares decomposes into a sum of two sums of squares with different "*sources*" – the variation of group means inside the classification A and the variation of measurements within the groups. If our hypothesis is true then Q is a sum of squares from a normal distribution, i.e. Q/σ^2 follows a χ^2-distribution with $n - 1$ degrees of freedom. Similarly for each group the expression

$$\frac{Q_i}{\sigma^2} = \frac{1}{\sigma^2} \sum_{j=1}^{n_i} (x_{ij} - \bar{x}_i)^2$$

follows a χ^2-distribution with $n_i - 1$ degrees of freedom. The sum

$$\frac{Q_W}{\sigma^2} = \sum_{i=1}^{t} \frac{Q_i}{\sigma^2}$$

is then described by a χ^2-distribution with $\Sigma_i(n_i - 1) = n - t$ degrees of freedom (cf. ch. 6, §5). Finally Q_A/σ^2 corresponds to χ^2 with $t - 1$ degrees of freedom.

The expressions

$$s^2 = \frac{Q}{n-1} = \frac{1}{n-1} \sum_i \sum_j (x_{ij} - \bar{x}),$$

$$s_A^2 = \frac{Q_A}{t-1} = \frac{1}{t-1} \sum_i n_i(\bar{x}_i - \bar{x})^2, \tag{1.4}$$

$$s_W^2 = \frac{Q_W}{n-t} = \frac{1}{n-t} \sum_i \sum_j (x_{ij} - \bar{x}_i)^2,$$

are unbiased estimates of the population variance (These expressions were called *mean squares* in ch. 6, §4.) The quotient

$$F = s_A^2/s_W^2 \tag{1.5}$$

can then be used to perform an F-test.

If the hypothesis of equal means is false then the $\bar{x}_i$ of the individual groups will be quite different. Therefore s_A^2 will be large while s_W^2, being an average over the sample variances within groups, will remain essentially unchanged. This means that the quotient (1.5) is large if the hypothesis is false. The one-tailed F-test is therefore appropriate. The hypothesis of equal means is rejected at a given level of significance, α, if

$$F = s_A^2/s_W^2 > F_{1-\alpha}(t-1, n-t). \tag{1.6}$$

The sums of squares can be computed using either of two equivalent formulae

$$Q = \sum_i \sum_j (x_{ij} - \bar{x})^2 = \sum_i \sum_j x_{ij}^2 - n\bar{x}^2,$$

$$Q_A = \sum_i n_i(\bar{x}_i - \bar{x})^2 = \sum_i n_i\bar{x}_i^2 - n\bar{x}^2, \tag{1.7}$$

$$Q_W = \sum_i \sum_j (x_{ij} - \bar{x}_i)^2 = \sum_i \sum_j x_{ij}^2 - \sum_i n_i\bar{x}_i^2.$$

The last expression of each line is usually easier to compute. Since each sum of squares is obtained by forming the difference of two rather large numbers, rounding errors can be important in this case. Although we only need Q_A and Q_W to compute F, the computation of Q allows a verification since, from (1.3), we have $Q = Q_A + Q_W$. The check is non-trivial only if the first expressions in (1.7) are used. Usually the results of an analysis of variance are summarized in a so-called *analysis of variance table* which has the form of table 11-1.

TABLE 11-1

Analysis of variance table for one-way classification

Source of variance	SS (sum of squares)	ND (number of degrees of freedom)	MS (mean square)	F
Between groups	$Q_A = \sum_{i=1}^{t} n_i (\bar{x}_i - \bar{x})^2$	$t - 1$	$s_A^2 = \frac{Q_A}{t - 1}$	$F = \frac{s_A^2}{s_W^2}$
Within groups	$Q_W = \sum_{i=1}^{t} \sum_{j=1}^{n_i} (x_{ij} - \bar{x}_i)^2$	$n - t$	$s_W^2 = \frac{Q_W}{n - t}$	
Total	$Q = \sum_{i=1}^{t} \sum_{j=1}^{n_i} (x_{ij} - \bar{x})^2$	$n - 1$	$s^2 = \frac{Q}{n - 1}$	

It is important to remember that an essential assumption was made in the derivation of the analysis of variance method namely the normal distribution of measurements within each group. This cannot always be assured. If, for example, the measured quantity is always positive (e.g. weight, length of an object), and has a rather large variance compared with its magnitude, the probability density can have a skew shape. If the original measurements – let us denote them by x' in this case – are transformed using a monotonous transformation like

$$x = a \log(x' + b), \tag{1.8}$$

where a and b are suitably chosen constants, a normal distribution can often be reasonably well approached. Sometimes other transformations like $x = \sqrt{x'}$, $x = 1/x'$ are used.

TABLE 11-2
Data of example 11-1

Experiment number	Groups treated with different medicaments I	II	III	
1	19	40	32	
2	45	28	26	
3	26	26	30	
4	23	15	17	
5	36	24	23	
6	23	26	24	
7	26	36	29	
8	33	27	20	
9	22	28	–	
10	–	19	–	$\sum_i \sum_j x_{ij}^2 = 20607$
$\sum_j x_j$	253	269	201	$\sum_i \sum_j x_{ij} = 723$
n_i	9	10	8	$n = 27$
				$n\bar{x}^2 = 19364$
$\bar{x}_i$	28.11	26.90	25.13	$\bar{x} = 26.78$
$\bar{x}_i^2$	790.17	723.61	631.52	$\sum_i n_i \bar{x}_i^2 = 19398$

Example 11-1: *Analysis of variance (one-way classification) on the influence of different medicaments*

The spleen of mice carrying a tumor is usually especially affected by the disease. The weight of the spleen can therefore be used as an indication of the response to different medicaments. Three test series (I, II and III) are performed. Table 11-2 contains the results that are already transformed according to $x = \log x'$, with x' being the weight in g. Most of the calculations are done within the framework of table 11-2. The analysis of variance table is given in table 11-3.

Since even at a 50% level of significance $F_{0.5}$ (2,24) = 0.72, the hypothesis of equal means cannot be rejected. The experiment gave no significant difference between the test series.

TABLE 11-3
Analysis of variance table for example 11-1

Source	SS	ND	MS	F
Between groups	34	2	17.0	0.337
Within groups	1209	24	50.4	
Total	1243	26	47.8	

11-2. Some aspects of two-way classification

Before we discuss the case of two external variables, we want to review the one-way classification somewhat more rigorously. The jth measurement of the quantity x in the group i was denoted by x_{ij}. Let us now assume that each group contains the same number of measurements, i.e. $n_i = J$ for all groups. For simplicity we now denote the number of groups by I. The classification into groups or classes was done following some criterion A (for example, the production number of a particular microscope) in which the groups are different. This situation is shown in table 11-4.

TABLE 11-4
One-way classification

Number of measurement	Classification A						
	A_1	A_2	...	A_i	...	A_I
1	x_{11}	x_{21}		x_{i1}		x_{I1}
2	x_{12}	x_{22}		x_{i2}		x_{I2}
⋮						
j	x_{1j}	x_{2j}		x_{ij}		x_{Ij}
⋮						
J	x_{1J}	x_{2J}		x_{iJ}		x_{IJ}

We write the different sample means in the form

$$\bar{x}_{..} = \bar{x} = \frac{1}{IJ} \sum_i \sum_j x_{ij},$$

$$\bar{x}_{i.} = \frac{1}{J} \sum_j x_{ij}, \tag{2.1}$$

$$\bar{x}_{.j} = \frac{1}{I} \sum_i x_{ij}.$$

The dot indicates summation over the subscript which has been replaced by the dot. This notation allows easy extension to more than two subscripts. The one-way classification was based on the assumption that the measurements within one class differ only by a measurement error which is normally distributed about zero with variance σ^2, i.e. we considered the *model*

$$x_{ij} = \mu_i + \varepsilon_{ij}. \tag{2.2}$$

The object of the analysis of variance was to test the hypothesis

$$H_0(\mu_1 = \mu_2 = \ldots = \mu_I = \mu). \tag{2.3}$$

Selecting only measurements of a particular group i and applying the maximum likelihood method to (2.2) we obtain the estimates

$$\tilde{\mu}_i = \bar{x}_{i.} = \frac{1}{J} \sum_j x_{ij}. \tag{2.4}$$

If H_0 is true, then

$$\tilde{\mu} = \bar{x} = \frac{1}{IJ} \sum_i \sum_j x_{ij} = \frac{1}{I} \sum_i \tilde{\mu}_i. \tag{2.5}$$

The (composite) alternative hypothesis is that not all the μ_i are equal. However we want to retain the concept of the total mean and write

$$\mu_i = \mu + a_i.$$

Then the model (2.2) reads

$$x_{ij} = \mu + a_i + \varepsilon_{ij}. \tag{2.6}$$

The a_i, which are a measure for the deviation of the mean of group i from the total mean, satisfy the condition

$$\sum_i a_i = 0. \tag{2.7}$$

Their maximum likelihood estimates are

$$\tilde{a}_i = \bar{x}_{i.} - \bar{x}. \tag{2.8}$$

The analysis of variance method was derived in §1 by using the identity

$$x_{ij} - \bar{x} = (\bar{x}_{i.} - \bar{x}) + (x_{ij} - \bar{x}_{i.}), \tag{2.9}$$

to describe the deviation of individual measurements from the total mean, and then forming the sum of squares Q of these deviations which decomposed into the two terms Q_A and Q_W (eq. (1.3)).

We can now study the case of two-way classification, in which the measurements are grouped into classes according to two criteria A and B. The measurement x_{ijk} then belongs to the class A_i, which denotes the classification according to A, and also to the class B_j. The subscript k indicates the measurement number within the group that belongs to both class A_i and class B_j. The measurements can be arranged in a rectangular block as in fig. 11-1. Again we restrict ourselves to the case in which all cells of the block are occupied by

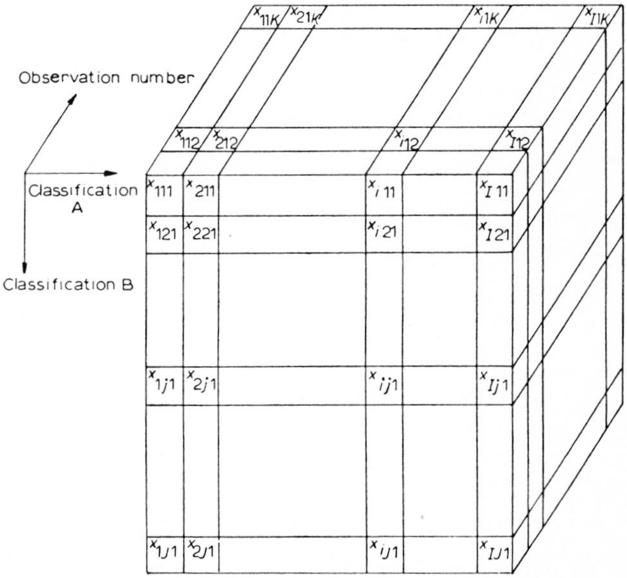

Fig. 11-1. Two-way classification with several observations.

measurements, i.e. $i = 1, 2, ..., I; j = 1, 2, ..., J; k = 1, 2, ..., K$ throughout. A two-way classification is called *crossed* if a particular classification B_j has the same meaning to all classes of the A classification. Therefore, if microscopes are classified by A and observers by B, and if every observer performs a measurement on every microscope, the classification is crossed. If however a survey is performed of microscopes by observers in distant laboratories and therefore in every laboratory another group of J observers performs measurements on the particular microscope i, the classification B is said to be *nested* in the classification A. The index j merely enumerates the classification B inside a particular class A.

The simplest case is the crossed two-way classification with only *one observation*. Since $k = 1$ for all observations x_{ijk} we can drop the subscript k altogether. The obvious model to be used is

$$x_{ij} = \mu + a_i + b_j + \varepsilon_{ij}, \quad \sum_i a_i = 0, \quad \sum_j b_j = 0, \tag{2.10}$$

with ε distributed normally with zero mean and variance σ^2. The null hypothesis states that there are no deviations from the total mean due to the grouping into classes according to the A or B classifications. We write it in the form of two separate hypotheses

$$H_0^{(A)}(a_1 = a_2 = \ldots = a_I = 0), \qquad H_0^{(B)}(b_1 = b_2 = \ldots = b_J = 0). \tag{2.11}$$

The least squares estimates for the a_i and b_j are

$$\tilde{a}_i = \bar{x}_{i.} - \bar{x}, \qquad \tilde{b}_j = \bar{x}_{.j} = \bar{x}.$$

By analogy with eq. (2.9) we can write

$$x_{ij} - \bar{x} = (\bar{x}_{i.} - \bar{x}) + (\bar{x}_{.j} - \bar{x}) + (x_{ij} - \bar{x}_{i.} - \bar{x}_{.j} + \bar{x}). \tag{2.12}$$

In a similar way the sum of squares can be written as

$$\sum_i \sum_j (x_{ij} - \bar{x})^2 = Q = Q_A + Q_B + Q_W, \tag{2.13}$$

with

$$\begin{aligned} Q_A &= J \sum_i (\bar{x}_{i.} - \bar{x})^2 = J \sum_i \bar{x}_{i.}^2 - IJ\,\bar{x}^2, \\ Q_B &= I \sum_j (\bar{x}_{.j} - \bar{x})^2 = I \sum_j \bar{x}_{.j}^2 - IJ\,\bar{x}^2, \\ Q_W &= \sum_i \sum_j (x_{ij} - \bar{x}_{i.} - \bar{x}_{.j} + \bar{x})^2 = \\ &= \sum_i \sum_j x_{ij}^2 - J \sum_i \bar{x}_{i.}^2 - I \sum_j \bar{x}_{.j}^2 + IJ\,\bar{x}^2. \end{aligned} \tag{2.14}$$

Divided by their respective numbers of degrees of freedom these sums of squares are estimates of σ^2, provided the hypotheses (2.11) are true. The hypotheses $H_0^{(A)}$ and $H_0^{(B)}$ can be tested separately by constructing the quotients

$$F^{(A)} = s_A/s_W, \qquad F^{(B)} = s_B/s_W, \tag{2.15}$$

and performing the one-tailed F-test as in §1. The situation is summarized in table 11-5.

If *several observations* are made the crossed classification can be generalized in many respects. The most important extension is the inclusion of *interaction* between classes. We then have the model

$$x_{ijk} = \mu + a_i + b_j + (ab)_{ij} + \varepsilon_{ijk}. \tag{2.16}$$

The quantity $(ab)_{ij}$, which should be read as a single symbol, is called the interaction between class A_i and B_j. It describes the deviation from the mean which is specific to the measurements belonging to both A_i and B_j. The parameters a_i, b_j, $(ab)_{ij}$ have to satisfy the conditions

$$\sum_i a_i = \sum_j b_j = \sum_i \sum_j (ab)_{ij} = 0. \tag{2.17}$$

Their least squares estimates are

$$\begin{aligned} \tilde{a}_i &= \bar{x}_{i..} - \bar{x}, \qquad \tilde{b}_j = \bar{x}_{.j.} - \bar{x}, \\ (\widetilde{ab})_{ij} &= \bar{x}_{ij.} + \bar{x} - \bar{x}_{i..} - \bar{x}_{.j.} \end{aligned} \tag{2.18}$$

The null hypothesis can be understood as 3 separate hypotheses

$$\begin{aligned} &H_0^{(A)}(a_i = 0; i = 1, 2, \dots, I), \qquad H_0^{(B)}(b_j = 0; j = 1, 2, \dots, J), \\ &H_0^{(AB)}((ab)_{ij} = 0; i = 1, 2, \dots, I; j = 1, 2, \dots, J), \end{aligned} \tag{2.19}$$

which can again be tested separately. The analysis of variance is based on the identity

$$\begin{aligned} x_{ijk} - \bar{x} = {} & (\bar{x}_{i..} - \bar{x}) + (\bar{x}_{.j.} - \bar{x}) \\ & + (\bar{x}_{ij.} + \bar{x} - \bar{x}_{i..} - \bar{x}_{.j.}) + (x_{ijk} - \bar{x}_{ij.}), \end{aligned} \tag{2.20}$$

which allows the decomposition of the sum of squares of deviations into four terms

$$Q = \sum_i \sum_j \sum_k (x_{ijk} - \bar{x})^2 = Q_A + Q_B + Q_{AB} + Q_W. \tag{2.21}$$

The degrees of freedom and mean squares, as well as the F-quotients that can be used to test the hypotheses (2.19) are given in table 11-6.

TABLE 11-5

Analysis of variance table for crossed two-way classification with single observations

Source	SS	ND	MS	F
Classification A	$Q_A = J \sum_i (\bar{x}_{i.} - \bar{x})^2$	$I - 1$	$s_A^2 = \dfrac{Q_A}{I - 1}$	$F^{(A)} = \dfrac{s_A^2}{s_W^2}$
Classification B	$Q_B = I \sum_j (\bar{x}_{.j} - \bar{x})^2$	$J - 1$	$s_B^2 = \dfrac{Q_B}{J - 1}$	$F^{(B)} = \dfrac{s_B^2}{s_W^2}$
Within groups	$Q_W = \sum_i \sum_j (x_{ij} - x_{i.} - x_{.j} + \bar{x})^2$	$(I - 1)(J - 1)$	$s_W^2 = \dfrac{Q_W}{(I - 1)(J - 1)}$	
Total	$Q = \sum_i \sum_j (x_{ij} - \bar{x})^2$	$IJ - 1$	$s^2 = \dfrac{Q}{IJ - 1}$	

TABLE 11-6

Analysis of variance table for crossed two-way classification

Source	SS	ND	MS	F
Classification A	$Q_A = JK \sum_i (\bar{x}_{i..} - \bar{x})^2$	$I - 1$	$s_A^2 = \frac{Q_A}{I - 1}$	$F^{(A)} = \frac{s_A^2}{s_W^2}$
Classification B	$Q_B = IK \sum_j (\bar{x}_{.j.} - \bar{x})^2$	$J - 1$	$s_B^2 = \frac{Q_B}{J - 1}$	$F^{(B)} = \frac{s_B^2}{s_W^2}$
Interaction	$Q_{AB} = K \sum_i \sum_j (\bar{x}_{ij.} + \bar{x} - \bar{x}_{i..} - \bar{x}_{.j.})^2$	$(I - 1)(J - 1)$	$s_{AB}^2 = \frac{Q_{AB}}{(I - 1)(J - 1)}$	$F^{(AB)} = \frac{s_{AB}^2}{s_W^2}$
Within groups	$Q_W = \sum_i \sum_j \sum_k (x_{ijk} - \bar{x}_{ij.})^2$	$IJ(K - 1)$	$s_W^2 = \frac{Q_W}{IJ(K - 1)}$	
Total	$Q = \sum_i \sum_j \sum_k (x_{ijk} - \bar{x})^2$	$IJK - 1$	$s^2 = \frac{Q}{IJK - 1}$	

Finally, we outline the simplest case of nested two-way classification. Since the classification B is defined only within each individual class A, terms like b_j and $(ab)_{ij}$ in eq. (2.10) have no meaning as they involve summation over i for fixed j. The model used instead is

$$x_{ijk} = \mu + a_i + b_{ij} + \varepsilon_{ijk}, \tag{2.22}$$

with

$$\sum_i a_i = 0, \qquad \sum_i \sum_j b_{ij} = 0,$$

$$\tilde{a}_i = \bar{x}_{i..} - \bar{x},$$

$$\tilde{b}_{ij} = \bar{x}_{ij.} - \bar{x}_{i..}.$$

The term b_{ij} is a measure of the deviation of the measurements belonging to class B_j within class A_i from the total mean of class A_i. The null hypothesis consists of

$$\begin{aligned} &H_0^{(A)}(a_i = 0; i = 1, 2, \ldots, I) \\ &H_0^{(B(A))}(b_{ij} = 0; i = 1, 2, \ldots, I; j = 1, 2, \ldots, J) \end{aligned} \tag{2.23}$$

The analysis of variance to test these hypotheses can be performed using table 11-7.

Different models of two-way or higher classification can be constructed in an analogous manner. According to the chosen model the total sum of squares decomposes into a sum of sums of squares which, if divided by their respective numbers of degrees of freedom, can be used to perform F-tests on hypotheses suggested by the model. Some models are, at least formally, contained in others. As an example comparison of tables 11-6 and 11-7 reveals the relation

$$Q_{B(A)} = Q_B + Q_{AB}. \tag{2.24}$$

A similar relation holds for the corresponding degrees of freedom

$$f_{B(A)} = f_B + f_{AB}. \tag{2.25}$$

These relations will be used in the program of the next section.

TABLE 11-7
Analysis of variance table for nested two-way classification

Source	SS	ND	MS	F
Classification A	$Q_A = JK \sum_i (\bar{x}_{i..} - \bar{x})^2$	$I - 1$	$s_A^2 = \dfrac{Q_A}{I - 1}$	$F^{(A)} = \dfrac{s_A^2}{s_W^2}$
Classification B within A	$Q_{B(A)} = K \sum_i \sum_j (\bar{x}_{ij.} - \bar{x}_{i..})^2$	$I(J - 1)$	$s_{B(A)}^2 = \dfrac{Q_{B(A)}}{I(J - 1)}$	$F^{(B(A))} = \dfrac{s_{B(A)}^2}{s_W^2}$
Within groups	$Q_W = \sum_i \sum_j \sum_k (x_{ijk} - \bar{x}_{ij.})^2$	$IJ(K - 1)$	$s_W^2 = \dfrac{Q_W}{IJ(K - 1)}$	
Total	$Q = \sum_i \sum_j \sum_k (x_{ijk} - \bar{x})^2$	$IJK - 1$	$s^2 = \dfrac{Q}{IJK - 1}$	

11-3. A FORTRAN program for two-way classification

Although the calculations required to perform an analysis of variance are simple, the number of operations involved makes the two-way classification already quite cumbersome. We therefore discuss a simple FORTRAN program to perform the analysis.

The program actually performs a crossed two-way classification. Using the relations (2.24) and (2.25), it also computes the relevant quantities for the nested case. Depending on an input parameter the analysis of variance table for crossed or nested classification is printed. To allow easy programming the measurements x_{ijk} are stored in a FORTRAN array X with 3 subscripts. This is uneconomical if I, J, K are smaller than the corresponding numbers in the DIMENSION statement defining the size of X. The FORTRAN names assigned to the different variables of §2 are listed in table 11-8.

TABLE 11-8
FORTRAN names of variables

Variable	FORTRAN name
I, J, K	NI, NJ, NK
x_{ijk}	X(I,J,K)
$\bar{x}$	XB
$\bar{x}_{i..}$	XBI(I)
$\bar{x}_{.j.}$	XBJ(J)
$\bar{x}_{ij.}$	XBIJ(I,J)
$Q_A, Q_B, Q_{AB}, Q_{B(A)}, Q_W, Q$	Q(1), Q(2),...,Q(6)
Degrees of freedom for above sums	NDF(1),...,NDF(6)
Text for analysis of variance table	SOURCE(1),...,SOURCE(6)

In the initial stage the program reads one card containing the numbers NI, NJ, NK, i.e. the number of classes A, classes B and repetitions, and the integer NTYPE. The analysis of the variance table for crossed or nested classification will be printed depending on whether NTYPE is zero or non-zero, respectively. As a next step the measurements are read and stored into X(I, J, K). The measurements belonging to a particular subgroup A_i, B_j are expected as a string of numbers, up to 8 floating point numbers per card. A new card is begun for each subgroup. The order of subgroups is

$$A_1B_1,\ A_1B_2,\ \ldots,\ A_1B_J,\ A_2B_1,\ A_2B_2,\ \ldots,\ A_2B_J,\ \ldots,\ A_IB_1,\ A_IB_2,\ \ldots,\ A_IB_J.$$

The data are printed again as they are read.

In the next section a complete list of all means is established. They are used to compute the different sums of squares. The corresponding degrees of freedom and mean squares are computed and the F-ratios are evaluated. Finally the analysis of variance table is printed using the number NTYPE to select the quantities appropriate to crossed or nested classification, respectively. The first word in each line of the output is taken from the array SOURCE. It contains words of explanatory text, defined by a DATA statement at the beginning of the program.

Program 11-1: *Main program* ANVAR *performing analysis of variance for crossed or nested two-way classification*

```
      DIMENSION X(20,20,20),XBI(20),XBJ(20),XBIJ(20,20),Q(6),DF(6),
     *S(6),F(4),SOURCE(6)
      DATA SOURCE/1HA,1HB,4HINT.,4HB(A),1HW,4HTTL./
C
C     READING OF INFORMATION ON INPUT DATA
C     NTYPE =0 FOR CROSSED CLASSIFICATION, NON ZERO FOR NESTED CLASSIFICATION.
C
      READ(5,1000) NI,NJ,NK,NTYPE
      AI=NI
      AK=NK
      AJ=NJ
C
C     READING OF INPUT DATA
C
      WRITE (6,1500)
      DO 10 I=1,NI
      DO 10 J=1,NJ
      READ (5,1100) (X(I,J,K),K=1,NK )
   10 WRITE (6,1200)I,J,(X(I,J,K),K=1,NK)
C
C     COMPUTATION OF ALL MEANS
C
      XB=0.
      DO 20 I=1,NI
   20 XBI(I)=0.
      DO 30 J=1,NJ
      XBJ(J)=0.
      DO 30 I=1,NI
   30 XBIJ(I,J)=0.
      DO 50 I=1,NI
      DO 50 J=1,NJ
      DO 40 K=1,NK
   40 XBIJ(I,J) =XBIJ(I,J)+X(I,J,K)
   50 XBIJ(I,J) =XBIJ(I,J)/AK
      DO 70 I=1,NI
      DO 60 J=1,NJ
   60 XBI(I)=XBI(I)+XBIJ(I,J)
   70 XBI(I)=XBI(I)/AJ
      DO 90 J=1,NJ
      DO 80 I=1,NI
   80 XBJ(J)=XBJ(J)+XBIJ(I,J)
      XBJ(J)=XBJ(J)/AI
   90 XB=XB+XBJ(J)
      XB=XB/AJ
C
C     SUMS OF  SQUARES
```

```
      DO  95 L=1,6
   95 Q(L)=0.
      DO 100 I=1,NI
  100 Q(1)=Q(1)+(XBI(I)-XB)**2
      Q(1)=AJ*AK*Q(1)
      DO 110 J=1,NJ
  110 Q(2)=Q(2)+(XBJ(J)-XB)**2
      Q(2)=AI*AK*Q(2)
      DO 120 I=1,NI
      DO 120 J=1,NJ
      Q(3)=Q(3)+(XBIJ(I,J)+XB-XBI(I)-XBJ(J))**2
      DO  120 K=1,NK
      Q(5)=Q(5)+(X(I,J,K)-XBIJ(I,J))**2
  120 Q(6)=Q(6)+(X(I,J,K)-XB)**2
      Q(3) =AK*Q(3)
      Q(4)=Q(2)+Q(3)
C
C     DEGREES OF FREEDOM
C
      DF(1)=AI-1.
      DF(2)=AJ-1.
      DF(3)=(AI-1.)*(AJ-1.)
      DF(4)=DF(2)+DF(3)
      DF(5)=AI*AJ*(AK-1.)
      DF(6)=AI*AJ*AK-1.
C
C     MEAN SQUARES
C
      DO 130 L=1,6
  130 S(L)=Q(L)/DF(L)
C
C     F-RATIONS
C
      DO 135 L=1,4
  135 F(L)=S(L)/S(5)
C
C     OUTPUT OF RESULTS
C
      WRITE (6,1300)
      WRITE (6,1400) SOURCE(1),Q(1),DF(1),S(1),F(1)
      IF(NTYPE)160,140,160
  140 DO 150 L=2,3
  150 WRITE (6,1400) SOURCE(L),Q(L),DF(L),S(L),F(L)
      GO TO 170
  160 WRITE (6,1400) SOURCE(4),Q(4),DF(4),S(4),F(4)
  170 DO 180 L=5,6
  180 WRITE (6,1400) SOURCE(L),Q(L),DF(L),S(L)
      STOP
C
C     FORMAT STATEMENTS
C
 1000 FORMAT(4I10)
 1100 FORMAT(10F10.5)
 1200 FORMAT(3H A(I2,5H), B(I2,2H) ,(10F10.2))
 1300 FORMAT(///13X,27H ANALYSIS OF VARIANCE TABLE //
     *52H SOURCE       SUM OF     DEGREES OF     MEAN          F-RATIO /
     *41H              SQUARES    FREEDOM        SQUARES //)
 1400 FORMAT(3X,A4,2X,F10.2,5X,F3.0,5X,F10.2,F10.5)
 1500 FORMAT(26H1                   OBSERVATIONS/)
      END
```

Example 11-2: *Analysis of variance (crossed two-way classification) in cancer research*

Two groups of rats are given injections of an amino acid (thymidine) containing traces of tritium, a radioactive isotope of hydrogen.

TABLE 11-9
Data of example 11-2

Observation number	Injection of	Time after injection (hours) 4	8	12	16	20	24	28	32	36	48
1		34	54	44	51	62	61	59	66	52	52
2	Thymidine	40	57	52	46	61	70	67	59	63	50
3		38	40	53	51	54	64	58	67	60	44
4		36	43	51	49	60	68	66	58	59	52
1		28	23	42	43	31	32	25	24	26	26
2	Thymidine	32	23	41	48	45	38	27	26	31	27
3	and	34	29	34	36	41	32	27	32	25	27
4	carcinogen	27	30	39	43	37	34	28	30	26	30

TABLE 11-10
Print-out for example 11-2

```
                      OBSERVATIONS

A( 1), B( 1)        34.00       40.00       38.00       36.00
A( 1), B( 2)        54.00       57.00       40.00       43.00
A( 1), B( 3)        44.00       52.00       53.00       51.00
A( 1), B( 4)        51.00       46.00       51.00       49.00
A( 1), B( 5)        62.00       61.00       54.00       60.00
A( 1), B( 6)        61.00       70.00       64.00       68.00
A( 1), B( 7)        59.00       67.00       58.00       66.00
A( 1), B( 8)        66.00       59.00       67.00       58.00
A( 1), B( 9)        52.00       63.00       60.00       59.00
A( 1), B(10)        52.00       50.00       44.00       52.00
A( 2), B( 1)        28.00       32.00       34.00       27.00
A( 2), B( 2)        23.00       23.00       29.00       30.00
A( 2), B( 3)        42.00       41.00       34.00       39.00
A( 2), B( 4)        43.00       48.00       36.00       43.00
A( 2), B( 5)        31.00       45.00       41.00       37.00
A( 2), B( 6)        32.00       38.00       32.00       34.00
A( 2), B( 7)        25.00       27.00       27.00       28.00
A( 2), B( 8)        24.00       26.00       32.00       30.00
A( 2), B( 9)        26.00       31.00       25.00       26.00
A( 2), B(10)        26.00       27.00       27.00       30.00


                   ANALYSIS OF VARIANCE TABLE

SOURCE     SUM OF      DEGREES OF    MEAN         F-RATIO
           SQUARES     FREEDOM       SQUARES


  A         9945.80        1.        9945.80  590.54736
  B         1917.50        9.         213.06   12.65051
  INT.      2234.94        9.         248.33   14.74481
  W         1010.50       60.          16.84
  TTL.     15108.75       79.         191.25
```

In addition a certain carcinogen is administered to the animals of one group. The assimilation of thymidine by the skin is studied as a function of time by counting the number of tritium decays per unit area of skin. The classifications are crossed since the time dependence is controlled in the same way for both series of test animals. The measurements are collected in table 11-9. The numbers have been obtained by transforming the original counting rates x' such that $x = 50 \log x' - 100$. The output of the program is reproduced in table 11-10. Since the quotient $F^{(A)}$ is very large there is no doubt that the presence or absence of carcinogen (classification A) influences the result. We want to test the existence of a time influence (classification B) and of an interaction between A and B at a level $\alpha = 0.01$. From table F-6 we find $F_{0.99}\,(9,60) = 2.72$. Therefore the hypotheses of time independence and of vanishing interactions have to be rejected.

Usually an analysis of variance program will include the computation of the critical values $F_{1-\alpha}$ for various levels α. However, this involves numerical integration or other methods which are outside the scope of this book. It would also be convenient to give a list of the various means or of the maximum likelihood estimates $\tilde{a}_i$, $\tilde{b}_j$, $(\widetilde{ab})_{ij}$ (eq. (2.18)). An inspection of the latter quantities is interesting if one of the null hypotheses fails.

Exercises

11-1: *One-way classification*

(The following agricultural example was taken from the book by MOORE et al. [1972]). The weight gain (in pounds) for pigs from five different litters is given in table 11-11. The hypothesis that the weight gain does not depend on the individual litter, i.e. that the animal stock of the farm is of uniform quality is to be tested for $\alpha = 5\%$ and 1%. Compute the complete analysis of variance table. Start by computing the equivalent of the lower part of table 11-1.

TABLE 11-11
Data for exercise 11-1

	Litter number				
	1	2	3	4	5
Number of pigs in litter	5	8	6	8	8
Weight gain (pounds)	23	29	38	30	31
	27	25	31	27	33
	26	33	28	28	31
	19	36	35	22	28
	30	32	33	33	30
		28	36	34	24
		30		34	29
		31		32	30
Litter totals	125	244	201	240	236

11-2: *Two-way classification*

Table 11-12 taken from the book by GILBERT [1973] contains sprinting speeds for three species of fast animals in feet per second.

(a) Is the classification crossed or nested?

(b) Set up the analysis of variance table and perform the *F*-tests at 10% significance level. Identify classification A with the different species and B with the sex. Start by specifying I, J and K and by forming a rectangular table of means $\bar{x}_{ij.}$ which by horizontal and vertical averaging also yields the $\bar{x}_{i..}$, $\bar{x}_{.j.}$.

TABLE 11-12
Data for exercise 11-2

	Cheetah	Greyhound	Kangaroo
Males	56, 52, 55	37, 33, 40	44, 48, 47
Females	53, 50, 51	38, 42, 41	39, 41, 36

CHAPTER 12

LINEAR REGRESSION

12-1. Linear regression as a simple case of least squares

In chapter 11 we have studied how a random variable (measurement) is influenced by one or several other variables which are not random but whose variation is exactly determined. This is a typical feature of most experiments. All variables except one are varied step by step in a controlled fashion. By measuring the last variable we try to establish an interdependence between the variables. So far we have used the analysis of variance to test whether the random variable was independent of the *controlled variables*. In this section we analyse a genuine dependence.

We denote the random variable by y and the controlled (non-random) variable by t. For example y could be the yield of a chemical reaction and t the temperature at which it takes place. In the simplest case of linear regression the following model is assumed for the dependence of the measurements y on the controlled variable t

$$y_i = \alpha + \beta t_i + \varepsilon_i, \qquad i = 1, 2, \ldots, n. \tag{1.1}$$

Here α and β are (unknown) constant parameters and ε_i denotes the measurement error of the ith measurement. Moreover all the measurements y_i are assumed to be *independent*, and the ε_i are assumed to be normally distributed about zero with equal variance σ_y^2. In this case the parameters α and β can be estimated using the simple least squares procedure of example 9-2. Although nothing essentially new will be added, we want to discuss the problem again in order to familiarize the reader with some of the terms used in regression analysis.

In example 9-2 the parameters α and β were grouped into a two-component vector

$$\boldsymbol{x} = \begin{pmatrix} \alpha \\ \beta \end{pmatrix}, \tag{1.2}$$

and the measurements into an n-component vector

$$\boldsymbol{y} = \begin{pmatrix} y_1 \\ y_2 \\ \vdots \\ y_n \end{pmatrix}. \tag{1.3}$$

The least squares solution of (1.1) was given by

$$\tilde{\boldsymbol{x}} = -(A^{\mathrm{T}}A)^{-1}A^{\mathrm{T}}\boldsymbol{y}, \tag{1.4}$$

with

$$A = \begin{pmatrix} -1 & -t_1 \\ -1 & -t_2 \\ \vdots & \\ -1 & -t_n \end{pmatrix}. \tag{1.5}$$

Substituting (1.5) into (1.4) we have

$$\tilde{\boldsymbol{x}} = \begin{pmatrix} n & \Sigma t_i \\ \Sigma t_i & \Sigma t_i^2 \end{pmatrix}^{-1} \begin{pmatrix} \Sigma y_i \\ \Sigma t_i\, y_i \end{pmatrix}.$$

Inversion of the (2×2)-matrix (eq. (B-3.13)) gives

$$\tilde{\boldsymbol{x}} = \frac{1}{n\,\Sigma t_i^2 - (\Sigma t_i)^2} \begin{pmatrix} \Sigma t_i^2 & -\Sigma t_i \\ -\Sigma t_i & n \end{pmatrix} \begin{pmatrix} \Sigma y_i \\ \Sigma t_i\, y_i \end{pmatrix}$$

$$= \frac{1}{n\,\Sigma t_i^2 - (\Sigma t_i)^2} \begin{pmatrix} (\Sigma t_i^2)(\Sigma y_i) - (\Sigma t_i)(\Sigma t_i\, y_i) \\ n\,\Sigma t_i\, y_i - (\Sigma t_i)(\Sigma y_i) \end{pmatrix},$$

i.e.

$$\tilde{\alpha} = \frac{(\Sigma t_i^2)(\Sigma y_i) - (\Sigma t_i)(\Sigma t_i\, y_i)}{n\,\Sigma t_i^2 - (\Sigma t_i)^2}, \tag{1.6}$$

$$\tilde{\beta} = \frac{n\,\Sigma t_i\, y_i - (\Sigma t_i)(\Sigma y_i)}{n\,\Sigma t_i^2 - (\Sigma t_i)^2}. \tag{1.7}$$

This solution is shown in fig. 12-1. The straight line with intercept $\tilde{\alpha}$ and slope $\tilde{\beta}$ is called the *empirical regression line*. Introducing

$$\bar{t} = \frac{1}{n}\,\Sigma t_i, \qquad \bar{y} = \frac{1}{n}\,\Sigma y_i, \tag{1.8}$$

we can transform to a coordinate system

$$t' = t - \bar{t}, \qquad y' = y - \bar{y}, \tag{1.9}$$

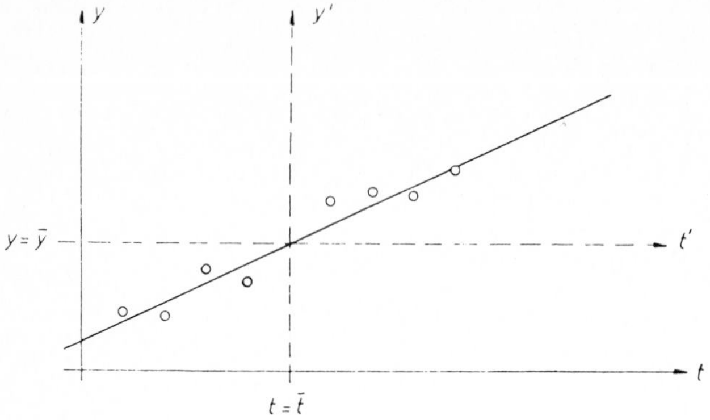

Fig. 12-1. Linear regression.

whose origin is at $(\bar{t}, \bar{y})$. It should be noted that $\bar{t}$ and $\bar{y}$ have only a formal similarity to sample means since they depend on the particular choice of values from the controlled variable t. Substituting (1.9) into (1.7) and (1.6) we get

$$\tilde{\beta} = \frac{\Sigma t_i' y_i'}{\Sigma t_i'^2}, \tag{1.10}$$

and

$$\tilde{\alpha} = \bar{y} - \bar{t}\tilde{\beta}. \tag{1.11}$$

The latter equation can also be easily read off from fig. 12-1.

The idea of any regression procedure is to attribute a part of the variation of the individual measurements y_i to the influence of external variables. In our simple model (1.1) there is only one such variable. If no influence had been suspected then the sum of squares

$$\Sigma(y_i - \bar{y})^2 = \Sigma y_i'^2$$

could have been used as a measure of this variation. By the use of regression it has been reduced to the sum of squares of deviations from the regression line

$$\Sigma\tilde{\varepsilon}_i^2 = \Sigma(y_i - \tilde{\alpha} - \tilde{\beta}t_i)^2 = \Sigma(y_i' - \tilde{\beta}t_i')^2. \tag{1.12}$$

The relative reduction is

$$r^2 = \frac{\Sigma y_i'^2 - \Sigma(y_i' - \tilde{\beta}t_i')^2}{\Sigma y_i'^2} = \frac{2\tilde{\beta}\,\Sigma t_i' y_i' - \tilde{\beta}^2\,\Sigma t_i'^2}{\Sigma y_i'^2}.$$

Using (1.10) we get

$$r^2 = \frac{2\tilde{\beta}\,\Sigma t_i'\,y_i' - \tilde{\beta}\,\Sigma t_i'\,y_i'}{\Sigma y_i'^2} = \tilde{\beta}\,\frac{\Sigma t_i'\,y_i'}{\Sigma y_i'^2},$$

or

$$r^2 = \frac{(\Sigma t_i'\,y_i')^2}{(\Sigma t_i'^2)(\Sigma y_i'^2)}. \tag{1.13}$$

Comparison with eq. (7-6.13) shows that r^2 is formally identical with the square of the sample correlation coefficient. It should be noted, however, that in the present case only one of the variables, y, is a true random variable.

According to (9-2.24) the covariance matrix of the $\tilde{x}$ is given by

$$C_{\tilde{x}} = \begin{pmatrix} \sigma^2(\tilde{\alpha}) & \operatorname{cov}(\tilde{\alpha}, \tilde{\beta}) \\ \operatorname{cov}(\tilde{\alpha}, \tilde{\beta}) & \sigma^2(\tilde{\beta}) \end{pmatrix} = \sigma_y^2(A^{\mathrm{T}}A)^{-1}$$

$$= \frac{\sigma_y^2}{n\,\Sigma t_i^2 - (\Sigma t_i)^2}\begin{pmatrix} \Sigma t_i^2 & -\Sigma t_i \\ -\Sigma t_i & n \end{pmatrix} = \sigma_y^2\begin{pmatrix} 1 - \dfrac{\bar{t}^2}{\Sigma t_i'^2} & -\dfrac{\bar{t}}{\Sigma t_i'^2} \\ -\dfrac{\bar{t}}{\Sigma t_i'^2} & \dfrac{1}{\Sigma t_i'^2} \end{pmatrix}. \tag{1.14}$$

Often we have an experimental situation in which the restrictions required by linear regression seem fulfilled, i.e. the deviations ε_j stem from a normal distribution with mean 0 and variance σ_y^2. The quantity σ_y^2 is however unknown. This can be the case, for example, when measurements are performed by the same instrument without systematic errors, but whose accuracy is not exactly known.

The variance σ_y^2 can be estimated from the measurements using

$$s_y^2 = \frac{\Sigma(y_i - \tilde{\alpha} - \tilde{\beta}t_i)^2}{n-2} = \frac{\Sigma\tilde{\varepsilon}_i^2}{n-2}. \tag{1.15}$$

From ch. 9, §2.3 we know that

$$M = \frac{1}{\sigma_y^2}\,\Sigma\tilde{\varepsilon}_i^2$$

follows a χ^2-distribution with $n - 2$ degrees of freedom since we have n measurements and 2 unknowns. Therefore

$$E\{\Sigma\tilde{\varepsilon}_i^2\} = \sigma_y^2 E(M) = \sigma_y^2(n - 2).$$

12-2. Confidence intervals

The uncertainty of the estimates $\tilde{\alpha}$ and $\tilde{\beta}$ of the regression parameters has already been derived in the covariance matrix (1.14). We now want to discuss it in greater detail. This point was already touched on in example 9-2, especially fig. 9-2d.

We consider the distribution of

$$\alpha + \beta t,$$

i.e. a point on the true regression line for a given value of the controlled variable t with respect to the corresponding point $\tilde{\alpha} + \tilde{\beta}t$ on the empirical regression line. Since $\tilde{\alpha}$ and $\tilde{\beta}$ are unbiased estimates we have

$$E(\tilde{\alpha} + \tilde{\beta}t) = E(\tilde{\alpha}) + tE(\tilde{\beta}) = \alpha + \beta t. \tag{2.1}$$

The variance of $\tilde{\alpha} + \tilde{\beta}t$ can be obtained from eq. (1.14)

$$\sigma^2(\tilde{\alpha} + \tilde{\beta}t) = \sigma^2(\tilde{\alpha}) + t^2\,\sigma^2(\tilde{\beta}) + 2t\,\mathrm{cov}(\tilde{\alpha}, \tilde{\beta})$$

$$= \sigma_y^2\left(\frac{1}{n} + \frac{\bar{t}^2 + t^2 - 2t\bar{t}}{\Sigma t_i'^2}\right) = \sigma_y^2\left(\frac{1}{n} + \frac{t'^2}{\Sigma t_i'^2}\right). \tag{2.2}$$

We can therefore construct a reduced random variable

$$u = \frac{(\tilde{\alpha} + \tilde{\beta}t) - (\alpha + \beta t)}{\sigma(\tilde{\alpha} + \tilde{\beta}t)},$$

which follows the standardized Gaussian distribution. From table F-3 we can read off confidence limits for the true regression line. For example with 95% confidence we have

$$\frac{|(\tilde{\alpha} + \tilde{\beta}t) - (\alpha + \beta t)|}{\sigma(\tilde{\alpha} + \tilde{\beta}t)} \leqslant \Omega'(0.95) = 1.96. \tag{2.3}$$

The limits of the 95% confidence region are then given by

$$\alpha + \beta t = \tilde{\alpha} + \tilde{\beta}t \pm \sigma(\tilde{\alpha} + \tilde{\beta}t)\,\Omega'(0.95). \tag{2.4}$$

This can be calculated as a function of t or, better, t' by substituting (1.10), (1.11) and (2.2). Fig. 12-2 shows these limits for the case of example 9-2.

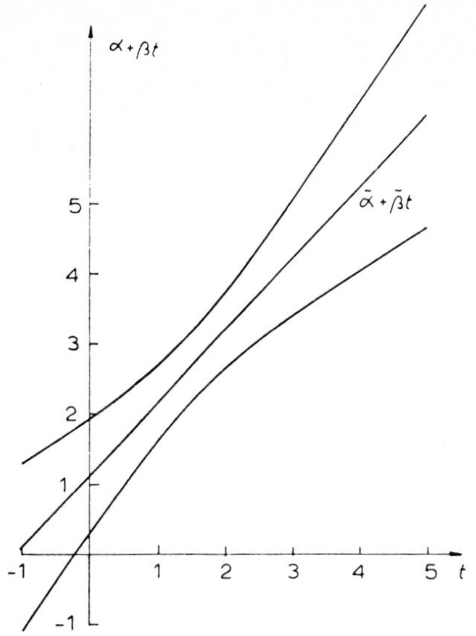

Fig. 12-2. Empirical regression line and 95% confidence limits for example 9-2.

If σ_y^2 is unknown, eq. (2.4) cannot be evaluated since $\sigma^2(\tilde{\alpha} + \tilde{\beta}t)$ is also unknown. But, using (1.15) and (2.2) we can construct an estimator

$$s^2_{\tilde{\alpha}+\tilde{\beta}t} = s_y^2\left(\frac{1}{n} + \frac{t'^2}{\Sigma t_i'^2}\right). \tag{2.5}$$

The quantity

$$v = \frac{(\tilde{\alpha} + \tilde{\beta}t) - (\alpha + \beta t)}{s_{\tilde{\alpha}+\tilde{\beta}t}} \tag{2.6}$$

then follows Student's t distribution with $n - 2$ degrees of freedom (cf. ch. 8, §2), i.e. for a confidence level of $1 - \alpha$ we have

$$\frac{|(\tilde{\alpha} + \tilde{\beta}t) - (\alpha + \beta t)|}{s_{\tilde{\alpha}+\tilde{\beta}t}} \leqslant t_{1-\frac{1}{2}\alpha}. \tag{2.7}$$

The right-hand side denotes the fractile of Student's distribution, which can be obtained from table F-7. (Neither t nor α on the right-hand side should be confused with the same symbols on the left-hand side.)

12-3. Testing of hypotheses

Often experimental data are used to check a certain hypothesis on regression parameters. To test the simple hypothesis $H_0(\beta = \beta_0)$ considerations similar to those in §2 can be used. From §1 we know that the estimator $\tilde{\beta}$ is normally distributed about the true value β with variance $\sigma^2(\tilde{\beta}) = \sigma_y^2/\Sigma t_i'^2$ therefore if H_0 is true, the variable

$$u = \frac{\tilde{\beta} - \beta_0}{\sigma_y/\sqrt{(\Sigma t_i'^2)}} \tag{3.1}$$

follows the standardized Gaussian distribution. It could be used to test H_0 if σ_y^2 were known. Substituting s_y (eq. (1.15)) for σ_y we obtain

$$v = \frac{\tilde{\beta} - \beta_0}{s_y/\sqrt{(\Sigma t_i'^2)}}. \tag{3.2}$$

The variable v is distributed like Student's t with $n - 2$ degrees of freedom. Therefore if at a given level of significance, α, the experimental data yield

$$|v| > t_{1-\frac{1}{2}\alpha},$$

the hypothesis H_0 has to be rejected.

12-4. Linear regression and analysis of variance

The starting point of linear regression was the assumption that

$$y_i = \alpha + \beta t_i + \varepsilon_i \tag{4.1}$$

described the inter-dependence of the random variable y and the controlled variable t. We could use (3.2) to test the simplest hypothesis on β, $H_0(\beta = \beta_0 = 0)$, namely that there is no such dependence. On the other hand eq. (4.1) can be regarded as a very simple model in analysis of variance.

The sum of squares of deviations of the measurement from their mean is

$$\Sigma(y_i - \bar{y})^2 = \Sigma y_i'^2.$$

The sum of squares of deviations from the regression line was given in (1.12) as

$$\begin{aligned}\Sigma\tilde{\varepsilon}_i^2 &= \Sigma(y_i - \tilde{\alpha} - \tilde{\beta}t_i)^2 = \Sigma(y_i' - \tilde{\beta}t_i')^2 \\ &= \Sigma y_i'^2 + \tilde{\beta}^2\,\Sigma t_i'^2 - 2\tilde{\beta}\,\Sigma y_i'\,t_i' \\ &= \Sigma y_i'^2 - \tilde{\beta}^2\,\Sigma t_i'^2,\end{aligned}$$

i.e.

$$\Sigma y_i'^2 = \tilde{\beta}^2\,\Sigma t_i'^2 + \Sigma(y_i' - \tilde{\beta}t_i')^2 = Q_{\text{REG}} + Q_{\text{RES}}. \tag{4.2}$$

Q_{REG} is the sum of squares due to regression, Q_{RES} the residual sum of squares. They have one and $n - 2$ degrees of freedom respectively. If H_0 is true the corresponding mean squares

$$s_{REG}^2 = \tilde{\beta}^2 \Sigma t_i'^2 = \frac{(\Sigma t_i' y_i')^2}{\Sigma t_i'^2},$$

$$s_{RES}^2 = \frac{\Sigma(y_i' - \tilde{\beta} t_i')^2}{n-2} = \frac{1}{n-2}\left\{\Sigma y_i'^2 - \frac{(\Sigma t_i' y_i')^2}{\Sigma t_i'^2}\right\},$$

are estimators of the variance σ^2. If, however, $\beta \neq 0$ then in general, s_{REG}^2 will be larger and s_{RES}^2 smaller than for H_0 true. A one-tailed F-test can therefore be used to test the null hypothesis. The situation is summarized in an analysis of variance table (table 12-1).

TABLE 12-1

Analysis of variance table for a simple regression model

Source	SS	ND	MS	F
Regression	$Q_{REG} = \dfrac{(\sum t_i' y_i')^2}{\sum t_i'^2}$	1	$s_{REG}^2 = Q_{REG}$	$F = \dfrac{s_{REG}^2}{s_{RES}^2}$
Residual	$Q_{RES} = \sum y_i'^2 - \dfrac{(\sum t_i' y_i')^2}{\sum t_i'^2}$	$n - 2$	$s_{RES}^2 = \dfrac{Q_{RES}}{n-2}$	
Total	$Q = \sum y_i'^2$	$n - 1$	$s^2 = \dfrac{Q}{n-1}$	

12-5. A general FORTRAN program for linear regression

Although the computing work to perform a linear regression is rather simple it can be quite time consuming in the case of many data points or if confidence intervals are to be computed. We therefore present a program which can perform all the calculations discussed in this chapter and which also gives a rough graphical display of data, regression line and confidence intervals.

Program 12-1: *Main program* LINREG *performing a linear regression including computation of* χ^2, *confidence intervals and analysis of variance table*

```
C
C      MAIN PROGRAM LINREG
C
       DIMENSION T(100),Y(100),X(2),TEXT(20),A(200),ATA1(4),HELP(200),
      *TP(500),YP(500),U(3,500),ATA1AT(200),CX(4),ETA(100),TVECT(2),
      *TALPHA(20)
       DATA TALPHA/6.31,2.92,2.35,2.13,2.02,1.94,1.89,1.86,1.83,1.81,
      *1.80,1.78,1.77,1.76,1.75,1.75,1.74,1.73,1.73,1.73/
       DATA OMEGA/1.645/
C
C      READ EXPLANATORY TEXT AND PARAMETERS
C
       READ(5,1000) TEXT
       READ(5,1100)N,TBEG,DELT,NT,YBEG,DELY,NY,SIGMA
C
C      READ INPUT DATA AND COMPUTE YBAR
C
       YBAR=0.
       DO 10 J=1,N
       READ(5,1200) T(J),Y(J)
 10    YBAR=YBAR+Y(J)
       YBAR=YBAR/N
C
C      COMPUTE X,ETA AND SUMS OF SQUARES
C
       DO 20 I=1,N
       K=(I-1)*2+1
       A(K)=-1.
 20    A(K+1)=-T(I)
       CALL MTXMAT(A,A,HELP,2,N,2)
       CALL MTXINV(HELP,ATA1,2,2)
       CALL MTXMBT(ATA1,A,ATA1AT,2,2,N)
       CALL MTXMSC(ATA1AT,ATA1AT,-1.,2,N)
       CALL MTXMLT(ATA1AT,Y,X,2,N,1)
       CALL MTXMLT(A,X,ETA,N,2,1)
       CALL MTXADD(Y,ETA,HELP,N,1)
       CALL MTXMAT(HELP,HELP,SE2,1,N,1)
       SY2=SE2/(N-2)
       SIGMA2=SIGMA**2
       IF(SIGMA)30,30,40
 30    SIGMA2=SY2
 40    CALL MTXMSC(ATA1,CX,SIGMA2,2,2)
       XM=SE2/SIGMA2
C
C      OUTPUT, FIRST PART
C
       WRITE(6,2000) TEXT,N,SIGMA
       WRITE(6,2200) X(1),X(2)
       CALL MTXWRT(CX,2,2)
       IF(SIGMA)60,60,50
 50       WRITE(6,2100)XM
C
C      SET UP GRAPHICAL OUTPUT
C
 60    DO 110 I=1,NT
       YP(I)=-10000.
 110   TP(I)=TBEG+(I-1)*DELT
       DO 140 J=1,N
       DIFF=1000000.
       DO 130 I=1,NT
       D=T(J)-(TBEG+(I-1)*DELT)
```

```
      AD=ABS(D)
      IF(DIFF-AD)130,130,120
  120 DIFF=AD
      I1=I
  130 CONTINUE
      TP(I1)=T(J)
  140 YP(I1)=Y(J)
C
C     COMPUTE REGRESSION LINE AND CONFIDENCE LIMITS
C     POINTWISE FOR GRAPHICAL OUTPUT
C
      TVECT(1)=1.
      DO 180 I=1,NT
      TVECT(2)=TP(I)
      CALL MTXMAT(X,TVECT,U(1,I),1,2,1)
      CALL MTXMLT(TVECT,ATA1,HELP,1,2,2)
      CALL MTXMBT(HELP,TVECT,SETA2,1,2,1)
      SETA=SQRT(SIGMA2*SETA2)
      FACT=OMEGA
      IF(SIGMA)150,170,150
  150 IF(N-22)160,160,170
  160 FACT=TALPHA(N-2)
  170 DIFF=SETA*FACT
      U(2,I)=U(1,I)-DIFF
  180 U(3,I)=U(1,I)+DIFF
      CALL PRGRAP(YBEG,DELY,NY,TP,YP,U,NT)
C
C     PERFORM ANALYSIS OF VARIANCE
C
      QRES=SE2
      CALL MTXMAT(Y,Y,Q,1,N,1)
      Q=Q-N*YBAR**2
      QREG=Q-QRES
      NDREG=1
      NDRES=N-2
      NDTOT=N-1
      SRES2=QRES/NDRES
      S2=Q/NDTOT
      F=QREG/SRES2
      WRITE(6,1300)
      WRITE(6,1400) QREG,NDREG,QREG,F,QRES,NDRES,SRES2,Q,NDTOT,S2
C
C     FORMAT STATEMENTS
C
 1000 FORMAT (20A4)
 1100 FORMAT(I10,2F10.5,I10,2F10.5,I10,F10.5)
 1200 FORMAT(2F10.5)
 1300 FORMAT(///13X,27H ANALYSIS OF VARIANCE TABLE //
     *52H SOURCE     SUM OF     DEGREES OF    MEAN         F-RATIO /
     *41H            SQUARES    FREEDOM       SQUARES //)
 1400 FORMAT(3X,3HREG,3X,F10.2,I8,5X,2F10.2/
     *3X,3HRES,3X,F10.2,I8,5X,F10.2/
     *3X,3HTTL,3X,F10.2,I8,5X,F10.2/)
 2000 FORMAT(/20A4,//18H LINEAR REGRESSION/
     *27H NUMBER OF MEASUREMENTS N =I5/
     *43H STANDARD DEVIATION OF MEASUREMENTS SIGMA =F10.5/19X,
     *19H (ZERO, IF UNKNOWN)/)
 2200 FORMAT(23H REGRESSION PARAMETERS:/
     * 9H ALPHA = F10.5/
     * 9H BETA  = F10.5//
     *19 H COVARIANCE MATRIX:)
 2100 FORMAT(/19H MINIMUM FUNCTION =F10.5//)
      END
```

The program consists of 4 parts:

1. *Computation of the regression parameters and the minimum function*

To begin with 2 punched cards are read containing a line of explanatory text which is reproduced as the heading line of the output and a card with the parameters N (the number of data points), TBEG, DELT, NT, YBEG, DELY, NY (lower limit, interval width and number of intervals of the variables t and y for the display, see program 12-2) and the standard deviation SIGMA (σ_y) of the data. SIGMA is left zero if it is unknown. The data are then read in pairs t_j, y_j, $j = 1, ..., n$ (each pair on one card) and stored in the arrays T and Y. Simultaneously $\bar{y} = (1/n) \sum y_i$ is computed and stored in YBAR. In the following section the regression parameters $\tilde{\alpha}$ and $\tilde{\beta}$ are found. Rather than eqs. (1.6) and (1.7) eq. (9-2.3) is used since it is written in matrix form so that the matrix subroutines of appendix B, § 4 can be applied. Use is made of the fact that σ_y is a multiple of the unit matrix and therefore does not appear in (9-2.3). The parameters $\tilde{\alpha}$ and $\tilde{\beta}$ form the elements x_1 and x_2 of the vector $\boldsymbol{x}$, which is stored in the array X. Intermediate results $(A^{\mathrm{T}}A)^{-1}$ and $(A^{\mathrm{T}}A)^{-1}A^{\mathrm{T}}$ are kept and stored in the arrays ATA1 and ATA1AT, respectively. Following (9-2.6) the improved measurements

$$\tilde{\boldsymbol{\eta}} = A(A^{\mathrm{T}}A)^{-1}\boldsymbol{y} = A\boldsymbol{x} \tag{5.1}$$

are computed. Finally $\sum \tilde{\varepsilon}_i^2 = \sum (y_i - \tilde{\eta}_i)^2$ stored in SE2 (note that $-\tilde{\boldsymbol{\eta}}$ is contained in ETA). According to (1.15) s_y^2 is found and placed in SY2. If σ_y is unknown SY2 is also copied into the word SIGMA2. If σ_y is known its square is stored in SIGMA2. The covariance matrix of the regression parameters and the minimum function M are computed according to eqs. (1.14) and (1.15) and stored in CX and XM, respectively. In the first output section the parameters $\tilde{\boldsymbol{x}}$, their covariance matrix and the minimum function are printed (the latter only if σ_y is known, otherwise the computed value for M is just $n - 2$ and does not provide information for a χ^2-test). If only these quantities are desired the program can be terminated by an END statement after statement 50.

2. *Preparation of graphical output*

A rough two-dimensional diagram t vs y is prepared for output on the printer. On the second input card the lower limits of t and y, the interval widths and the number of intervals of both variables were specified. The plot is prepared in the arrays TP and YP. First the words TP(1) ... TP(NT) are filled with numbers corresponding to the lower limit of each t-interval. The corresponding words YP(1) ... YP(NT) contain a large negative number (−10000) which falls outside the range of y and is therefore ignored by the printing subroutine PRGRAP (program 12-2). Next the data points T(J), Y(J) are scanned and for each data point the value of TP is found which is nearest to T(J). The corresponding index of TP, called I1 in the program is used to replace TP(I1) and YP(I1) by T(J) and Y(J), respectively. In this way TP contains a total of NT ordered values which are more or less equidistant; among them are the t-values of the data. YP contains the y-values of the data.

3. *Computation of regression line and confidence limits for graphical output*

For all values of TP the corresponding point on the regression line (L=1) and the lower (L=2) and upper (L=3) confidence limits are now computed and stored in the array U(L,I), I=1 ... NT. To facilitate the calculations a matrix T is formed (which in our case is (2×1) and therefore is called TVECT which performs the transformation of $\boldsymbol{x}$ to $\tilde{\eta}$ (cf. ch. 4, § 4.5)

$$\tilde{\eta}_j = T\tilde{x}, \quad T_1 = 1, \quad T_j = t_j, \tag{5.2}$$

$$\sigma^2_{\tilde{\eta}_j} = TC_xT^{\mathrm{T}}. \tag{5.3}$$

Eq. (5.3) is identical to eq. (2.2) but written in matrix form. In this way $\sigma_{\tilde{\eta}}$ is computed for every value t_j and stored in SETA. According to (2.3) and (2.7) it is multiplied by Ω' (0.95) or $t_{0.95}$ to obtain the difference (stored in DIFF) between $\tilde{\eta} = \tilde{\alpha} + \tilde{\beta}t_j$ and the confidence limits. $t_{0.95}$ is the fractile of Student's distribution for $f = n - 2$ degrees of freedom which were stored by a DATA statement for $f = 1, \ldots, 20$ in the array TALPHA. Subroutine PRGRAP performs the graphical output.

4. *Analysis of variance table*

Rather than transforming from y_i, t_i to y'_i, t'_i we make use of the results obtained so far and the relations:

$$\begin{aligned} Q &= \sum y_i'^2 = \sum (y_i - \bar{y})^2 = \sum y_i^2 - n\bar{y}^2, \\ Q_{\mathrm{RES}} &= \sum (y'_i - \tilde{\beta}t'_i)^2 = \sum \tilde{\varepsilon}_i, \\ Q_{\mathrm{REG}} &= Q - Q_{\mathrm{RES}}. \end{aligned}$$

The computations are straightforward and can be read off from the program. Their results are printed in the form of an analysis of variance table.

Program 12-2: *Subroutine* PRGRAP, *plotting a two-dimensional diagram of data points y versus a controlled variable t together with fitted curve and confidence limits*

```
      SUBROUTINE PRGRAP(YBEG,DELY,NY,T,Y,U,LENGTH)
      DIMENSION U(3,500),Y(1),YLINE(100),ZLINE(101),YSCALE(10),SYMBOL(3)
      DIMENSION T(500)
      DATA BLANK,CROSS,HOR,VERT/1H ,1HX,1H-,1HI/
      DATA SYMBOL/1H*,1H+,1H+/
C
C     WRITE SCALE ON Y-AXIS
C
      DO 10 I=1,100
 10   YLINE(I)=BLANK
      DO 20 I=1,NY
 20   YLINE(I)=HOR
      NMARKS=(NY-1)/10+1
      DO 30 I=1,NMARKS
      YSCALE(I)=YBEG+((I-1)*10+0.5)*DELY
 30   YLINE((I-1)*10+1)=VERT
      WRITE(6,1000) (YSCALE(J),J=1,NMARKS)
      WRITE(6,1100)YLINE
C
C     PREPARE ONE LINE OF OUTPUT FOR EVERY J
C
      DO 200 J=1,LENGTH
 100  DO 110 I1=1,101
 110  ZLINE(I1)=BLANK
 135  M=(Y(J)-YBEG)/DELY+1.5
      IF(M)170,170,140
 140  IF(M-NY)  150,150,170
 150  M2=M-1
      M3=M+1
      IF(M2) 151,151,152
 151  M2=M
 152  IF(M2-NY) 154,154,153
 153  M3=M
 154  DO 160 M1=M2,M3
 160  ZLINE(M1)=CROSS
 170  DO 190 M2=1,3
      M=(U(M2,J)-YBEG)/DELY+1.5
      IF(M)190,190,180
 180  IF(M-NY) 185,185,190
 185  ZLINE(M)=SYMBOL(M2)
 190  CONTINUE
      ZLINE(NY+1)=VERT
      WRITE(6,1200) T(J),Y(J),U(1,J),ZLINE
 200  CONTINUE
C
C     REPEAT SCALE ON Y-AXIS
C
      WRITE(6,1100)YLINE
      WRITE(6,1300) (YSCALE(J),J=1,NMARKS)
      RETURN
 1000 FORMAT(50X,1HY/5X,1HT,8X,1HY,7X,3HETA,10F10.4)
 1100 FORMAT(30X,100A1)
 1200 FORMAT(1X,3(F8.3,1X),1X,1HI,101A1)
 1300 FORMAT(25X,10F10.4)
      END
```

The print-out is arranged in such a way that the variable y runs horizontally from left to right and the variable t vertically from top to bottom of the paper. Each printed line corresponds to one value of t. A total of l lines is printed. The variable l is contained in the argument LENGTH. The scale in y is defined by the arguments YBEG, DELY and NY which characterize the lower limit, the interval width and the number of intervals in y, respectively. The first part of the program corresponds to the beginning of the subroutine PRSCAT (program 6-5) and prepares a y-axis and its scale for later printing. In the following DO-loop for every t-value one line of output is prepared in the array ZLINE. First the complete array is filled with blanks. Next an index M is computed in statement 135. It is equal to 1 if Y(J) corresponds to the first interval on the y-axis of the plot, equal to 2 for the second interval etc. The three adjacent words ZLINE(M2), ZLINE(M), ZLINE(M3) with M2=M—1, M3=M+1 are filled with crosses, if their indices are in the permissible range between 1 and NY. Similarly the word corresponding to the position of U(1,J) is marked by an asterisk (*) and those of U(2,J) and U(3,J) by plus signs (+). The original data Y(J) were marked by three crosses – the middle one giving the exact position – in order to avoid complete overwriting in case one or several of the U-values fall into the same interval. The word ZLINE(NY+1) is filled with a vertical line (I) for the right-hand margin of the plot. Finally each line is printed. It is preceded by the numerical values of T(J), Y(J) and U(1,J) and a vertical line for the left-hand margin. The FORMAT for the printing of these numbers is chosen as F 8.3. Thus if one of them is very large in absolute value it cannot be printed in this FORMAT. Automatically a row of asterisks is printed instead. This feature was used in the program LINREG to mark that for some t-values no measurements y were given. Just before the end of the program the y-axis and its scale are repeated. A print-out resulting from the program is shown in table 12-3.

Example 12-1: *Price of building land as a function of public expenditure*

The price of building land has been steadily increasing during the past few years. But so have other economic indices like the public expenditure (being more or less equal to the total tax paid) per citizen. For Germany the situation is summarized in table 12-2. We might now assume that the price of building land is a linear function of the public expenditure and perform a linear regression. The output from the program is shown in table 12-3. One observes that our crude model describes the general tendency rather well, although of course the "measurements", marked by (XXX), do not in general fall directly on the regression line (*). The confidence limits (+ +) reflect the fluctuation of the measurements about the regression line. The analysis of

TABLE 12-2

xpenditure and price of building land in Germany for the years 1962–1971

Year	Public expenditure (DM/citizen)	Price of building land (DM/m²)
1962	1883	14.83
1963	2033	16.92
1964	2198	18.46
1965	2381	21.89
1966	2559	23.61
1967	2605	25.71
1968	2646	28.37
1969	2872	29.86
1970	3191	30.74
1971	3677	33.56

variance table shows that the residual variance is much smaller than the one due to the linear tendency. Their ratio F is much larger than $F_{0.95}(1.8) = 5.33$ therefore the hypothesis of independence of the two variables is certainly false, as is also quite obvious from the plot in table 12-3. Closer inspection of the plot indicates that if only the last four points had been used the fit of a straight line would have been even better. (Moreover the numerical value of $\tilde{\beta}$ would have been smaller.) This is the reflection of the general rule that linear models can usually only be applied over a restricted range.

12-6. Interpretation of results from linear regression

The results from linear regression (as from every other method which implies a model – in our case a linear dependence between y and t) have to be interpreted with care. We demonstrate this using the following example.

Example 12-2: *Linear regression using the same measurements but different assumptions about their errors*

A regression is performed for the short set of data in table 12-4. However three different assumptions are made about the standard deviation of the measurements y:

(a) $\sigma_y = 1$,

(b) $\sigma_y = 8$,

(c) σ_y unknown.

TABLE 12-3

Print-out for example 12-1

```
PUBLIC EXPENSES (DM/CITIZEN)  VS. PRICE OF BUILDING LAND (DM/SQUARE METER)

LINEAR REGRESSION
NUMBER OF MEASUREMENTS N =   10
STANDARD DEVIATION OF MEASUREMENTS SIGMA =    0.00000
                    (ZERO, IF UNKNOWN)

REGRESSION PARAMETERS:
ALPHA =    -4.65831
BETA  =     0.01116

COVARIANCE MATRIX:
 11.25323   -0.00416
 -0.00416    0.00000
                                                 Y
    I        Y      ETA    10.5000   20.5000   30.5000   40.5000
                             I---------I---------I---------I---------
1600.000 ********   13.190  I + *  +                                 I
1700.000 ********   14.305  I  + * +                                 I
1800.000 ********   15.421  I   + * +                                I
1883.000   14.830   16.347  I    X+* +                               I
2033.000   16.920   18.020  I      +X* +                             I
2100.000 ********   18.767  I       + *+                             I
2198.000   18.460   19.860  I       X+X*+                            I
2300.000 ********   20.998  I          +*+                           I
2381.000   21.890   21.902  I           +*+                          I
2500.000 ********   23.229  I            +*+                         I
2646.000   28.370   24.858  I              +*+XXX                    I
2700.000 ********   25.460  I              +* +                      I
2800.000 ********   26.576  I               + *+                     I
2872.000   29.860   27.379  I                +* +XX                  I
3000.000 ********   28.807  I                 + *+                   I
3100.000 ********   29.922  I                  + *+                  I
3191.000   30.740   30.937  I                   +X*X+                I
3300.000 ********   32.153  I                    + * +               I
3400.000 ********   33.269  I                     + * +              I
3500.000 ********   34.384  I                      + *  +            I
3600.000 ********   35.500  I                       + *  +           I
3677.000   33.560   36.359  I                       X+X*  +          I
3800.000 ********   37.731  I                         +  * +         I
3900.000 ********   38.846  I                          +  *  +       I
                             I---------I---------I---------I---------
                           10.5000   20.5000   30.5000   40.5000


             ANALYSIS OF VARIANCE TABLE

SOURCE      SUM OF     DEGREES OF     MEAN        F-RATIO
            SQUARES    FREEDOM        SQUARES


  REG        327.51         1          327.51      77.92
  RES         33.63         8            4.20
  TTL        361.13         9           40.13
```

TABLE 12-4

Data for example 12-2

t	0.5	11.2	19.3	30.7	51.0
y	5.0	8.0	23.0	36.0	43.0

TABLE 12-5

Print-out for example 12-2

```
                                                 Y
    T        Y        ETA  -9.0000   11.0000   31.0000   51.0000   71.0000
                             I---------I---------I---------I---------I----
-10.000 ********    -3.920  I  +*+                                        I
 -5.000 ********     0.216  I    +*+                                      I
  0.500    5.000     4.767  I       ++X                                   I
  5.000 ********     8.489  I        +*+                                  I
 11.200    8.000    13.619  I        XXX++                                I
 15.000 ********    16.762  I             ++                              I
 19.300   23.000    20.320  I               ++XX                          I
 25.000 ********    25.035  I                 ++                          I
 30.700   36.000    29.751  I                   ++ XXX                    I
 35.000 ********    33.308  I                     ++                      I
 40.000 ********    37.444  I                       ++                    I
 45.000 ********    41.581  I                         +*+                 I
 51.000   43.000    46.545  I                          X+*+               I
 55.000 ********    49.854  I                             +*+             I
 60.000 ********    53.990  I                               +*+           I
                             I---------I---------I---------I---------I----
                           -9.0000   11.0000   31.0000   51.0000   71.0000
```

```
                                                 Y
    T        Y        ETA  -9.0000   11.0000   31.0000   51.0000   71.0000
                             I---------I---------I---------I---------I----
-10.000 ********    -3.920  I   *        +                                I
 -5.000 ********     0.216  I     *       +                               I
  0.500    5.000     4.767  I +     *XX    +                              I
  5.000 ********     8.489  I   +     *     +                             I
 11.200    8.000    13.619  I       +XXX *    +                           I
 15.000 ********    16.762  I         +   *    +                          I
 19.300   23.000    20.320  I           +   *XXX+                         I
 25.000 ********    25.035  I             +    *   +                      I
 30.700   36.000    29.751  I               +    * XXX+                   I
 35.000 ********    33.308  I                +     *    +                 I
 40.000 ********    37.444  I                  +     *     +              I
 45.000 ********    41.581  I                   +      *      +           I
 51.000   43.000    46.545  I                    +    XX*        +        I
 55.000 ********    49.854  I                     +        *        +     I
 60.000 ********    53.990  I                      +         *         +  I
                             I---------I---------I---------I---------I----
                           -9.0000   11.0000   31.0000   51.0000   71.0000
```

```
                                                 Y
    T        Y        ETA  -9.0000   11.0000   31.0000   51.0000   71.0000
                             I---------I---------I---------I---------I----
-10.000 ********    -3.920  I   *   +                                     I
 -5.000 ********     0.216  I +   *   +                                   I
  0.500    5.000     4.767  I    +  *XX +                                 I
  5.000 ********     8.489  I      +  *  +                                I
 11.200    8.000    13.619  I        X+X * +                              I
 15.000 ********    16.762  I           + *  +                            I
 19.300   23.000    20.320  I             + *X+X                          I
 25.000 ********    25.035  I               +  * +                        I
 30.700   36.000    29.751  I                  + * +XX                    I
 35.000 ********    33.308  I                   +  * +                    I
 40.000 ********    37.444  I                     +  *  +                 I
 45.000 ********    41.581  I                      +   *  +               I
 51.000   43.000    46.545  I                        + XX*   +            I
 55.000 ********    49.854  I                          +   *   +          I
 60.000 ********    53.990  I                           +    *    +       I
                             I---------I---------I---------I---------I----
                           -9.0000   11.0000   31.0000   51.0000   71.0000
```

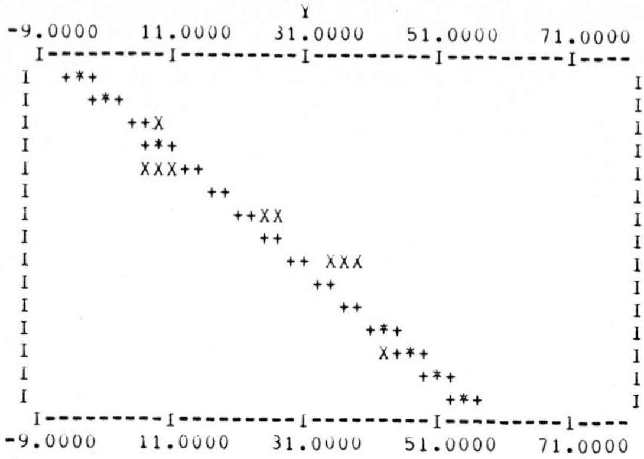

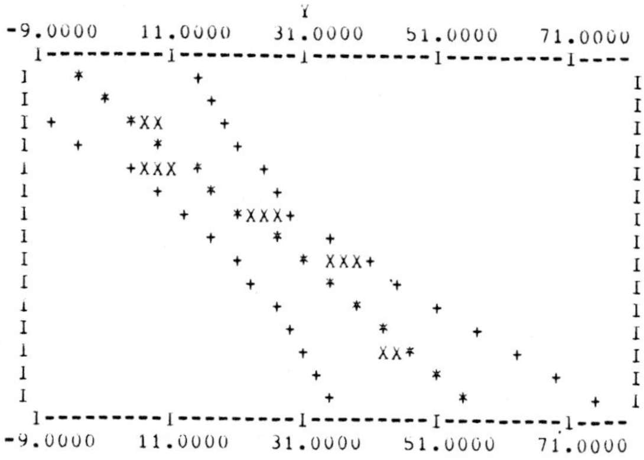

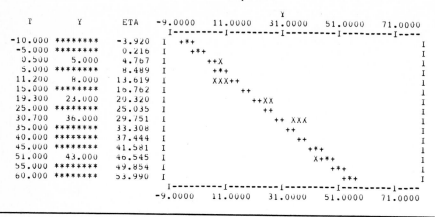

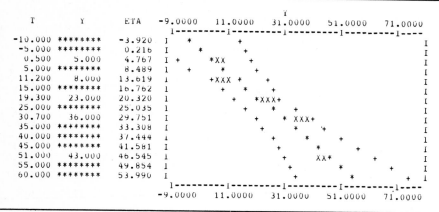

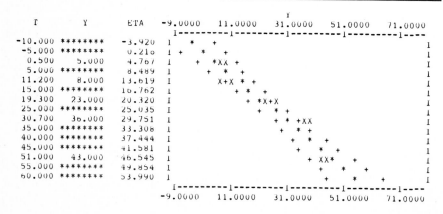

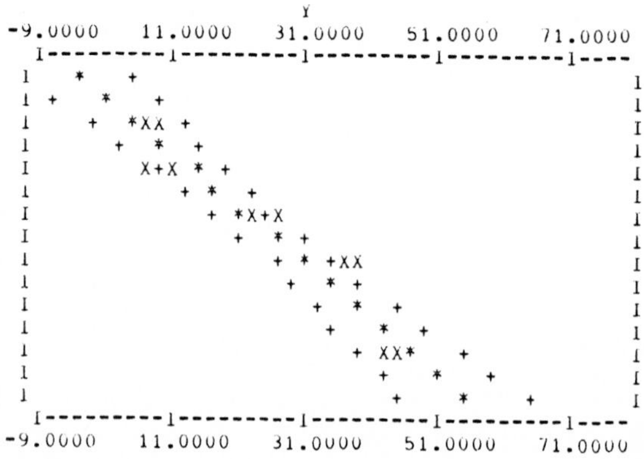

```
DEMONSTRATION DATA

LINEAR REGRESSION
NUMBER OF MEASUREMENTS N =     5
STANDARD DEVIATION OF MEASUREMENTS SIGMA =    1.00000
                      (ZERO, IF UNKNOWN)

REGRESSION PARAMETERS:
ALPHA =     4.35292
BETA  =     0.82729

COVARIANCE MATRIX:
  0.53838  -0.01501
 -0.01501   0.00067

MINIMUM FUNCTION =  90.42569
```

```
DEMONSTRATION DATA

LINEAR REGRESSION
NUMBER OF MEASUREMENTS N =     5
STANDARD DEVIATION OF MEASUREMENTS SIGMA =    8.00000
                      (ZERO, IF UNKNOWN)

REGRESSION PARAMETERS:
ALPHA =     4.35292
BETA  =     0.82729

COVARIANCE MATRIX:
 34.45648  -0.96080
 -0.96080   0.04263

MINIMUM FUNCTION =   1.41290
```

```
DEMONSTRATION DATA

LINEAR REGRESSION
NUMBER OF MEASUREMENTS N =     5
STANDARD DEVIATION OF MEASUREMENTS SIGMA =    0.00000
                      (ZERO, IF UNKNOWN)

REGRESSION PARAMETERS:
ALPHA =     4.35292
BETA  =     0.82729

COVARIANCE MATRIX:
 16.22787  -0.45251
 -0.45251   0.02008
```

Although the regression parameters $\tilde{\alpha}$ and $\tilde{\beta}$ are not influenced by these assumptions there is a strong influence on their covariance matrix, on the minimum function and on the confidence limits.

This is demonstrated in table 12-5, which shows the results for the three cases (The numerical results are printed to the right rather than on top of the graphical output for reasons of space.) We observe:

– For case (a): The confidence limits and the errors of the regression parameters are very small. However the minimum function is large, so that the χ^2-test on the fit of a straight line fails ($\chi^2_{0.99}(3) = 11.34$).
– For case (b): Errors and confidence limits are larger, but the minimum function is small so that the linear fit seems reasonable.
– For case (c): No minimum function can be computed. The confidence limits seem reasonable compared to the fluctuation of the data about the regression line.

From this example we can abstract the following rules for the interpretation of results from linear regression:

(1) Only the χ^2-test (and not the errors of the regression parameters) signifies whether the model implied in linear regression, i.e. a linear dependence of y and t is justified. It is therefore very important to obtain at least a rough estimate of the measurement error σ_y since otherwise the minimum function necessary for the χ^2-test cannot be computed.

(2) If σ_y is unknown linear regression alone does not allow one to decide whether a linear model is justified. The simplest (qualitative) way out is the visual inspection of a t–y-plot and a search for a systematic deviation of the data points from a straight line. If such a deviation seems to exist other (non-linear) models have to be constructed. They can be fitted to the data by least squares (cf. ch. 9). A comparison between the regression model R yielding the improved measurements

$$\tilde{\eta}_i^{\mathrm{R}} = \tilde{\alpha} + \tilde{\beta} t_i, \quad i = 1, \cdots, n$$

and an alternative model A yielding $\tilde{\eta}_i^{\mathrm{A}}$ can be performed by constructing the sums of squares

$$Q_{\mathrm{R}} = \sum_i (y_i - \tilde{\eta}_i)^2, \quad Q_{\mathrm{A}} = \sum_i (y_i - \tilde{\eta}_i)^2 \tag{6.1}$$

and the mean squares

$$s_{\mathrm{R}}^2 = Q_{\mathrm{R}}/(n-2), \quad s_{\mathrm{A}}^2 = Q_{\mathrm{A}}/(n-p), \tag{6.2}$$

p being the number of parameters in model A. An F-test on the quotient

$$F = s_R^2/s_A^2 \tag{6.3}$$

can be used to test that A is not significantly better than R. The hypothesis has to be rejected at the level α, i.e. A is better at this level of significance, if

$$F > F_{1-\alpha}(n - 2, n - p) \tag{6.4}$$

(for the fractiles $F_{1-\alpha}$ see table F-6).

(3) If the χ^2-test fails (1) or if a much better model is found (2), the results from linear regression are meaningless. This is not changed by the fact that the errors of the regression variables may be very small as in case (a) of example 12-2. On the contrary: since the errors of measurement are much smaller than the mean deviation of the measurements from the regression line only a small variation of the position and direction of the line (i.e. a small variation of the parameters) is possible within the limits given by the measurement errors.

However, sometimes no better model can be found since no systematic deviation of the measurements from the line is apparent. In this case the assumptions about the measurement errors should be reconsidered. The errors may have been underestimated.

Exercises

12-1: *Variance of measurements unknown*
The number of people employed in farming and related professions in Germany is given in table 12-6 for various years.
(a) Draw a diagram of the numbers as a function of time to check whether linear regression can be applied.
(b) Compute the regression parameters $\tilde{\alpha}$ and $\tilde{\beta}$ using eqs. (1.10) and (1.11).
(c) Predict the number for 1980 and its confidence limits at the 95% level.
(d) Why can linear regression not be applied to this problem over a very long time interval?

TABLE 12-6
Farmers and farm-workers in Germany 1962–1973 (in millions)

year t	1962	1964	1966	1967	1968	1969	1970	1971	1972	1973
number y	3.307	3.002	2.790	2.638	2.523	2.395	2.262	2.101	1.953	1.812

12-2: *Variance of measurements known*
Perform the calculations of table 12-5 by hand, i.e. find the numbers given on the left-hand side of the graphical output.

12-3: *Analysis of variance*

Rather than the simple null hypothesis H_0 ($\beta = \beta_0 = 0$) one may assume more generally H_0 ($\beta = \beta_0 = b$). Show that with $Q_{\text{PRED}} = \sum_i (y_i' - bt_i')^2$, $Q_{\text{RES}} = \sum_i (y_i' - \tilde{\beta} t_i')^2$ and $S^2_{\text{PRED}} = Q_{\text{PRED}}/n$, $S^2_{\text{RES}} = Q_{\text{RES}}/(n-2)$ an F-test can be performed on $F = S^2_{\text{PRED}}/S^2_{\text{RES}}$ with (n − 1, $n - 2$) degrees of freedom.

CHAPTER 13

TIME SERIES ANALYSIS

13-1. Time series. Trend

In the last chapter we have discussed a random variable y and its dependence on a controlled variable t. As in that chapter we assume that y is composed of two parts

$$y_i = \eta_i + \varepsilon_i, \quad i = 1, 2, \ldots, n. \tag{1.1}$$

In chapter 12 η_i was assumed to be a linear function of t, i.e. $\eta_i = \alpha + \beta t_i$, and ε_i was identified with a normally distributed measurement error of zero mean.

We shall now be less specific with our assumptions about η. The controlled variable t will be called the time in this chapter although the method to be described can also be applied to other variables. The method, called *time series analysis*, is widely used in economic applications. It can be used in all cases where there is little or no knowledge on the dependence of η on t. It is customary in the discussion of time series to assume that the y_i are observed at equidistant time intervals

$$t_i - t_{i-1} = \Delta t = \text{const.} \tag{1.2}$$

since by this convention the formulae become much simpler.

An example of a time series is given in fig. 13-1. It shows the mean number of sunspots observed each month over a period of several years. It is clear from the figure that although the values y_i fluctuate largely the number of sunspots tends to vary smoothly as a function of time. The smooth variation could be obtained qualitatively by drawing a free-hand line through the data. But since such drawings are not reproducible we shall discuss more objective methods.

Let us use the notation of eq. (1.1) and call η_i the *trend*, ε_i the *erratic component* of the *measurement* y_i. One way to obtain a smoother function of t would be to form

$$u_i = \frac{1}{2k+1} \sum_{j=i-k}^{i+k} y_i \tag{1.3}$$

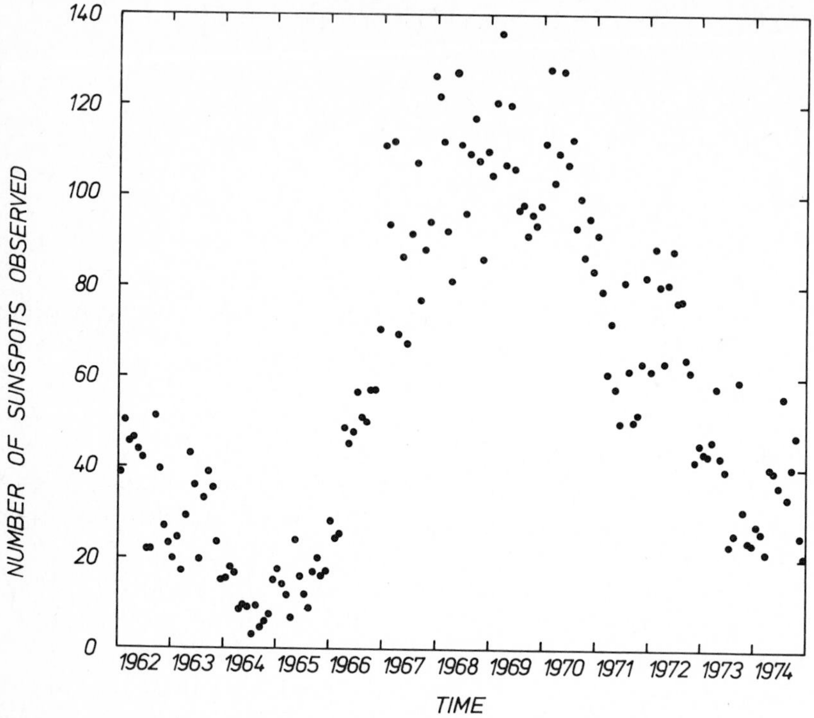

Fig. 13-1. Number of sunspots observed per month from January 1962 to December 1974.

for every i, i.e. the unweighted average of the measurements at the times

$$t_{i-k}, t_{i-k+1}, ..., t_{i-1}, t_i, t_{i+1}, ..., t_{i+k}.$$

Eq. (1.3) is called a *moving average* of y.

13-2. Moving averages

Of course the moving average (1.3) is a very simple construction. We shall show (in example 13-1) that use of the moving average (1.3) implies that η is a linear function of t in the interval used

$$\eta_j = \alpha + \beta t_j, \quad j = -k, -k+1, ..., k. \tag{2.1}$$

Here α and β are constants. They can be estimated from the data by linear regression.

Rather than restricting ourselves to the assumption that the trend η can be described as a linear function of t we shall assume that η can be a polynomial of degree l.

Within the interval of averaging t assumes the values

$$t_j = t_i + j\Delta t, \quad j = -k, -k+1, ..., k. \tag{2.2}$$

Thus if η is a polynomial in t

$$\eta_j = a_1 + a_2 t_j + a_3 t_j^2 + \cdots + a_{l+1} t_j^l \tag{2.3}$$

it is also a polynomial in j

$$\eta_j = x_1 + x_2 j + x_3 j^2 + \cdots + x_{l+1} j^l \tag{2.4}$$

since (2.2) describes a linear transformation between t_j and j (i.e. just a change of scale).

Let us now obtain the coefficients $x_1, x_2, ..., x_{l+1}$ by a least squares fit from the data. The problem is that of example 9-6, where all measurements (for lack of better knowledge) are assumed to be of equal accuracy. In that case the matrix $G_y = aI$ is a multiple of the unit matrix I. Since B is the unit matrix we have $G_B = G_y = aI$. The vector of coefficients is then given by (cf. eq. (9-4.14))

$$\tilde{\boldsymbol{x}} = -(A^{\mathrm{T}}A)^{-1}A^{\mathrm{T}}\boldsymbol{y} \tag{2.5}$$

with A being a $((2k+1) \times (l+1))$ matrix

$$A = -\begin{pmatrix} 1 & -k & (-k)^2 & \dots & (-k)^l \\ 1 & -k+1 & (-k+1)^2 & \dots & (-k+1)^l \\ \vdots & & & & \\ 1 & k & k^2 & \dots & k^l \end{pmatrix}. \tag{2.6}$$

The least squares estimate of the trend $\tilde{\eta}_0$ for the middle of the averaging interval ($j = 0$) is given by (2.4) as

$$\tilde{\eta}_0 = \tilde{x}_1. \tag{2.7}$$

It is equal to the first coefficient of the polynomial. From (2.5) we see that $\tilde{x}_1$ is obtained by multiplying the column vector of measurements $\boldsymbol{y}$ from the left by the row vector

$$\boldsymbol{a} = (-(A^{\mathrm{T}}A)^{-1}A^{\mathrm{T}})_1, \tag{2.8}$$

i.e. by the first row of the matrix $-(A^{\mathrm{T}}A)^{-1}A^{\mathrm{T}}$.

We have

$$\tilde{\eta}_0 = \boldsymbol{a}\boldsymbol{y} = a_{-k}y_{-k} + a_{-k+1}y_{-k+1} + \dots + a_0 y_0 + \dots + a_k y_k. \tag{2.9}$$

This is a linear function of the measurements within the averaging interval.

The vector $\boldsymbol{a}$ does not depend on the measurements but only on l and k, i.e. on the degree of the polynomial and on the length of the interval. Of course one has to choose

$$l < 2k + 1$$

since otherwise there are no degrees of freedom in the least squares fit.

Eq. (2.9) describes the moving average belonging to the assumption of the polynomial (2.4). Once the vector $\boldsymbol{a}$ is determined the moving average

$$u_i = \tilde{\eta}_0(i) = \boldsymbol{a}\boldsymbol{y}(i) = a_1 y_{i-k} + a_2 y_{i-k+1} + \dots + a_{2k+1} y_{i+k} \tag{2.10}$$

for every value of i is easily computed.

Example 13-1: *Moving average for linear trend*

In the case of a linear trend function

$$\eta_j = x_1 + x_2 j$$

the matrix A simply becomes

$$A = -\begin{pmatrix} 1 & -k \\ 1 & -k+1 \\ \vdots & \\ 1 & k \end{pmatrix}.$$

Then

$$A^{\mathrm{T}}A = \begin{pmatrix} 1 & 1 & \dots & 1 \\ -k & -k+1 & \dots & k \end{pmatrix} \begin{pmatrix} 1 & -k \\ 1 & -k+1 \\ \vdots & \vdots \\ 1 & k \end{pmatrix} = \begin{pmatrix} (2k+1) & 0 \\ 0 & k(k+1) \end{pmatrix},$$

$$(A^{\mathrm{T}}A)^{-1} = \frac{1}{k(k+1)(2k+1)} \begin{pmatrix} k(k+1) & 0 \\ 0 & (2k+1) \end{pmatrix} = \begin{pmatrix} \frac{1}{2k+1} & 0 \\ 0 & \frac{1}{k(k+1)} \end{pmatrix},$$

$$\boldsymbol{a} = (-(A^{\mathrm{T}}A)^{-1}A^{\mathrm{T}})_1 = \frac{1}{2k+1}(1, 1, \cdots, 1).$$

Therefore we obtain for the moving average the unweighted average (1.3).

For more complicated models the vectors $\boldsymbol{a}$ can be found in table 13-1

TABLE 13-1

Coefficients of the vector $\boldsymbol{a}$ for the construction of moving averages

$$a = (a_{-k}, a_{-k+1}, \ldots, a_0, a_1, \ldots, a_k) = \frac{1}{A}(\alpha_{-k}, \alpha_{-k+1}, \ldots, \alpha_0, \alpha_1, \ldots, \alpha_k)$$

$$\alpha_{-j} = \alpha_j$$

$l = 2$ and $l = 3$

k	$\frac{1}{A}$	α_{-7}	α_{-6}	α_{-5}	α_{-4}	α_{-3}	α_{-2}	α_{-1}	α_0
2	$\frac{1}{35}$						−3	12	17
3	$\frac{1}{21}$					−2	3	6	7
4	$\frac{1}{231}$				−21	14	39	54	59
5	$\frac{1}{429}$			−36	9	44	69	84	89
6	$\frac{1}{143}$		−11	0	9	16	21	24	25
7	$\frac{1}{1105}$	−78	−13	42	87	122	147	162	167

$l = 4$ and $l = 5$

k	$\frac{1}{A}$	α_{-7}	α_{-6}	α_{-5}	α_{-4}	α_{-3}	α_{-2}	α_{-1}	α_0
3	$\frac{1}{231}$					5	−30	75	131
4	$\frac{1}{429}$				15	−55	30	135	179
5	$\frac{1}{429}$			18	−45	−10	60	120	143
6	$\frac{1}{2431}$		110	−198	−135	110	390	600	677
7	$\frac{1}{46189}$	2145	−2860	−2937	−165	3755	7500	10125	11063

or obtained by solving eq. (2.8). From the symmetry of A it can be shown that for odd polynomials (i.e. $l = 2n$, n integer) the same values of $\boldsymbol{a}$ are obtained as for $l = 2n - 1$. It is also easy to show that $\boldsymbol{a}$ is symmetric with

$$a_j = a_{-j}, \quad j = 1, 2, \ldots, k. \tag{2.11}$$

13-3. End effects

It is clear that the moving average (2.10) as an estimator of the trend can be formed only for those points i which have a neighbourhood of length $(2k+1)$ of measured points around them. That means that for the first k and the last k points of a time series another estimator has to be used. The most natural estimator is constructed by using the polynomial (2.4) not only for the middle point of the interval. One then obtains the estimators.

$$\begin{aligned}
\tilde{\eta}_i = u_i &= \tilde{x}_1^{k+1} + \tilde{x}_2^{k+1}(i-k-1) + \tilde{x}_3^{k+1}(i-k-1)^2 + \cdots \\
&\quad + \tilde{x}_{l+1}^{k+1}(i-k-1)^l, \quad i \leqslant k, \\
\tilde{\eta}_i = u_i &= \tilde{x}_1^{n-k} + \tilde{x}_2^{n-k}(i+k-n) + \tilde{x}_3^{n-k}(i+k-n)^2 + \cdots \\
&\quad + \tilde{x}_{l+1}^{n-k}(i+k-n), \quad i > n-k.
\end{aligned} \tag{3.1}$$

Here $\tilde{x}^{k+1}$ and $\tilde{x}^{n-k}$ signify that the coefficients $\tilde{x}$ were determined for the first and last interval centered at $(k+1)$ and $(n-k)$ respectively.

The estimators (3.1) are defined even for $i < 1$ and $i > n$, i.e. they offer the possibility of continuing the time series (e.g. into the future). Such extrapolations have to be interpreted with great caution for two reasons.

(i) There is usually no theoretical justification to assume that the trend is given by a polynomial, which is just an easy means of forming a moving average. Without theoretical understanding of the trend the meaning of an extrapolation usually is not very clear.

(ii) Even if the trend were described by a polynomial with theoretical justification the confidence in the fitted polynomial decreases in the region with no data.

While the applicability of (i) cannot be discussed in general but depends on the particular problem under investigation the question (ii) which is familiar to us already from linear regression (cf. fig. 12-2) is studied in § 4.

13-4. Confidence interval

Let us first consider the confidence interval belonging to a moving average u_i obtained by eq. (2.8). Since the errors of the measurement y_j are unknown they have to be estimated. By a simple generalization of eq. (12-1.5) one obtains as the empirical variance of the y_j within the interval of length $2k + 1$

$$s_y^2 = \frac{1}{2k - l} \sum_{j=-k}^{k} (y_j - \tilde{\eta}_j)^2, \tag{4.1}$$

where $\tilde{\eta}_j$ is given by

$$\tilde{\eta}_j = \tilde{x}_1 + \tilde{x}_2 j + \tilde{x}_3 j^2 + \cdots + \tilde{x}_{l+1} j^l. \tag{4.2}$$

Then the covariance matrix of measurements can be estimated by

$$G_B^{-1} = G_y^{-1} \approx s_y^2 I. \tag{4.3}$$

The covariance matrix of the coefficients $\boldsymbol{x}$ is then given by eq. (9-4.19)

$$G_{\tilde{x}}^{-1} \approx (A^{\mathrm{T}} G_B A)^{-1} = s_y^2 (A^{\mathrm{T}} A)^{-1}. \tag{4.4}$$

Now since $u_i = \tilde{\eta}_0 = \tilde{x}_1$ we have as an estimate of the variance of u_i

$$s_{\tilde{x}_1}^2 = (G_{\tilde{x}}^{-1})_{11} = s_y^2 ((A^{\mathrm{T}} A)^{-1})_{11} = s_y^2 a_0. \tag{4.5}$$

From eqs. (2.6), (2.7) and (2.8) one easily sees that $((A^{\mathrm{T}} A)^{-1})_{11} = a_0$, since the central row of A is $-(1, 0, 0, \ldots, 0)$.

With the same reasoning as in ch. 12, § 2 we then have at a confidence level $1 - \alpha$

$$\frac{|\tilde{\eta}_0(i) - \eta_0(i)|}{s_y^2 a_0} \leqslant t_{1 - \frac{1}{2}\alpha}. \tag{4.6}$$

For given α we can therefore quote confidence limits

$$\eta_0^{\pm}(i) = \tilde{\eta}_0(i) \pm a_0 s_y t_{1 - \frac{1}{2}\alpha}, \tag{4.7}$$

where $t_{1-\frac{1}{2}\alpha}$ is a quantile of Student's distribution for $2k - l$ degrees of freedom (cf. table F-7). The true value of the trend lies within these limits at confidence level $1 - \alpha$.

The determination of confidence limits at the ends of the time series is more involved although conceptually identical. The moving average is now given by eq. (3.1). If we denote the argument of the expressions (3.1) by $j = i - k + 1$ or $j = i + k - n$, respectively we obtain

$$\tilde{\eta} = T\boldsymbol{x}, \tag{4.8}$$

where T is a row vector of length $l + 1$

$$T = (1, j, j^2, \ldots, j^l). \tag{4.9}$$

According to the law of transformation of errors (eq. (4-5.4)) we obtain

$$G_{\tilde{\eta}}^{-1} = TG_x^{-1}T^{\mathrm{T}}. \tag{4.10}$$

Using (4.6) we have

$$G_{\tilde{\eta}}^{-1} \approx s_{\tilde{\eta}}^2 = s_y^2 T(A^{\mathrm{T}}A)^{-1}T^{\mathrm{T}}, \tag{4.11}$$

where s_y^2 again is given by (4.1).

Now $s_{\tilde{\eta}}^2$ is a number which can be computed for every value of j even outside the region of the time series. With it we obtain the confidence limits

$$\eta^{\pm}(i) = \tilde{\eta}(i) \pm s_{\tilde{\eta}} t_{1-\frac{1}{2}\alpha}. \tag{4.12}$$

An actual example of these confidence limits is given in table 13-2.

13-5. A FORTRAN program for time series analysis

The computation of moving averages (except arithmetic means) and especially of confidence limits involves a lot of arithmetic work and should always be done by computer. The FORTRAN program described here reads the parameters n, k and l and the data $y_i, j = 1, \ldots, n$. For the "inner points" of the series $j = k+1, k+2, \ldots, m-k$ the moving average $u_j = \tilde{\eta}_0(j)$ of eq. (2.10) and the confidence limits $\tilde{\eta}_0^{\pm}(j)$ according to eq. (4.7) are computed. For the "outer points" ($j = 1, \ldots, k$ and $j = n-k+1, \ldots, n$) the trend $\tilde{\eta}_j$ is found according to eq. (3.1) and extrapolated for k points on both sides of the time series, i.e. for $j = -k, -k+1, \ldots, -1, 0$ and $j = n+1, n+2, \ldots, n+k$. The confidence intervals for the end parts, i.e. for the outer and extrapolated point are found according to eq. (4.12). The results are listed and plotted using program 12-2.

Program 13-1: *Main program* TIMSER *performing a time series analysis*

```
C        MAIN PROGRAM TIMSER
C
      DIMENSION Y(500),U(3,500),A(200),ETA(21),TEXT(20),ATA1
     *(100),ATA1AT(200),X(10),TALPHA(20),T(500),HELP(200),TVECT(10)
      DATA TALPHA/6.31,2.92,2.35,2.13,2.02,1.94,1.89,1.86,1.83,1.81,
     *1.80,1.78,1.77,1.76,1.75,1.75,1.74,1.73,1.73,1.73/
      DATA OMEGA/1.645/
C
C     READ EXPLANATORY TEXT AND PARAMETERS
C
      READ(5,1000) TEXT
      READ(5,1100) N,K,L,YBEG,DELY,NY
C
C     READ DATA OF TIME SERIES AND STORE THEM IN Y
C
```

```
      J1=K+1
      J2=K+N
      READ (5,1200) (Y(J),J=J1,J2)
C
C      COMPUTE MATRICES DEPENDING ONLY ON K AND L
C
      K21=2*K+1
      L1=L+1
      DO 20 I=1,L1
      J1=2*K+1
      DO 20  J=1,J1
      I1=(J-1)*(L+1)+I
      IF(I-1) 10,10,15
 10   A(I1)=-1.
      GO TO 20
 15   A(I1)=(-1.)*(J-K-1)**(I-1)
 20   CONTINUE
      CALL MTXMAT(A,A,HELP,L1,K21,L1)
      CALL MTXINV(HELP,ATA1,L1)
      CALL MTXMBT(ATA1,A,ATA1AT,L1,L1,K21)
      CALL MTXMSC(ATA1AT,ATA1AT,-1.,L1,K21)
C
C     DO-LOOP OVER INNER PART OF TIME SERIES
C
      IA=2*K+1
      IB=N
      DO 90 I=IA,IB
C
C     COMPUTE MOVING AVERAGE AND CONFIDENCE LIMITS FOR INNER PART
C
      CALL MTXMLT(ATA1AT,Y(I-K),X,L1,K21,1)
      U(1,I)=X(1)
      CALL MTXMLT(A,X,ETA,K21,L1,1)
      CALL MTXADD(Y(I-K),ETA,ETA,K21,1)
      CALL MTXMAT(ETA,ETA,SY2,1,K21,1)
      SY2=SY2/(2*K-L)
      A0=ATA1AT(K+1)
      FACT=OMEGA
      IF(2*K-L-20) 22,22,25
 22   FACT=TALPHA(2*K-L)
 25   UDIFF=A0*SQRT(SY2)*FACT
      U(2,I)=U(1,I)-UDIFF
      U(3,I)=U(1,I)+UDIFF
      T(I)=I-K
C
C     COMPUTE MOVING AVERAGE AND CONFIDENCE LIMITS FOR ENDS
C
      IF(I-IA)  40,30,40
 30   IADD=IA
      IS=-1
      GO TO 60
 40   IF(I-IB)  90,50,90
 50   IADD=IB
      IS=1
 60   I3=2*K
      DO 80 I1=1,I3
      J=IS*I1
      DO 70 I2=1,L1
 70   TVECT(I2)=J**(I2-1)
      CALL MTXMBT(ATA1,TVECT,HELP,L1,L1,1)
      CALL MTXMLT(TVECT,HELP,SETA2,1,L1,1)
      SETA2=SY2*SETA2
      CALL MTXMLT(TVECT,X,ETA,1,L1,1)
      UDIFF=SQRT(SETA2)*TALPHA(2*K-L)
      U(1,IADD+J)=ETA
      U(2,IADD+J)=ETA-UDIFF
      U(3,IADD+J)=ETA+UDIFF
      T(IADD+J)=IADD+J-K
```

```
 80    CONTINUE
 90    CONTINUE
C
C      WRITE OUTPUT
C
       WRITE(6,1000) TEXT
       WRITE(6,1300) N,K21,L
       DO 100 I=1,K
       I1=I+K+N
       Y(I)= -10000.
 100   Y(I1)=-10000.
       LENGTH=N+2*K
       CALL PRGRAP (YBEG,DELY,NY,T,Y,U,LENGTH)
 1000  FORMAT (20A4)
 1100  FORMAT (3I10,2F10.5,I10)
 1200  FORMAT (8F 10.5)
 1300  FORMAT (/23H LENGTH OF TIME SERIES ,12X,4HN = ,I5,/
      *39H LENGTH OF AVERAGING INTERVAL  2*K+1 = ,I5/
      *22H DEGREE OF POLYNOMIAL ,13X,4HL = ,I5//)
       END
```

In the input part one card with alphanumeric text is read for use as a heading of the output followed by a card with the parameters n, k and l (stored in N, K and L) and 3 numbers to specify the lower limit, the interval size and the number of intervals of the y-axis of the graphical output (YBEG, DELY and NY). The actual data are then read from a series of cards. 8 of them are contained on one card. The data $y_1, y_2, \ldots, y_n$ are stored in the words Y(K+1), Y(K+2), ..., Y(K+N), thus leaving k words at the beginning of the array Y free for extrapolation. Next the two integers $l+1$ and $2k+1$ are formed and written into the words L1 and K21, respectively. They are used frequently in the program beginning with the computation of the matrices A, $(A^{\mathrm{T}}A)^{-1}$ and $(A^{\mathrm{T}}A)^{-1}A^{\mathrm{T}}$ which are stored in A, ATA1 and ATA1AT, respectively. In the following DO-loop, which extends over the inner part of the time series and in which the index I characterizes the middle point of the averaging interval the moving average and its confidence limits are computed and stored in U(1,I) and U(2,I), U(3,I). The vector X contains the coefficients $x_1, \ldots, x_{l+1}$ of eq. (2.4). The difference between the moving average and its confidence limits is computed at 90 % confidence level. The fractile $t_{1-\frac{1}{2}\alpha}$ is taken from the array TALPHA, where it is stored for 1 to 20 degrees of freedom. For higher degrees of freedom the fractile OMEGA of the normal distribution is taken instead.

After calculations for the first and the last point of the inner part of the time series given by I=IA or I=IB, respectively the program computes the values $\tilde{\eta}$ and $\tilde{\eta}^{\pm}$ according to eqs. (3.1) and (4.12) for the outer and extrapolated points. The indices IADD and IS are used to store the results in the correct locations of the array U. The array T(1), ..., T(N+2*K) is filled with the numbers —K, —K+1, ..., —1, 0, 1, ..., N, N+1, ..., N+K which correspond to the index i of eq. (1.1). This index assumes negative values for the extrapolation to the left thus starting with $i = 1$ for the first measurement.

Finally the numerical and graphical output is performed by the subroutine PRGRAP (program 12-2).

Example 13-2: *Time series analysis of the mean number of sun spots observed per month*

A print-out is shown in table 13-2 for the sun spot data of fig. 13-1. The time series analysis was performed with $k = 4$ and $l = 2$. The current index i for each data point is given in the first column (T). The second and third columns (Y) and ETA contain the data points and corresponding moving averages. The latter are extrapolated for k points before and after the data points. The data (marked by XXX) fluctuate around the smoother line of moving averages (marked by *). The interval between the confidence limits (marked by +) is wide in regions of large fluctuations and narrow where the fluctuations are small. Moreover the confidence interval widens rather fast if one goes to the regions of extrapolation.

13-6. A word of caution

We have stressed in ch. 12, § 6 that the results of a linear regression have to be interpreted with care. This is even more true for the results of time series analysis. The reasons are twofold:
(1) While it is often reasonable to construct a linear model, there is usually no *a priori* justification for the model underlying a time series analysis other than the wish to separate the trend from "statistical" fluctuations.
(2) There is great freedom for the user in choosing the parameters k and l which can strongly influence the results. To get a feeling for the last point we study the following example:

Example 13-3: *Time series analysis of the same data but for averaging intervals of different lengths and polynomials of different degrees*

Table 13-3 contains the graphical part of the print-out for a time series analysis of the number of people killed in road accidents in Germany in the years 1956 to 1973, where different values of k and l were used. (The numerical part is not reproduced for reasons of space.) The diagrams (a) to (d) were obtained for $l = 1$ (linear averaging) but

TABLE 13-2

Print out for example 13-2

```
MEAN NUMBER OF SUN SPOTS IN THE MONTHS JAN 1962 TO DEC 1964

LENGTH OF TIME SERIES              N  =      36
LENGTH OF AVERAGING INTERVAL   2*K+1 =       9
DEGREE OF POLYNOMIAL               L =       2


                                                       Y
     T         Y        ETA   -19.0000    1.0000   21.0000    41.0000    61.0000
                             I---------I---------I---------I---------I---------
   -3.000  ********   58.536  I                                            *           I
   -2.000  ********   55.320  I           +                                *           I
   -1.000  ********   52.339  I                 +                          *           I
    0.000  ********   49.593  I               +                       *                I
    1.000   38.700   47.082   I                    +      XXX      *          +         I
    2.000   50.300   44.807   I                     +       *  XXX    +                 I
    3.000   45.600   42.767   I                     +      *XXX   +                      I
    4.000   46.400   40.963   I                    +     *  XXX  +                       I
    5.000   43.700   39.394   I                   +   *XX+                               I
    6.000   42.000   35.262   I                  +  *  +XX                               I
    7.000   21.800   35.136   I       XXX      *  *  +                                   I
    8.000   21.600   34.366   I       XXX  +   *   +                                     I
    9.000   51.300   33.833   I              +   *   +         XXX                        I
   10.000   39.500   32.452   I            +   *   +XX                                   I
   11.000   26.900   32.795   I           XX+  *   +                                     I
   12.000   23.200   27.131   I          +XX*  +                                         I
   13.000   19.800   19.210   I       ++X                                               I
   14.000   24.400   21.103   I       +  *+X                                            I
   15.000   17.100   26.878   I       XXX+  *   +                                        I
   16.000   29.300   28.974   I          +  *XX+                                         I
   17.000   43.000   30.584   I            +  *  +        XXX                            I
   18.000   35.900   32.316   I            +  *X+X                                       I
   19.000   19.600   35.203   I        XXX      *   *  +                                 I
   20.000   33.200   34.156   I               +X*X+                                      I
   21.000   38.800   30.961   I            +  *   +XX                                    I
   22.000   35.300   28.919   I           +  *  +XXX                                     I
   23.000   23.400   26.648   I          +X*  +                                          I
   24.000   14.900   21.352   I       XXX+  *+                                          I
   25.000   15.300   16.535   I       ++*+                                              I
   26.000   17.700   13.880   I       +*+XX                                             I
   27.000   16.500   13.595   I       +*+X                                              I
   28.000    8.600   11.714   I     XX+*+                                               I
   29.000    9.500    9.691   I       ++*+                                              I
   30.000    9.100    7.413   I       ++XX                                              I
   31.000    3.100    6.209   I     X+*+                                                I
   32.000    9.300    5.635   I       +*+XX                                             I
   33.000    4.700    6.259   I     X+*+                                                I
   34.000    6.100    7.624   I     X+*+                                                I
   35.000    7.400    9.728   I     +X+*+                                               I
   36.000   15.100   12.572   I       +  *XX+                                           I
   37.000  ********   16.155   I         +   *    +                                      I
   38.000  ********   20.478   I         +     *    +                                    I
   39.000  ********   25.540   I           +      *       +                              I
   40.000  ********   31.342   I            +             *            +                 I
                             I---------I---------I---------I---------I---------
                           -19.0000    1.0000   21.0000    41.0000    61.0000
```

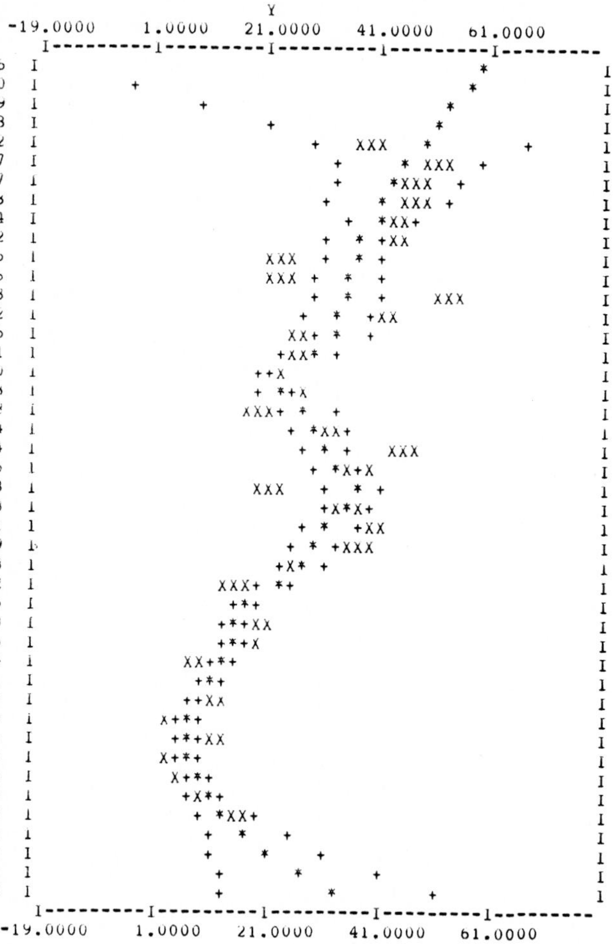

interval lengths $2k + 2 = 3, 5, 7$ and 9 respectively. One observes that
the line of moving average becomes smoother and the confidence
interval narrower for increasing k but also that the average deviation
of the individual observations from the line increases. The extrapolation

of course is a straight line. (For $l = 0$ we would have obtained the same moving average for the inner points. The outer and the extrapolated

TABLE 13-3
Print out for example 13-3

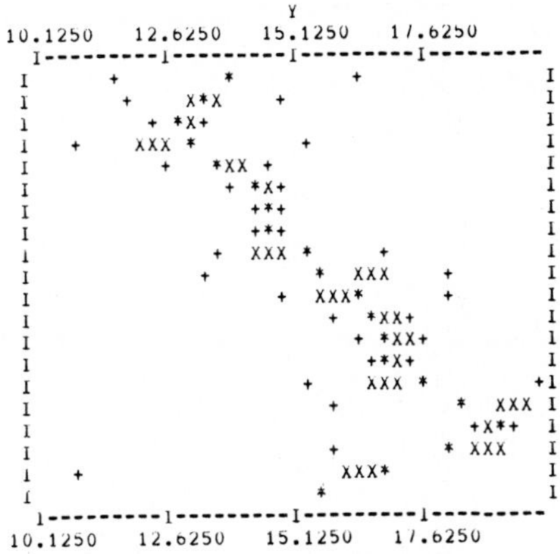

(a) $2k+1=3, \; l=1$

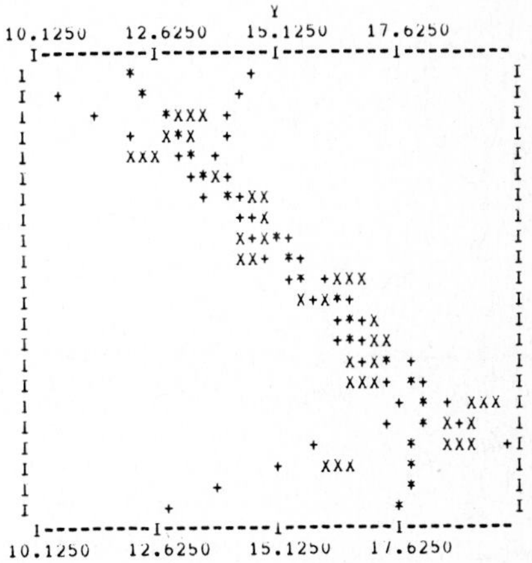

(b) $2k+1=5, \; l=1$

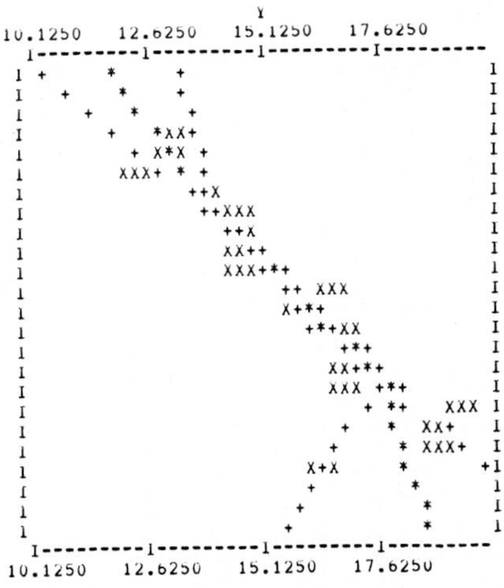

(c) $2k+1=7, \; l=1$

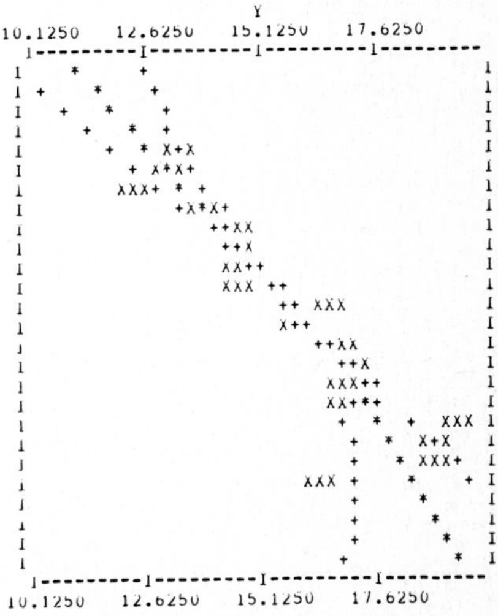

(d) $2k+1=9, \; l=1$

TABLE 13-3 (continued)

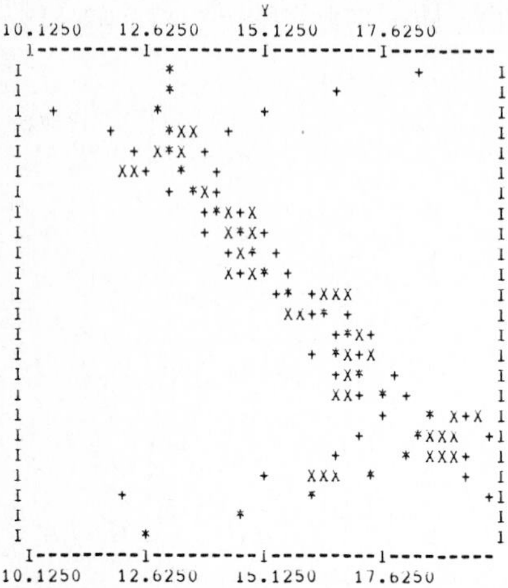

(e) 2k+1=7, l=2

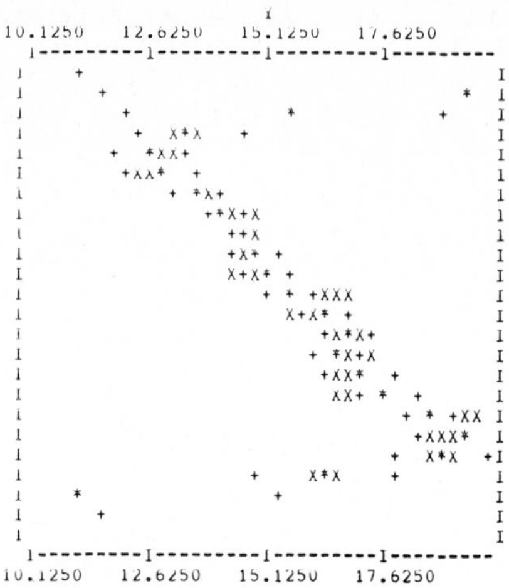

(f) 2k+1=7, l=3

points would however have been located on a vertical straight line since a polynomial of degree zero is a constant.) Diagrams (e) and (f) were obtained for $2k + 1 = 7$ and $l = 2$ and $l = 3$ respectively. The moving average is nearer to the data and the confidence interval is wider than for $l = 1$. The moving average is identical in the inner part of both diagrams. However the extrapolation is rather different. It is obvious that extrapolation becomes more and more dangerous for increasing l since a polynomial of higher power can change very rapidly.

From these observations we can set up the following qualitative rules:
(1) The averaging interval should not be chosen larger than a region in which the data can be expected to be described well by a polynomial of the chosen degree, i.e. for $l = 1$ the interval $2k + 1$ must be such that no large non-linear effects are expected.
(2) On the other hand the smoothing is more effective if the averaging interval is long. As a rule, the larger $2k + 1 - l$ the more effective the averaging.
(3) Great care is required if the time series is to be extrapolated, especially in the non-linear case.
The art of time series analysis is of course much more developed than can be described in a short chapter. The particularly interested reader is referred to the specialized literature, where the use of functions other than polynomials and the analysis of several variables are described.

Exercise

13-1: *Moving average*

Compute the moving average for $i = 5, 6, \ldots, 10$ of example 13-1 by hand for $k = 4$, $l = 2$, as in table 13-2. (Hint: Write the measurements $y_1, y_2, \ldots, y_{14}$ in a column. Write the coefficients $a_{-4}, a_{-3}, \ldots, a_4$ in a column of equal spacing on a separate piece of paper and move it along the first row to perform the calculations of eq. (2.7).)

SOLUTION AND DISCUSSION OF THE EXERCISES

Exercise 2-1

(a) $N = 365^n$.

(b) $N' = 365 \cdot 364 \ldots (365 - n + 1)$.

(c) $$P_{\text{diff}} = N'/N = \frac{364}{365} \cdot \frac{363}{365} \ldots \frac{365 - n + 1}{365}$$

$$= \left(1 - \frac{1}{365}\right)\left(1 - \frac{2}{365}\right) \ldots \left(1 - \frac{n + 1}{365}\right).$$

(d) $P = 1 - P_{\text{diff}}$.

Inserting numbers one obtains $P \approx 0.5$ for $n = 23$ and $P = 0.99$ for $n = 57$.

Exercise 2-2

$P(A + B) = P(A) + P(B) - P(AB)$,

$P(\text{ace or diamond}) = P(\text{ace}) + P(\text{diamond}) - P(\text{ace and diamond})$

$$= \frac{4}{52} + \frac{13}{52} - \frac{1}{52}.$$

Exercise 2-3

(a) $P(A) = \frac{4}{52}$, $P(B) = \frac{13}{52}$, $P(AB) = \frac{1}{52}$, i.e. $P(AB) = P(A)P(B)$.

(b) $P(A) = \frac{4}{53}$, $P(B) = \frac{13}{53}$, $P(AB) = \frac{1}{53}$, i.e. $P(AB) \neq P(A)P(B)$

Exercise 2-4

$P(\bar{A}\bar{B}) = 1 - P(A + B) = 1 - P(A) - P(B) + P(AB)$.

For A, B independent $P(AB) = P(A)P(B)$, therefore

$P(\bar{A}\bar{B}) = 1 - P(A) - P(B) + P(A)P(B) = (1 - P(A))(1 - P(B)) = P(\bar{A})P(\bar{B})$.

Exercise 2-5

(a) $P(A) = 0.2$, $P(B) = 0.3$, $P(C) = 0.5$, $P(AB) = 2 \cdot 0.2 \cdot 0.3$, $P(AC) = 2 \cdot 0.2 \cdot 0.5$, $P(BC) = 2 \cdot 0.3 \cdot 0.5$, $P(AA) = 0.2^2$, $P(BB) = 0.3^2$, $P(CC) = 0.5^2$.

(b) $P(A) = 2/10 = 0.2$, $P(B) = 3/10 = 0.3$, $P(C) = 5/10 = 0.5$,

$$P(AB) = \frac{2}{10} \cdot \frac{3}{9} + \frac{3}{10} \cdot \frac{2}{9}, \quad P(AC) = \frac{2}{10} \cdot \frac{5}{9} + \frac{5}{10} \cdot \frac{3}{9},$$

$$P(AA) = \frac{2}{10} \cdot \frac{1}{9}, \quad P(BB) = \frac{3}{10} \cdot \frac{2}{9}, \quad P(CC) = \frac{5}{10} \cdot \frac{4}{9}.$$

Valid for (a) and (b):
$P(2 \text{ equal coins}) = P(AA) + P(BB) + P(CC)$,
$P(2 \text{ different coins}) = P(AB) + P(AC) + P(BC) = 1 - P(2 \text{ equal coins})$.

Exercise 3-1

(a) $\hat{x} = \sum_{i=1}^{6} x_i p_i = \frac{1}{6} \sum_{i=1}^{6} i = \frac{21}{6} = 3.5,$

$$\sigma^2(x) = \sum_{i=1}^{6} (x_i - \hat{x})^2 p_i = \frac{1}{6}(2.5^2 + 1.5^2 + 0.5^2 + 0.5^2 + 1.5^2 + 2.5^2)$$

$$= \frac{2}{6}(6.25 + 2.25 + 0.25) = 2.92,$$

$$\mu_3 = \sum_{i=1}^{6} (x_i - \hat{x})^3 p_i = \frac{1}{6} \sum_{i=1}^{6} (i - 3.5)^3 = 0,$$

$\gamma = \mu_3/\sigma^3 = 0.$

(b) $\hat{x} = \frac{1}{12}(2 + 2 + 3 + 8 + 15 + 18) = 4,$

$$\sigma^2(x) = \frac{1}{12}(2 \cdot 3^2 + 1 \cdot 2^2 + 1 \cdot 1^2 + 2 \cdot 1^2 + 3 \cdot 2^2 + 3 \cdot 3^2) = \frac{64}{12} = 5.33,$$

$$\mu_3 = \frac{1}{12}(-2 \cdot 3^3 - 1 \cdot 2^3 - 1 \cdot 1^3 + 2 \cdot 1^3 + 3 \cdot 2^3 + 3 \cdot 3^3)$$

$$= \frac{1}{12}(3^3 + 2 \cdot 2^3 + 1^3) = \frac{44}{12} = 3.67,$$

$\gamma = \mu_3/\sigma^3 = 3.67/(5.33)^{\frac{3}{2}} = 0.27.$

Exercise 3-2

$$\hat{x} = \frac{2}{b - a}\left[\int_a^0 x \frac{a - x}{a} \,dx + \int_0^b x \frac{b - x}{b} \,dx\right] = \frac{2}{b - a}\left(\frac{b^2}{2} - \frac{a^2}{2} - \frac{b^2}{3} + \frac{a^2}{3}\right) = \frac{1}{3}(a + b),$$

$x_m = c = 0,$

$$x_{0.5} = \frac{2}{b - a} \int_a^{x_{0.5}} \frac{a - x}{a} \,dx = \frac{2}{b - a}\left[x_{0.5} - a - \frac{x_{0.5}^2}{2a} + \frac{a}{2}\right] \text{ yields } x_{0.5} = a + \frac{a(a - b)}{2}.$$

$$E(x^2) = \frac{2}{b - a}\left[\int_a^0 x^2 \frac{a - x}{a} \,dx + \int_0^b x^2 \frac{b - x}{b} \,dx\right] = \frac{2}{b - a}\left(\frac{b^3}{3} - \frac{a^3}{3} - \frac{b^3}{4} + \frac{a^3}{4}\right)$$

$$= \frac{b^3 - a^3}{6(b - a)}.$$

Using (3.15) one obtains

$$\sigma^2(x) = E(x^2) - \hat{x}^2 = \frac{b^3 - a^3}{6(b - a)} - \frac{1}{9}(a + b)^2,$$

For $a = -b : \hat{x} = 0,\ x_{0.5} = 0,\ \sigma^2(x) = \dfrac{b^2}{6}$.

For $a = -2b : \hat{x} = -\dfrac{b}{3} = -0.33\,b,\ x_{0.5} = b(\sqrt{3} - 2) = -0.27\,b,\ \sigma^2(x) = \dfrac{7}{18}b^2$.

Exercise 3-3

$\hat{x} = 0$ since $f(x)$ is symmetric about 0. Therefore with (3.15)

$$\sigma^2(x) = E(x^2) = \frac{1}{\pi}\int_{-\infty}^{\infty} \frac{x^2}{1 + x^2}\,\mathrm{d}x = \frac{1}{\pi}\int_{-\infty}^{\infty}\left(1 - \frac{1}{1 + x^2}\right)\mathrm{d}x = \infty - 1.$$

Exercise 4-1

(a) $\theta = \arctan y,\ \dfrac{\mathrm{d}\theta}{\mathrm{d}y} = \dfrac{1}{1 + y^2},\ g(y) = \dfrac{1}{\pi(1 + y^2)}$.

(b) $\mathrm{e}^{-\frac{1}{2}y^2} = \dfrac{\mathrm{d}x}{\mathrm{d}y}\mathrm{e}^{-\frac{1}{2}x^2} = \sigma\mathrm{e}^{-\frac{1}{2}x^2}$.

Exercise 4-2

(a) $f(x) = \dfrac{1}{\sqrt{(2\pi)}\sigma_x}\exp\left(-\dfrac{1}{2}\dfrac{x^2}{\sigma_x^2}\right),\ f(y) = \dfrac{1}{\sqrt{(2\pi)}\sigma_y}\exp\left(-\dfrac{1}{2}\dfrac{x^2}{\sigma_x^2}\right)$.

(b) Yes, since eq. (1.6) is fulfilled.

(c) The transformation can be written as

$$\begin{pmatrix} u \\ v \end{pmatrix} = R\begin{pmatrix} x \\ y \end{pmatrix},\quad R = \begin{pmatrix} \cos\phi & \sin\phi \\ -\sin\phi & \cos\phi \end{pmatrix}.$$

It is orthogonal, since $R^{\mathrm{T}}R = I$. Therefore

$$g(u, v) = f(x, y) = \frac{1}{2\pi\sigma_x\sigma_y}\exp\left(-\frac{x^2}{2\sigma_x^2} - \frac{y^2}{2\sigma_y^2}\right)$$

$$= \frac{1}{2\pi\sigma_x\sigma_y}\exp\left(-\frac{(u\cos\phi - v\sin\phi)^2}{2\sigma_x^2} - \frac{(u\sin\phi + v\cos\phi)^2}{2\sigma_y^2}\right)$$

$$= \frac{1}{2\pi\sigma_x\sigma_y}\exp\left(-\frac{u^2\cos^2\phi + v^2\sin^2\phi - 2uv\cos\phi\sin\phi}{2\sigma_x^2}\right.$$

$$\left. - \frac{u^2\sin^2\phi + v^2\cos^2\phi + 2\,uv\cos\phi\sin\phi}{2\sigma_y^2}\right).$$

(d) For $\phi = 90°$: $\cos\phi = 0$, $\sin\phi = 1$ etc. The expression $g(u, v)$ then factorizes, i.e. $g(u, v) = g(u)g(v)$.

For $\sigma_x = \sigma_y = \sigma$:

$$g(u, v) = \frac{1}{2\pi\sigma^2} \exp\left(-\frac{u^2 + v^2}{2\sigma^2}\right) = g(u)g(v).$$

(e)
$$J = \begin{vmatrix} \dfrac{\partial x}{\partial r} & \dfrac{\partial y}{\partial r} \\ \dfrac{\partial x}{\partial \phi} & \dfrac{\partial y}{\partial \phi} \end{vmatrix} = \begin{vmatrix} \cos\phi & \sin\phi \\ -r\sin\phi & r\cos\phi \end{vmatrix} = r,$$

$$g(r, \phi) = r f(x, y) = \frac{r}{2\pi\sigma^2} \exp\left(-\frac{x^2 + y^2}{2\sigma^2}\right) = \frac{r}{2\pi\sigma^2} \exp\left(-\frac{r^2}{2\sigma^2}\right);$$

$$g(r) = \int_0^{2\pi} g(r, \phi)\mathrm{d}\phi = \frac{r}{\sigma^2} \exp\left(-\frac{r^2}{2\sigma^2}\right),$$

$$g(\phi) = \frac{1}{2\pi\sigma^2} \int_0^\infty r \exp\left(-\frac{r^2}{2\sigma^2}\right) \mathrm{d}r = \frac{1}{4\pi\sigma^2} \int_0^\infty \exp\left(-\frac{u}{2\sigma^2}\right) du$$

$$= -\frac{1}{2\pi}\left[\exp\left(-\frac{u}{2\sigma^2}\right)\right]_0^\infty = \frac{1}{2\pi};$$

$g(r, \phi) = g(r)g(\phi)$, therefore r and ϕ are independent.

Exercise 4-3

$$g = 4\pi^2 \frac{l}{T^2} = 4\pi^2 \frac{99.8}{2.03^2} = 956.17,$$

$$\frac{\partial g}{\partial l} = \frac{4\pi^2}{T^2} = 9.58, \quad \frac{\partial g}{\partial T} = -\frac{8\pi^2 l}{T^3} = -942,$$

$$(\Delta g)^2 = \left(\frac{\partial g}{\partial l}\right)^2 (\Delta l)^2 + \left(\frac{\partial g}{\partial T}\right)^2 \Delta T^2 = 2310,$$

$\Delta g = 48.06.$

Exercise 4-4

(a)
$$\mathbf{y} = T\mathbf{x}, \; T = \begin{pmatrix} \dfrac{\partial y_1}{\partial x_1} & \dfrac{\partial y_1}{\partial x_2} \\ \dfrac{\partial y_2}{\partial x_1} & \dfrac{\partial y_2}{\partial x_2} \end{pmatrix} = \begin{pmatrix} \dfrac{\partial p}{\partial m} & \dfrac{\partial p}{\partial v} \\ \dfrac{\partial E}{\partial m} & \dfrac{\partial E}{\partial v} \end{pmatrix} = \begin{pmatrix} v & m \\ \frac{1}{2}v^2 & mv \end{pmatrix};$$

$$C_y = TC_x T^{\mathrm{T}} = \begin{pmatrix} v & m \\ \frac{1}{2}v^2 & mv \end{pmatrix} \begin{pmatrix} a^2m^2 & 0 \\ 0 & b^2v^2 \end{pmatrix} \begin{pmatrix} v & \frac{1}{2}v^2 \\ m & mv \end{pmatrix}$$

$$= \begin{pmatrix} (a^2 + b^2)m^2v^2 & (\frac{1}{2}a^2 + b^2)m^2v^3 \\ (\frac{1}{2}a^2 + b^2)m^2v^3 & (\frac{1}{4}a^2 + b^2)m^2v^4 \end{pmatrix};$$

$$\rho(p, E) = \frac{\mathrm{cov}\,(p, E)}{\sigma(p)\sigma(E)} = \frac{(\frac{1}{2}a^2 + b^2)m^2v^3}{\sqrt{[(a^2 + b^2)m^2v^2]}\sqrt{[(\frac{1}{4}a^2 + b^2)m^2v^4]}} = \frac{\frac{1}{2}a^2 + b^2}{\sqrt{[(a^2 + b^2)(\frac{1}{4}a^2 + b^2)]}}.$$

For $a = 0$ or $b = 0 : \rho = 1$(in this case either m or v are completely determined and there is a linear relation between E and p). However for $a, b \neq 0$ one has $\rho \neq 1$, e.g. for $a = b$ one obtains $\rho = 3/\sqrt{13}$.

(b) $m = \frac{1}{2}p^2/E,\ v = E/2p,\ C_y = \begin{pmatrix} (a^2 + b^2)p^2 & (a^2 + 2b^2)Ep \\ (a^2 + 2b^2)Ep & (a^2 + 4b^2)E^2 \end{pmatrix}$,

$$m = Ty,\ T = \left(\frac{\partial m}{\partial y_1}, \frac{\partial m}{\partial y_2}\right) = \left(\frac{\partial m}{\partial p}, \frac{\partial m}{\partial E}\right) = \left(\frac{p}{E}, -\frac{p^2}{2E^2}\right),$$

$$C_m = \sigma^2(m) = TC_yT^{\mathrm{T}} = \left(\frac{p}{E}, -\frac{p^2}{2E^2}\right)\begin{pmatrix} (a^2 + b^2)p^2 & (a^2 + 2b^2)Ep \\ (a^2 + 2b^2)Ep & (a^2 + 4b^2)E^2 \end{pmatrix}\begin{pmatrix} \frac{p}{E} \\ -\frac{p^2}{2E^2} \end{pmatrix}$$

$$= a^2\frac{p^4}{4E^2} = a^2m^2, \quad \text{q.e.d.}$$

Exercise 5-1

(a) $W_{k+1}^n = \binom{n}{k+1}p^{k+1}q^{n-k-1} = \frac{n!}{(k+1)!(n-k-1)!}p^kq^{n-k}\frac{p}{q}$

$$= \frac{n!}{k!(n-k)!}\frac{n-k}{k+1}p^kq^{n-k}\frac{p}{q} = W_k^n\frac{n-k}{k+1}\frac{p}{q}.$$

(b) $W_3^5 = \binom{5}{3}0.8^3 \cdot 0.2^2 = 10 \cdot 0.513 \cdot 0.04 = 0.2052,$

$$W_4^5 = W_3^3 \cdot \frac{2}{4} \cdot \frac{0.8}{0.2} = 0.2052 \cdot 2 = 0.4104,$$

$$W_5^5 = W_4^5 \cdot \frac{1}{5} \cdot \frac{0.8}{0.2} = 0.4014 \cdot 0.8 = 0.3211,$$

$$P_3 = 1 - W_4^5 - W_5^5 = 0.2685,$$
$$P_2 = P_3 - W_3^5 = 0.0633.$$

(c) Using the result of exercise 5.1(a) we have

$$W_k^n - W_{k-1}^n = W_{k-1}^n\left(\frac{n-k+1}{k}\frac{p}{q} - 1\right),$$

i.e. the probability W_k^n increases with k as long as the expression in brackets is positive, i.e.

$$\frac{(n-k+1)p}{kq} - 1 > 0.$$

Since k and q are positive, we have

$$(n-k+1)p > kq = k(1-p), \qquad k < (n+1)p.$$

The mode k_m is the largest value of k for which this inequality holds.

(d) $\varphi_{x_i}(t) = E\{e^{itx_i}\} = qe^{it\cdot 0} + pe^{it} = q + pe^{it}$;

$$\varphi_x = (q + pe^{it})^n = \sum_{k=0}^{m} \binom{n}{k} q^{n-k}p^k e^{itk} = \sum_{k=0}^{m} f(k)e^{itk}$$

$$= E\{e^{itk}\}, \text{ i.e. } f(k) = W_k^n.$$

Exercise 5-2

(a) From table F-1 we find $f(0) \approx 0.05$ for $\lambda = 3$.

(b) The probability of finding $x \leqslant x < x + \Delta x$ in a single experiment is $p = f(x)\Delta x$, the probability of finding it elsewhere is $q = 1 - p$. Therefore in n experiments the probability of finding k times x in this interval is

$$W_k^n = \binom{n}{k} p^k q^{n-k}.$$

Since for $n \to \infty$ and $np = \lambda = \text{const.}$

$$W_k^n \to f(k) = \frac{\lambda^k}{k!} e^{-\lambda},$$

the probability becomes Poisson distributed with $\lambda = nf(x)\Delta x$.

(c) According to (b) we have

$$\lambda = nf(x)\Delta x = \frac{n\Delta x}{\tau} e^{-x/\tau}.$$

Exercise 5-3

(a) With $g(y) = \dfrac{1}{b-a} = \text{const.}$, eq. (5.13) yields

$$x = G(y) = \frac{1}{b-a}\int_a^y dt = \frac{y-a}{b-a}, \text{ i.e. } y = a + x(b-a).$$

(b) $$x = G(y) = \frac{1}{\pi}\int_{-\infty}^{y} \frac{1}{1+t^2}\,dt = \frac{1}{\pi}[\arctan]_{-\infty}^{y} = \frac{1}{\pi}[\arctan y + \tfrac{1}{2}\pi],$$

$$y = \tan(\pi(x - \tfrac{1}{2})).$$

Exercise 5-4

(a) According to (9.5) the fraction

$$f_r = 2\psi_0\left[\frac{(R_0 - a_2R_0) - R_0}{bR_0}\right] = 2\psi_0\left(-\frac{a_2}{b}\right)$$

is rejected because $R < R_0 - \Delta_2$ or $R > R_0 + \Delta_2$. Similarly the fractions

$$f_2 = 2\psi_0\left(-\frac{a_1}{b}\right) - f_r \quad \text{and} \quad f_5 = 1 - f_2 - f_r$$

yield the prices $2C$ and $5C$ respectively. Therefore

$$P = 2f_2C + 5f_5C - C = C\{2f_2 + 5 - 5f_2 - 5f_r - 1\} = C\{4 - 3f_2 - 5f_r\}$$

$$= C\left\{4 - 6\psi_0\left(-\frac{a_1}{b}\right) + 3f_r - 5f_r\right\} = C\left\{4 - 6\psi_0\left(-\frac{a_1}{b}\right) - 4\psi_0\left(-\frac{a_2}{b}\right)\right\}.$$

(b) $\psi_0(-0.2) = 0.421;\ \psi_0(-1) = 0.159,$

i.e. $P = C\{4 - 2.526 - 0.636\} = 0.838C.$

(c) $$\frac{d^2 f}{dx^2} = \frac{1}{\sqrt{(2\pi)b^3}} \exp\left(-(x-a)^2/2b^2\right)\{(x-a)^2/b^2 - 1\} = 0$$

is fulfilled if the last bracket vanishes.

Exercise 5-5

(a) $$P(R) = \int_0^R g(r)\,dr = \frac{1}{\sigma^2}\int_0^R r e^{-r^2/2\sigma^2}\,dr = \frac{1}{2\sigma^2}\int_0^{R^2} e^{-u/2\sigma^2}\,du = [e^{-u/2\sigma^2}]_{R^2}^{0} = 1 - e^{-R^2/2\sigma^2}.$$

(b) $1 - P = \exp(-R^2/2\sigma^2),\ R^2/2\sigma^2 = -\ln(1-P).$
For $\sigma = 1,\ P = 0.9$ one obtains

$$R = \sqrt{(-2\ln 0.1)} = \sqrt{4.61} = 2.15.$$

Exercise 5-6

(a) The numbers x_i generated by RAND have mean $a = 0.5$ and variance $\sigma^2 = \frac{1}{12}$. The arithmetic mean of N such numbers then follows (in the limit $N \to \infty$) a Gaussian with mean 0.5 and variance $1/12N$. Using the sum $x = \sum_{i=1}^{N} x_i$ rather than the arithmetic mean one obtains normally distributed numbers with mean $0.5N$ and variance $N/12$ (cf. eq. (3-3.14)). Therefore for $N = 12$ one obtains unit variance. The program could have the form

```
   FUNCTION GAUSS (XX)
   A=0
   DO 10 J=1,12
10 A=A+RAND (YY)
   GAUSS=A—6.0
   RETURN
   END
```

(b) Whereas in the above program N random numbers are always picked, in program 5-2 the number of pairs y_r, g_r is equal to the ratio between the surface of the two-dimensional area, from which the pairs y_r, g_r can be picked, and the area under the probability density curve, the latter being 1, i.e.

$$N' = 2 \cdot 2d g_{max} = 4d/\sqrt{(2\pi)}.$$

For $d = 6$ (which would correspond to the program in the above exercise insofar as random numbers with absolute values up to 6 would be possible) one has $N' = 9.6$. Counting only the number of random numbers to be picked the methods are of comparative efficiency. However, since in program 5-2 the exponential has to be computed it is in fact much slower.

(c) For λ integer the Poisson distribution of the variable k can be regarded as the distribution of a sum of $n = \lambda$ independently Poisson distributed variables $x_1, x_2, ..., x_n$ stemming from a Poisson distribution with $\lambda_1 = 1$. According to the central limit theorem $\bar{x} = k/n = (x_1 + x_2 + ... + x_n)/n$ follows a normal distribution with mean $a_1 = \lambda_1 = 1$ and variance $\sigma_{\bar{x}}^2 = \lambda_1/n = 1/n$. The variable $k = n\bar{x}$ then follows a normal distribution with mean $a_k = n = \lambda$ and $\sigma_k^2 = n^2/n = \lambda$.

Exercise 5-7

(a) $0 \leqslant u < 1$:

$$f(u) = \int_{u-1}^{u} f_1(y)\,dy = \int_{0}^{u} y\,dy = \tfrac{1}{2}u^2.$$

$1 \leqslant u < 2$:

$$f(u) = \int_{u-1}^{u} f_1(y)\,dy + \int_{u-1}^{u} f_2(y)\,dy = \int_{u-1}^{1} y\,dy + \int_{1}^{u} (2-y)\,dy$$

$$= \tfrac{1}{2}(1-(u-1)^2) + \tfrac{1}{2}(1-(2-u)^2) = \tfrac{1}{2}(-3 + 6u - 2u^2).$$

$2 \leqslant u < 3$:

$$f(u) = \int_{u-1}^{2} f_2(y)\,dy = \int_{u-1}^{2} (2-y)\,dy = -\int_{3-u}^{0} z\,dz = \tfrac{1}{2}(3-u)^2.$$

(b) $$f(u) = \frac{1}{2\pi\sigma_x\sigma_y} \int_{-\infty}^{\infty} \exp\left(-\frac{x^2}{2\sigma_x^2} - \frac{(u-x)^2}{2\sigma_y^2}\right) dx.$$

Using quadratic extension the exponent can be written in the form

$$-\frac{\sigma^2}{2\sigma_x^2\sigma_y^2}\left\{\left(x - \frac{\sigma_x^2}{\sigma^2}u\right)^2 - \frac{\sigma_x^4}{\sigma^4}u^2 + \frac{\sigma_x^2}{\sigma^2}u^2\right\}.$$

With $v = (\sigma/\sigma_x\sigma_y)(x - \sigma_x^2 u/\sigma^2)$ as integration variable one obtains

$$f(u) = \frac{1}{2\pi\sigma_x\sigma_y} \exp\left(\frac{\sigma_x^4 - \sigma^2\sigma_x^2}{2\sigma^2\sigma_x^2\sigma_y^2}u^2\right) \frac{\sigma_x\sigma_y}{\sigma} \int_{-\infty}^{\infty} \exp(-\tfrac{1}{2}v^2)\,dv$$

$$= \frac{1}{\sqrt{2\pi}\,\sigma} \exp\left(-\frac{u^2}{2\sigma^2}\right),$$

since $\sigma_x^4 = \sigma_x^2(\sigma^2 - \sigma_y^2)$.

Exercise 5-8

(a) $f(a) = 2/\pi\Gamma, \quad \dfrac{1}{\pi\Gamma} = \dfrac{2}{\pi\Gamma}\,\dfrac{\Gamma^2}{4(x_{1,2} - a)^2 + \Gamma^2},$

$$4(x_{1,2} - a)^2 = \Gamma^2, \quad x_{1,2} = a \pm \tfrac{1}{2}\Gamma.$$

(b) $f(x) = \dfrac{1}{\sqrt{(2\pi)}\sigma} \exp(-(x-a)^2/2\sigma^2), \quad f(a) = \dfrac{1}{\sqrt{(2\pi)}\sigma},$

$$\frac{1}{2\sqrt{(2\pi)}\sigma} = \frac{1}{\sqrt{(2\pi)}\sigma} \exp(-(x_{1,2} - a)^2/2\sigma^2),$$

$$\ln 0.5 = -(x_{1,2} - a)^2/2\sigma^2,$$
$$(x_{1,2} - a)^2 = \sigma^2(-2 \ln 0.5),$$
$$x_{1,2} = a \pm \sigma\sqrt{(-2 \ln 0.5)},$$
$$\text{FWHM} = |x_2 - x_1| = \sigma \cdot (2\sqrt{[-2 \ln 0.5]}) = 1.177\sigma.$$

Exercise 6-1

(a) $E(S) = E\{\sum a_i x_i\} = \sum a_i E(x_i) = \hat{x} \sum a_i = \hat{x}$ if and only if $\sum a_i = 1$.

(b) $\sigma^2 = \sum \left(\dfrac{\partial S}{\partial x_i}\right)^2 \sigma^2 = \sigma^2 \sum a_i^2,$

$$\sigma^2(S_1) = \sigma^2\left(\frac{1}{16} + \frac{1}{16} + \frac{1}{4}\right) = \sigma^2 \cdot \frac{3}{8} = 0.375\sigma^2,$$

$$\sigma^2(S_2) = \sigma^2\left(\frac{1}{25} + \frac{4}{25} + \frac{4}{25}\right) = \sigma^2 \cdot \frac{9}{25} = 0.360\sigma^2,$$

$$\sigma^2(S_3) = \sigma^2\left(\frac{1}{36} + \frac{1}{9} + \frac{1}{4}\right) = \sigma^2 \cdot \frac{7}{18} = 0.389\sigma^2.$$

(c) $\sigma^2(S) = [a_1^2 + a_2^2 + (1 - (a_1 + a_2))^2]\sigma^2,$

$$\frac{\partial\sigma^2(S)}{\partial a_1} = [2a_1 - 2(1 - (a_1 + a_2))]\sigma^2 = 0,$$

$$\frac{\partial\sigma^2(S)}{\partial a_2} = [2a_2 - 2(1 - (a_1 + a_2))]\sigma^2 = 0.$$

$$a_1 = 1 - (a_1 + a_2),\ a_2 = 1 - (a_1 + a_2),\ a_1 = a_2 = \tfrac{1}{3};$$
$$\sigma^2(S) = \tfrac{3}{9}\sigma^2 = 0.333\sigma^2.$$

Exercise 6-2

$$a = 20,\ \Delta = \frac{1}{n}\sum \delta_i = \frac{1}{10}(-2 + 1 + 3 - 1 + 0 + 1 + 0 - 1 + 0 - 3) = -0.2.$$

$\bar{x} = 19.8.$

$$s^2 = \tfrac{1}{9}(1.8^2 + 1.2^2 + 3.2^2 + 0.8^2 + 0.2^2 + 1.2^2 + 0.2^2 + 0.8^2 + 0.2^2 + 2.8^2)$$
$$= \tfrac{1}{9} \cdot 25.60 = 2.84.$$

$s_{\bar{x}}^2 = 0.284.$ Therefore $\bar{x} = 19.8 \pm 0.53.$

Exercise 6-3

(a) $\bar{x}_1 = 0.2$, $\bar{x}_2 = 0.5$, $\bar{x}_3 = 0.8$, $\tilde{x} = 0.02 + 0.35 + 0.16 = 0.53$,

$$s_1^2 = \tfrac{1}{9}(2 - 10 \cdot 0.2^2) = 0.178, \; s_2^2 = \tfrac{1}{9}(5 - 10 \cdot 0.5^2) = 0.278,$$

$$s_3^2 = \tfrac{1}{9}(8 - 10 \cdot 0.8^2) = 0.178,$$

$$s_{\tilde{x}}^2 = \frac{0.1^2}{10} 0.178 + \frac{0.7^2}{10} 0.278 + \frac{0.2^2}{10} 0.178 = 0.005 \cdot 0.178 + 0.049 \cdot 0.278 = 0.0145,$$

$$s_{\tilde{x}} = +\sqrt{(s_{\tilde{x}}^2)} = 0.12.$$

Therefore $\tilde{x} = 0.53 \pm 0.12$ is not significantly in favour of one party.

(b) $s_1 = 0.422$, $\quad s_2 = 0.527$, $\quad s_3 = 0.422$,

$$p_1 s_1 = 0.0422, \quad p_2 s_2 = 0.369, \quad p_3 s_3 = 0.0844,$$

$$\sum p_i s_i = 0.496,$$

$$\frac{n_1}{n} = 0.085, \quad \frac{n_2}{n} = 0.744, \quad \frac{n_3}{n} = 0.170.$$

Exercise 6-4

(a) For convenience we use $\chi^2 = u$.

$$\begin{aligned} \mu_3 &= E\{(u - \hat{u})^3\} = E\{u^3 - 3u^2\hat{u} + 3u\hat{u}^2 - \hat{u}^3\} \\ &= E(u^3) - 3\hat{u}E(u^2) + 3\hat{u}^2 E(u) - \hat{u}^3 = \lambda_3 - 3\hat{u}E(u^2) + 2\hat{u}^3, \end{aligned}$$

$$E(u^3) = \lambda_3 = \frac{1}{i^3}\varphi'''(0) = i\varphi'''(0) = i(-\lambda)(-\lambda - 1)(-\lambda - 2)(-2i)^3$$

$$= 8\lambda(\lambda + 1)(\lambda + 2) = 8\lambda^3 + 24\lambda^2 + 16\lambda,$$

$$3\hat{u}E(u^2) = 6\lambda(4\lambda^2 + 4\lambda) = 24\lambda^3 + 24\lambda^2,$$

$$2\hat{u}^3 = 2 \cdot (2\lambda)^3 = 16\lambda^3,$$

$$\mu_3 = 16\lambda,$$

$$\gamma = \mu_3/\sigma^3 = 16\lambda/8\lambda^{\frac{3}{2}} = 2\lambda^{-\frac{1}{2}}.$$

(b) Obviously $\gamma = 2/\sqrt{\lambda} \to 0$ for $\lambda = \frac{1}{2}u \to \infty$.

Exercise 6-5

$$\Delta k_+ = 4.30, \; \Delta k_- = 1.65, \; \Delta k = 2.33.$$

$$\Delta k_+/\Delta k = 1.85, \; \Delta k_-/\Delta k = 0.71, \; (\Delta k_+ + \Delta k_-)/2\Delta k = 1.28.$$

Exercise 6-6

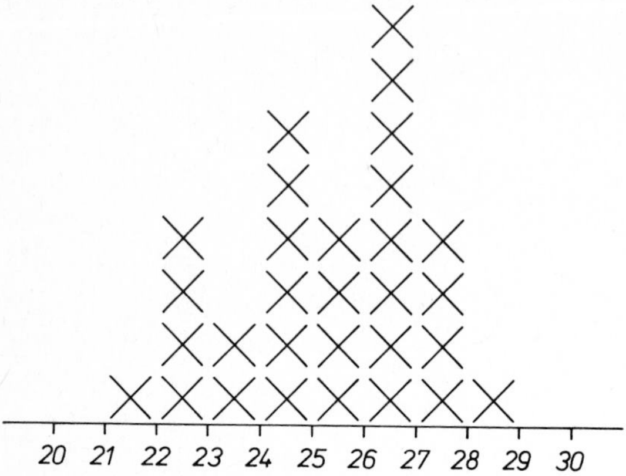

Histogram from data of exercise 6.6

Exercise 7-1

(a) $L = \prod_{j=1}^{N} \frac{1}{b} = \frac{1}{(b)^N}$; obviously $L = \max$ for $b = x_{\max}$.

(b) $\lambda_1 = a$, $\lambda_2 = \Gamma$,

$$L = \prod_{j=1}^{N} \frac{2}{\pi\lambda_2} \frac{\lambda_2^2}{4(x^{(j)} - \lambda_1)^2 + \lambda_2^2} = \left(\frac{2\lambda_2}{\pi}\right)^N \prod_{j=1}^{N} \frac{1}{4(x^{(j)} - \lambda_1)^2 + \lambda_2^2},$$

$$l = N(\ln 2 - \ln \pi + \ln \lambda_2) - \sum_{j=1}^{N} \ln [4(x^{(j)} - \lambda_1)^2 + \lambda_2^2],$$

$$\frac{\partial l}{\partial \lambda_1} = \sum_{j=1}^{N} \frac{8(x^{(j)} - \lambda_1)}{4(x^{(j)} - \lambda_1)^2 + \lambda_2^2} = 0,$$

$$\frac{\partial l}{\partial \lambda_2} = \frac{N}{\lambda_2} - 2\lambda_2 \sum_{j=1} \frac{1}{4(x^{(j)} - \lambda_1)^2 + \lambda_2^2} = 0.$$

There is no unique solution since the equations are nonlinear in λ_1 and λ_2. However for $|x^{(j)} - \lambda_1| \ll \lambda_2$ we have

$$\frac{8}{\lambda_2^2} \sum (x^{(j)} - \lambda_1) = 0, \quad N\lambda_1 = \sum x^{(j)}, \quad \lambda_1 = a = \bar{x}.$$

Exercise 7-2

(a) $L = \prod_{j=1}^{N} \frac{1}{\sqrt{2\pi}\sigma} \exp [-(x^{(j)} - \lambda)^2/2\sigma^2]$

$$= (\sqrt{2\pi}\sigma)^{-N} \prod \exp [-(x^{(j)} - \lambda)^2/2\sigma^2],$$

$$l = -N \ln (\sqrt{2\pi}\sigma) - \frac{1}{2\sigma^2} \sum (x^{(j)} - \lambda)^2,$$

$$l' = \frac{1}{\sigma^2} \sum (x^{(j)} - \lambda),$$

$$l'' = -N/\sigma^2,$$

$$I(\lambda) = -E(l'') = N/\sigma^2.$$

(b) $$L = \prod_{j=1}^{N} \frac{1}{\sqrt{(2\pi)}\sqrt{\lambda}} \exp\left(-\frac{(x^{(j)} - a)^2}{2\lambda}\right)$$

$$= (2\pi)^{-N/2}\lambda^{-N/2} \prod_{j=1}^{N} \exp\left(-\frac{(x^{(j)} - a)^2}{2\lambda}\right),$$

$$l = -\frac{N}{2}\ln(2\pi) - \frac{N}{2}\ln\lambda - \frac{1}{2\lambda}\sum_{j=1}^{N}(x^{(j)} - a)^2,$$

$$l' = -\frac{N}{2\lambda} + \frac{1}{2\lambda^2}\sum (x^{(j)} - a)^2,$$

$$l'' = \frac{N}{2\lambda^2} - \frac{1}{\lambda^3}\sum (x^{(j)} - a)^2,$$

$$I(\lambda) = -E(l'') = -\frac{N}{2\lambda^2} + \frac{1}{\lambda^3}E\{\sum (x^{(j)} - a)^2\} = -\frac{N}{2\lambda^2} + \frac{1}{\lambda^3}N\lambda,$$

$$I(\lambda) = \frac{1}{2\lambda^2}(-N + 2N) = \frac{N}{2\lambda^2},$$

$$E(S) = \frac{1}{N}E\{\sum (x^{(j)} - a)^2\} = \sigma^2 = \lambda, \text{ since } \sigma^2 = E\{(x^{(j)} - a)^2\} \text{ by definition.}$$

Exercise 7-3

(a) $\sigma^2(S) \geqslant \dfrac{1}{I(\lambda)} = \dfrac{\sigma^2}{N}$.

In exercise 6-1 (c) for $N = 3$ we had $\sigma^2(\bar{x}) = \frac{1}{3}\sigma^2$.

(b) $\sigma^2(S) \geqslant \dfrac{1}{I(\lambda)} = \dfrac{2\lambda^2}{N}$.

(c) From exercise 7-2 we have

$$l' = -\frac{N}{2\lambda} + \frac{1}{2\lambda^2}NS = \frac{N}{2\lambda^2}(S - \lambda) = \frac{N}{2\lambda^2}(S - E(S)).$$

Therefore $\sigma^2(S) = \dfrac{2\lambda^2}{N}$.

Exercise 8-1

$\bar{x}_1 = 21.11$, $\bar{x}_2 = 21.00$,

$s_1^2 = 14.91$, $s_2^2 = 10.00$,

$T = s_1^2/s_2^2 = 1.491$, $F_{0.95}(8, 6) = 4.15$.

The variance of sample (2) is not significantly smaller.

Exercise 8-2

$\bar{x} = 25.142, \quad s^2 = 82.69/29 = 2.85, \quad s = 1.69,$

$$T = \frac{\bar{x} - 25.5}{s/\sqrt{30}} = \frac{25.142 - 25.5}{1.69/5.48} = -\frac{0.358}{0.289} = -1.24,$$

$|T| < t_{0.95}(f = 29) = 1.70,$

therefore the hypothesis cannot be rejected.

Exercise 8-3

(a) $$f(x^{(1)}, x^{(2)}, \ldots, x^{(N)}, \tilde{\lambda}^{(\Omega)}) = \prod_{j=1}^{N} \frac{1}{\sqrt{(2\pi)}\sigma_1} \exp - \frac{(x^{(j)} - \bar{x})^2}{2\sigma_1^2}$$

$$= (\sqrt{(2\pi)}\sigma_1)^{-N} \exp\left[-\frac{1}{2\sigma_1^2}\sum (x^{(j)} - \bar{x})^2\right],$$

$$f(x^{(2)}, x^{(3)}, \ldots, x^{(N)}, \tilde{\lambda}^{(\omega)}) = (\sqrt{(2\pi)}\sigma_0)^{-N} \exp\left[-\frac{1}{2\sigma_0^2}\sum (x^{(j)} - \bar{x})^2\right],$$

$$T = \left(\frac{\sigma_0}{s}\right)^N \exp\left[\tfrac{1}{2}(N - 1)s^2\left(\frac{1}{\sigma_0^2} - \frac{1}{s^2}\right)\right],$$

$$\ln T = N(\ln \sigma_0 - \ln s) + \tfrac{1}{2}(N - 1)s^2\left(\frac{1}{\sigma_0^2} - \frac{1}{s^2}\right)$$

$$= N(\ln \sigma_0 - \ln s) + \tfrac{1}{2}(N - 1)\frac{s^2}{\sigma_0^2} - \tfrac{1}{2}(N - 1).$$

(b) $T' = (N - 1)s^2/\sigma_0^2$ is a monotonic function in T if $\sigma_1 > \sigma_2$; otherwise it is negative monotonic.

Therefore the test is

$T' > T'_{1-\frac{1}{2}\alpha}, \; T' < T'_{\frac{1}{2}\alpha}$ if $H_0(\sigma = \sigma_0)$,

$T' > T'_{1-\alpha}$ if $H_0(\sigma < \sigma_0)$,

$T' < T'_{\alpha}$ if $H_0(\sigma > \sigma_0)$.

(c) follows from eq. (6-6.2).

Exercise 8-4

(a) $$\tilde{a} = \frac{1}{n}\sum_{i=1}^{t} n_i x_i = 202.36.$$

$$\tilde{\sigma}^2 = \frac{1}{n - 1}\sum_{i=1}^{t} n_i(x_i - \tilde{a})^2 = 13.40, \quad \tilde{\sigma} = 3.66.$$

(b) $p_k(x_k) = \psi_0\left(\frac{x_k + \frac{1}{2}\Delta x - \tilde{a}}{\tilde{\sigma}}\right) - \psi_0\left(\frac{x_k - \frac{1}{2}\Delta x - \tilde{a}}{\tilde{\sigma}}\right),$

x_k	n_k	$\psi_0\left(\frac{x_k + \frac{1}{2}\Delta x - \tilde{a}}{\tilde{\sigma}}\right)$	$\psi_0\left(\frac{x_k - \frac{1}{2}\Delta x - \tilde{a}}{\tilde{\sigma}}\right)$	np_k	$\frac{(n_k - np_k)^2}{np_k}$
193	1	0.011	0.002	(0.9)	–
195	2	0.041	0.011	(3.0)	–
197	9	0.117	0.041	7.6	0.258
199	12	0.261	0.117	14.4	0.400
201	23	0.460	0.261	19.9	0.483
203	25	0.674	0.460	21.4	0.606
205	11	0.813	0.674	13.9	0.605
207	9	0.938	0.813	12.5	0.980
209	6	0.982	0.938	4.4	0.582
211	2	0.996	0.982	(1.4)	–
					$X^2 = 3.914$

There are $7-2 = 5$ degrees of freedom. Since $\chi^2_{0.90}(5) = 9.24$ the test does not fail.

Exercise 8-5

(a) $\tilde{p}_1 = \frac{1}{111}(13 + 44) = \frac{57}{111}, \quad \tilde{p}_2 = \frac{1}{111}(25 + 29) = \frac{54}{111}.$

$$\tilde{q}_1 = \frac{1}{111}(13 + 25) = \frac{38}{111}, \quad \tilde{q}_2 = \frac{1}{111}(44 + 29) = \frac{73}{111},$$

$$X^2 = \frac{\left(13 - \frac{57 \cdot 38}{111}\right)^2}{\frac{57 \cdot 38}{111}} + \frac{\left(44 - \frac{57 \cdot 73}{111}\right)^2}{\frac{57 \cdot 73}{111}} + \frac{\left(25 - \frac{54 \cdot 38}{111}\right)^2}{\frac{54 \cdot 38}{111}} + \frac{\left(29 - \frac{54 \cdot 73}{111}\right)^2}{\frac{54 \cdot 73}{111}}$$

$$= \frac{42.43}{19.51} + \frac{42.43}{37.49} + \frac{42.43}{18.49} + \frac{42.43}{35.51} = 6.78.$$

Since $\chi^2_{0.90} = 4.61$ for $f = 2$ the hypothesis of independence must be rejected.

(b) $n_{ij} - \frac{1}{n}(n_{i1} + n_{i2})(n_{1j} + n_{2j})$

$$= \frac{1}{n}[n_{ij}(n_{11} + n_{12} + n_{21} + n_{22}) - (n_{i1} + n_{i2})(n_{1j} + n_{2j})].$$

It is easy to show that the expression in brackets is $n_{11}n_{22} - n_{12}n_{21}$ for every i, j.

Exercise 9-1

y_j	σ_j	$1/\sigma_j^2 = g_j$	$y_j g_j$	$y_j - \tilde{x}$	$(y_j - \tilde{x})^2 g_j$
10.23	0.3	11.11	113.66	0.22	0.54
11.14	0.4	6.25	69.63	1.13	7.98
9.82	0.2	25.00	245.50	−0.19	0.90
9.71	0.2	25.00	242.75	−0.30	2.25
10.78	0.5	4.00	43.12	0.77	2.37
		$\sum g_j = 71.36$	$\sum y_i g_j = 714.66$		$M = 14.04$
		$\tilde{x} = \sum y_i g_j / \sum g_j$	$= 10.01$		
		$\Delta\tilde{x} = (\sum g_j)^{-\frac{1}{2}}$	$= 0.12$		

The χ^2-test fails since $M = 14.04$ and $\chi^2_{0.95} = 9.49$ for 4 degrees of freedom.

$\Delta\tilde{x}' = 0.12\sqrt{(14.04/4)} = 0.22.$

Exercise 9-2

(a) The straight line through the origin is defined by

$$\eta_j = x t_j,$$

$$\text{i.e. } \boldsymbol{\eta} + Ax = 0, \quad A = -\begin{pmatrix}0\\1\\2\\3\end{pmatrix}.$$

As in example 9-2 we have $G_y = 4I$, therefore, with eq. (2.20)

$$\tilde{x} = -(A^{\mathrm{T}}A)^{-1}A^{\mathrm{T}}y = \left[(0\;1\;2\;3)\begin{pmatrix}0\\1\\2\\3\end{pmatrix}\right]^{-1}(0\;1\;2\;3)\begin{pmatrix}1.4\\1.5\\3.7\\4.1\end{pmatrix} = \frac{1}{14}\cdot 21.2 = 1.514,$$

$$G_{\tilde{x}}^{-1} = (\Delta\tilde{x})^2 = (A^{\mathrm{T}}G_y A)^{-1} = \frac{1}{4}\cdot\frac{1}{14} = 0.0179,$$

$$\Delta\tilde{x} = 0.134.$$

(b) In example 9-2 we had $\tilde{x}_2 = 1.03$, $(\Delta\tilde{x}_2)^2 = 0.175$, $\Delta\tilde{x}_2 = 0.418$.

(c) $$M = (y + A\tilde{x})^{\mathrm{T}}G_y(y + A\tilde{x}) = \left[\begin{pmatrix}1.4\\1.5\\3.7\\4.1\end{pmatrix} - 1.514\begin{pmatrix}0\\1\\2\\3\end{pmatrix}\right]^{\mathrm{T}}\cdot 4I\left[\begin{pmatrix}1.4\\1.5\\3.7\\4.1\end{pmatrix} - 1.514\begin{pmatrix}0\\1\\2\\3\end{pmatrix}\right]$$

$$= 4(1.4,\; -0.014,\; 0.628,\; -0.442)\begin{pmatrix}1.4\\-0.014\\0.628\\-0.442\end{pmatrix}.$$

$M = 10.20$, $\chi^2_{0.90}(f = 3) = 6.25$: test fails.
For example 9-2 one obtains

$$M = \left[\begin{pmatrix}1.4\\1.5\\3.7\\4.1\end{pmatrix} - \begin{pmatrix}1 & 0\\1 & 1\\1 & 2\\1 & 3\end{pmatrix}\begin{pmatrix}1.13\\1.03\end{pmatrix}\right]^{\mathrm{T}} \cdot 4I \left[\begin{pmatrix}1.4\\1.5\\3.7\\4.1\end{pmatrix} - \begin{pmatrix}1 & 0\\1 & 1\\1 & 2\\1 & 3\end{pmatrix}\begin{pmatrix}1.13\\1.03\end{pmatrix}\right]$$

$$= \left[\begin{pmatrix}1.4\\1.5\\3.7\\4.1\end{pmatrix} - \begin{pmatrix}1.13\\2.26\\3.19\\4.22\end{pmatrix}\right]^{\mathrm{T}} \cdot 4I \left[\begin{pmatrix}1.4\\1.5\\3.7\\4.1\end{pmatrix} - \begin{pmatrix}1.13\\2.26\\3.19\\4.22\end{pmatrix}\right]$$

$$= 4(0.27,\ -0.76,\ 0.51,\ -0.12)\begin{pmatrix}0.27\\-0.76\\0.51\\-0.12\end{pmatrix},$$

$M = 3.70$, $\chi^2_{0.90}(f = 2) = 4.61$: test does not fail.

(d) A drawing of a line through the origin with slope $\tilde{x} = 1.514$ in fig. 9-2a shows that the line is systematically too steep compared with the two-parameter fit of example 9-2. Therefore M is high and the χ^2-test fails. The error obtained for a parameter in the fit is a measure of how much the parameter can be varied while it still fits the data reasonably well within their errors (compare fig. 9-2d). Since in our one-parameter fit (1) the line is forced to go through the origin and (2) the error bars of the first and last points do not even touch the straight line the allowed variation is very small. We therefore find that the error is small because the fit is poor! The exercise shows that it is important to inspect the minimum function M.

Exercise 9-3

$$\frac{\partial f_j}{\partial x_1} = -t_j^{x_2}, \quad \frac{\partial f_j}{\partial x_2} = -(x_1 \ln t_j)t_j^{x_2},$$

$$A = \begin{pmatrix} -t_1^{x_{20}} & -(x_1 \ln t_1)t_1^{x_{20}} \\ -t_2^{x_{20}} & -(x_1 \ln t_2)t_2^{x_{20}} \\ \vdots & \vdots \\ -t_5^{x_{20}} & -(x_1 \ln t_5)t_5^{x_{20}} \end{pmatrix} = \begin{pmatrix} -\ 0.250 & 0.087 \\ -\ 4.000 & -\ 1.383 \\ -16.000 & -11.090 \\ -25.000 & -20.118 \\ -49.000 & -47.674 \end{pmatrix},$$

$B = I$, $G_y = I$,

$$\boldsymbol{c} = \boldsymbol{f}(\boldsymbol{x}_0, \boldsymbol{y}) = \begin{pmatrix} -2.125 \\ 1.000 \\ -2.000 \\ -1.500 \\ -2.500 \end{pmatrix},$$

$$\tilde{\xi} = -(A^{\mathrm{T}}G_yA)^{-1}AG_yc = -(A^{\mathrm{T}}A)^{-1}Ac = \begin{pmatrix} -0.133 \\ 0.083 \end{pmatrix},$$

$$\tilde{x}_{11} = x_{10} + \tilde{\xi}_1 = 0.367, \qquad \tilde{x}_{21} = x_{20} + \tilde{\xi}_2 = 2.083.$$

Exercise 9-4

```
      SUBROUTINE INVAL (Y,CY,N)
      DIMENSION Y(1),CY(1),A(5)
      DATA A /—2.0,3.0,6.0,11.0,22.0/
      N=5
      CALL MTXTRA(A,Y,5,1)
      CALL MTXUNT(CY,5)
      RETURN
      END

      SUBROUTINE DERIV(X,Y,A,B,C,NR,N,M,NHYP,NSTEP,MAXSTP)
      DIMENSION X(1),Y(1),A(1),B(1),C(1),T(5)
      DATA T /0.5,2.0,4.0,5.0,7.0/
      IF(NHYP—1)60,10,60
   10 IF(NSTEP—1)30,20,30
C
C     SET UP FIRST APPROXIMATION
C
   20 X(1)=0.5
      X(2)=2.
      NR=2
      M=N
      MAXSTP=5
C
C     CALCULATE A,B AND C
C
   30 CALL MTXUNT(B,N)
      DO 40 I=1,N
      IJ=(I—1)*NR+1
      A(IJ)=(—1.)*T(I)**X(2)
      A(IJ+1)=(—1.)*X(1)*ALOG(T(I))*T(I)**X(2)
   40 C(I)=Y(I)—X(1)*T(I)**X(2)
   50 RETURN
   60 NHYP=—1
      GO TO 50
      END
```

Exercise 9-5

$$\eta_1 + \eta_2 + \eta_3 + \eta_4 - 100 = 0,$$

$$\boldsymbol{B} = (1, 1, 1, 1), \; \boldsymbol{b}_0 = -100,$$

$$c = \boldsymbol{By} + b_0 = 42 + 28 + 8 + 10 - 100 = -7,$$

$$G_y^{-1} = \begin{pmatrix} 9 & 0 & 0 & 0 \\ 0 & 16 & 0 & 0 \\ 0 & 0 & 25 & 0 \\ 0 & 0 & 0 & 9 \end{pmatrix},$$

$$G_B = (BG_y^{-1}B^T)^{-1} = \left[(1, 1, 1, 1) \begin{pmatrix} 9 \\ 16 \\ 25 \\ 9 \end{pmatrix} \right]^{-1} = 0.017,$$

$$\tilde{\eta} = y - G_y^{-1}B^T G_B c,$$

$$= \begin{pmatrix} 42 \\ 28 \\ 8 \\ 15 \end{pmatrix} - \begin{pmatrix} 9 & 0 & 0 & 0 \\ 0 & 16 & 0 & 0 \\ 0 & 0 & 25 & 0 \\ 0 & 0 & 0 & 9 \end{pmatrix} \begin{pmatrix} 1 \\ 1 \\ 1 \\ 1 \end{pmatrix} 0.017 \cdot (-7)$$

$$= \begin{pmatrix} 42 + 1.071 \\ 28 + 1.904 \\ 8 + 2.975 \\ 15 + 1.071 \end{pmatrix} = \begin{pmatrix} 43.07 \\ 29.90 \\ 10.98 \\ 16.07 \end{pmatrix}.$$

$$G_\eta = G_y^{-1} - G_y^{-1}B^T G_B B G_y^{-1}$$

$$= G_y^{-1} - 0.017 \begin{pmatrix} 9 & 0 & 0 & 0 \\ 0 & 16 & 0 & 0 \\ 0 & 0 & 25 & 0 \\ 0 & 0 & 0 & 9 \end{pmatrix} \begin{pmatrix} 1 & 1 & 1 & 1 \\ 1 & 1 & 1 & 1 \\ 1 & 1 & 1 & 1 \\ 1 & 1 & 1 & 1 \end{pmatrix} \begin{pmatrix} 9 & 0 & 0 & 0 \\ 0 & 16 & 0 & 0 \\ 0 & 0 & 25 & 0 \\ 0 & 0 & 0 & 9 \end{pmatrix}$$

$$= G_y^{-1} - 0.017 \begin{pmatrix} 9 & 0 & 0 & 0 \\ 0 & 16 & 0 & 0 \\ 0 & 0 & 25 & 0 \\ 0 & 0 & 0 & 9 \end{pmatrix} \begin{pmatrix} 9 & 16 & 25 & 9 \\ 9 & 16 & 25 & 9 \\ 9 & 16 & 25 & 9 \\ 9 & 16 & 25 & 9 \end{pmatrix}$$

$$= \begin{pmatrix} 9 & 0 & 0 & 0 \\ 0 & 16 & 0 & 0 \\ 0 & 0 & 25 & 0 \\ 0 & 0 & 0 & 9 \end{pmatrix} - \begin{pmatrix} 1.38 & 2.45 & 3.83 & 1.38 \\ 2.45 & 4.35 & 6.80 & 2.45 \\ 3.83 & 6.80 & 10.63 & 3.83 \\ 1.38 & 2.45 & 3.83 & 1.38 \end{pmatrix} = \begin{pmatrix} 7.62 & -2.45 & -3.83 & -1.38 \\ -2.45 & 11.65 & -6.80 & -2.45 \\ -3.83 & -6.80 & 14.37 & -3.83 \\ -1.38 & -2.45 & -3.83 & 7.62 \end{pmatrix}.$$

Therefore $\eta_1 = (43.07 \pm 2.76)\%$,
$\eta_2 = (29.90 \pm 3.41)\%$,
$\eta_3 = (10.98 \pm 3.79)\%$,
$\eta_4 = (16.07 \pm 2.76)\%$.

Exercise 9-6

(a) With the constraint equation

$$\eta_1 + \eta_2 + \dots + \eta_n - b_0 = 0$$

we have

$$B = (1, 1, \dots, 1)$$

and

$$c = \sum_{i=1}^{n} y_i - b_0 = 0.$$

The covariance matrix of the measurements is

$$G_y^{-1} = \begin{pmatrix} \sigma_1^2 & 0 & \dots & 0 \\ 0 & \sigma_2^2 & & 0 \\ \vdots & & & \\ 0 & & & \sigma_n^2 \end{pmatrix}$$

Therefore

$$G_B = (BG_y^{-1}B^{\mathrm{T}})^{-1} = (\sum_{i=1}^{n} \sigma_i^2)^{-1}$$

and

$$\tilde{\boldsymbol{\delta}} = \tilde{\boldsymbol{\eta}} - \boldsymbol{y} = -G_y^{-1}B^{\mathrm{T}}G_B c = -\begin{pmatrix} \sigma_1^2 \\ \sigma_2^2 \\ \vdots \\ \sigma_n^2 \end{pmatrix} (\sum_{i=1}^{n} \sigma_i^2)^{-1} (\sum_{i=1}^{n} y_i - b_0).$$

(b) The covariance matrix of $\tilde{\eta}$ becomes

$$G_{\tilde{\eta}}^{-1} = G_y^{-1} - G_y^{-1}B^{\mathrm{T}}G_B BG_y^{-1}$$

$$= G_y^{-1} - (\sum_{i=1}^{n} \sigma_i^2)^{-1} \begin{pmatrix} \sigma_1^2 & 0 & \dots & 0 \\ 0 & \sigma_2^2 & \dots & 0 \\ \vdots & & & \\ 0 & & & \sigma_n^2 \end{pmatrix} \begin{pmatrix} 1 & 1 & \dots & 1 \\ 1 & 1 & \dots & 1 \\ \vdots & & & \\ 1 & 1 & \dots & 1 \end{pmatrix} \begin{pmatrix} \sigma_1^2 & 0 & \dots & 0 \\ 0 & \sigma_2^2 & \dots & 0 \\ \vdots & & & \\ 0 & & & \sigma_n^2 \end{pmatrix}$$

$$= \begin{pmatrix} \sigma_1^2 & 0 & \dots & 0 \\ 0 & \sigma_2^2 & & \\ \vdots & & & \\ 0 & & & \sigma_n^2 \end{pmatrix} - (\sum \sigma_i^2)^{-1} \begin{pmatrix} \sigma_1^4 & \sigma_1^2\sigma_2^2 & \dots & \sigma_1^2\sigma_n^2 \\ \sigma_2^2\sigma_1^2 & \sigma_2^4 & \dots & \sigma_2^2\sigma_n^2 \\ \vdots & & & \\ \sigma_n^2\sigma_1^2 & \sigma_n^2\sigma_2^2 & \dots & \sigma_n^4 \end{pmatrix},$$

or, in components,

$$(\sigma_{\tilde{\eta}}^2)_k = \sigma_k^2 - \frac{\sigma_k^4}{\sum \sigma_i^2},$$

$$\operatorname{cov}(\tilde{\eta}_k, \tilde{\eta}_l) = -\frac{\sigma_k^2 \sigma_l^2}{\sum \sigma_i^2}.$$

(c) If two measurements are allowed to vary the increase in one measurement leads to the decrease in the other.

Exercise 9-7

(a) The equations are

$$f_1 = x - \tfrac{1}{2}\eta_1 \eta_2^2 = 0,$$

$$f_2 = \eta_3 - \eta_1 \eta_2 = 0.$$

The matrices become

$$A = \begin{pmatrix} 1 \\ 0 \end{pmatrix}, \quad B = \begin{pmatrix} -\frac{1}{2}\eta_1 & -\eta_1\eta_2 & 0 \\ -\eta_2 & -\eta_1 & 1 \end{pmatrix}_{\boldsymbol{\eta} = \boldsymbol{\eta}_0},$$

$$G_B = \begin{pmatrix} \frac{1}{4}\eta_2 \sigma_1^2 + \eta_1^2 \eta_2^2 \sigma_2^2, & \frac{1}{2}\eta_2^2 \sigma_1^2 + \eta_1^2 \eta_2 \sigma_2^2 \\ \frac{1}{2}\eta_2^2 \sigma_1^2 + \eta_1^2 \eta_2 \sigma_2^2, & \eta_2^2 \sigma_1^2 + \eta_1^2 \sigma_2^2 + \sigma_3^2 \end{pmatrix}^{-1},$$

$$c = \begin{pmatrix} x - \frac{1}{2}\eta_1 \eta_2^2 \\ \eta_3 - \eta_1 \eta_2 \end{pmatrix}_{x = x_0, \boldsymbol{\eta} = \boldsymbol{\eta}_0}$$

(b) For $\boldsymbol{\eta}_0 = \mathbf{y}$, $x_0 = \frac{1}{2} y_1 y_2^2$ we obtain

$$B = \begin{pmatrix} -0.5 & -1 & 0 \\ -1 & -1 & 1 \end{pmatrix},$$

$$G_B = \begin{pmatrix} 0.0425 & 0.045 \\ 0.045 & 0.14 \end{pmatrix}^{-1} = \begin{pmatrix} 32.75 & -10.53 \\ -10.53 & 9.94 \end{pmatrix}.$$

Then

$$\tilde{\xi} = -(A^{\mathrm{T}} G_B A)^{-1} A^{\mathrm{T}} G_B c$$

$$= -\left[(1, 0)\begin{pmatrix} 32.75 & -10.53 \\ -10.53 & 9.94 \end{pmatrix}\begin{pmatrix} 1 \\ 0 \end{pmatrix}\right]^{-1} (1, 0)\begin{pmatrix} 32.75 & -10.53 \\ -10.53 & 9.94 \end{pmatrix}\begin{pmatrix} 0 \\ -0.2 \end{pmatrix}$$

$$= -\frac{2.106}{32.75} = -0.064,$$

and finally

$$\tilde{x}_1 = x_0 + \tilde{\xi} = 0.5 - 0.064 = 0.436,$$

$$\sigma_{\tilde{x}_1}^2 = G_{\tilde{x}}^{-1} = (A^{\mathrm{T}} G_B A)^{-1} = \frac{1}{32.75} = 0.0305,$$

$$\Delta \tilde{x}_1 = 0.175.$$

Exercise 11-1

Litter	1	2	3	4	5	Sums and averages
$x_i = \sum_j x_{ij}$	125	244	201	240	236	$\sum_{ij} x_{ij} = 1046$
n_i	5	8	6	8	8	$n = 35$
$\bar{x}_i$	25.00	30.50	33.50	30.00	29.50	$\bar{x} = 29.89$
$\bar{x}_i^2$	625.00	930.25	1122.25	900.00	870.25	$\sum_i n_i \bar{x}_i^2 = 31462.5$

$n\bar{x}^2 = 31260.46, \sum_{ij} x_{ij}^2 = 31848,$

$Q = \sum_{ij} x_{ij}^2 - n\bar{x}^2 = 587.54, \; Q_A = \sum_i n_i \bar{x}_i^2 - n\bar{x}^2 = 202.04,$

$Q_W = \sum_{ij} x_{ij}^2 - \sum_i n_i \bar{x}_i^2 = 385.50.$

Analysis of variance table

Source	SS	ND	MS	F
Between groups	202.04	4	50.51	3.93
Within groups	385.50	30	12.85	
Total	587.54	34		

At 5% significance level: $F_{0.95}(4.30) = 2.69.$
At 1% significance level: $F_{0.99}(4.30) = 4.02.$

Exercise 11-2
(a) The classification is crossed.
(b) $I = 3, \; J = 2, \; K = 3.$
Table of $x_{ij\cdot}$:

j \ i	1	2	3	
1	54.33	36.67	46.33	$\bar{x}_{\cdot 1 \cdot} = 45.78$
2	51.33	40.33	38.67	$\bar{x}_{\cdot 2 \cdot} = 43.44$
	$\bar{x}_{1\cdot\cdot} =$ 52.83	$\bar{x}_{2\cdot\cdot} =$ 38.50	$\bar{x}_{3\cdot\cdot} =$ 42.50	$\bar{x} = 44.61$

Analysis of variance table

	SS	ND	MS	F
Classification A (species)	656.44	2	328.22	57.92
Classification B (male/female)	24.50	1	24.50	4.32
Interaction	97.33	2	48.67	8.59
Within groups	68.00	12	5.67	
Total	846.28	17	49.78	

At 10% significance
- there is a dependence of speed on species since $F_{0.9}(2.12) = 2.81 < 57.92$,
- there is a dependence of speed on sex since $F_{0.9}(1.12) = 3.18 < 4.32$,
- there is an interaction since $F_{0.9}(2.12) = 2.81 < 8.59$.

Exercise 12-1
(a) The diagram (easy to draw and therefore not shown) indicates a linear dependence.
(b) $\bar{t} = 68.2,\ \bar{y} = 2.4783,$
$\sum t_i'^2 = 111.6, \quad \sum t_i' y_i' = -14.9306,$
$\tilde{\beta} = -0.1338, \quad \tilde{\alpha} = 11.6026.$
(c) $\tilde{\eta} = \alpha + \tilde{\beta} t = 11.6026 - 0.1338 \cdot 80 = 0.8986,$

$$\sum \tilde{\varepsilon}_i^2 = \sum (y_i' - \tilde{\beta} t_i')^2 = 0.0041, \ s_y^2 = 0.0041/8 = 0.00052,$$

$$s^2_{\tilde{\alpha}+\tilde{\beta}t} = 0.00052 \left(\frac{1}{10} + \frac{80^2}{111.6} \right) = 0.0299.$$

The limits are

$$\eta_{\pm} = \tilde{\eta} \pm t_{0.95}(8) \cdot s^2_{\tilde{\alpha}+\tilde{\beta}t} = 0.8986 \pm 1.86 \cdot 0.0299 = 0.8986 \pm 0.0556,$$

i.e. $\eta_+ = 0.954, \quad \eta_- = 0.843.$

(d) The linear model would predict negative numbers for very large times.

Exercise 12-2
Solutions can be checked by comparison with table 12-5.

Exercise 12-3
Q_{PRED} is the sum of squares of deviations of the measurements from the predicted regression line. It corresponds to Q in § 4. Q_{RES} specifies the deviation from the regression line. The quotient F will therefore be large if both lines are significantly different.

Exercise 13-1

i	y_i	j	a_j
1	38.7	−4	−0.0909
2	50.3	−3	0.0606
3	45.6	−2	0.1688
4	46.4	−1	0.2338
5	43.7	0	0.2554
6	42.0	1	0.2338
7	21.8	2	0.1688
8	21.8	3	0.0606
9	51.3	4	−0.0909
10	39.5		
11	26.9		
12	23.2		
13	19.8		
14	24.4		

$$\tilde{\eta}_0(i = 5) = \sum_{j=-4}^{4} a_j y_{5+j}$$

$$= -0.0905 \cdot 38.5 + 0.0606 \cdot 50.3 + \ldots$$

$$-0.0905 \cdot 51.3 = 39.39.$$

Similarly for $i = 6, \ldots, 10$.

APPENDIX A

SOME ELEMENTS OF THE FORTRAN PROGRAMMING LANGUAGE

The programming language most commonly employed by the users of large digital computers is the FORTRAN language originally developed by the IBM-Corporation. Programs for a number of applications have been given in this book using FORTRAN. In this section we attempt to give a short introduction to the most important elements of this language. It should be noted that only a very limited part of the language is discussed here, in particular the rather complex procedures for input or output of information are practically ignored. The selection of FORTRAN features discussed here refers to most FORTRAN variants, with the exception of the DATA statement which, in the form used below, refers only to IBM FORTRAN IV: It attempts to provide the reader with sufficient understanding to enable him to follow the examples, rather then to offer a complete course in the language.

A programming language like FORTRAN (FORmula TRANslation) allows the user of a computer to express his problem in terms similar to the formula language of mathematics. A program written in such a language is first translated by the use of a special program (compiler) into a string of machine instructions which are then executed one by one. A FORTRAN program is made up of a series of *statements*, contained in the columns 7 through 72 of a standard 80-column punched card; if a statement is longer it can be continued on further cards if column 6 of the continuation cards contains any symbol, e.g. 1 or *. The cards are represented in our examples by printed lines. Statements can be numbered in the first five columns of the corresponding card for reference in some other part of the program. The *statement numbers* need not be in consecutive order. If the first column contains a C the card is not considered a statement but simply as a *comment* describing the program. There are different classes of statements. The ones actually used for calculations are the arithmetic statements. An example is

```
F=3.14159 * R ** 2.+(A * B—C)/X.
```

It performs the calculation

$$3.14159\,R^2 + \frac{AB - C}{X}$$

and stores the result in a location to which the name F is assigned. Another example is

```
N=2 * J+K * L—M.
```

The right-hand side of an arithmetic statement is an *arithmetic expression.* It is composed of *variables* and *constants* associated by *arithmetic operators.* We introduce here only real and integer variables. *Real variables* can take any fractional value. They are denoted by names composed of up to 6 letters or numbers starting with any letter of the alphabet except I, J, K, L, M, N. Examples are A, X ALPHA, G17Y. *Integer variables* can only take integer values, e.g. -5, 0, 3. They are denoted by names starting with one of the six letters I, J, K, L, M, N. Such names could be M, N12, IXY7A. It is necessary to distinguish between integer and real variable names because most computers use different logic for integer and real arithmetic. Correspondingly there are *integer* and *real constants.* Examples are

```
— 5,0,+3,7,12078
```

and

```
— 5.0,— 2.72,0.0,+5.7,+137.35.
```

There are five *arithmetic operators* +, —, *, /, ** which signify addition, subtraction, multiplication, division and exponentiation, respectively. Corresponding to the conventions of arithmetic the exponentiation operator has the largest binding strength followed by multiplication and division operators whose binding is stronger than the one of addition and subtraction, i.e.

```
F=A*B+C*D**2
```

corresponds to

$$F = AB + C \cdot D^2.$$

The expression

$$F = A((B + C) \cdot D)^2$$

would have to be written

```
F=A*((B+C)*D)**2.
```

To any variable or constant a "*word*", i.e. a location of computer storage, is assigned. The value of a variable (i.e. the contents of the word referred to by the name of the variable) can be changed by placing its name on the left-hand side of an arithmetic statement. A statement such as

```
N=N+1
```

is thus meaningful. It should be interpreted as "increase the contents of the word with the name N by one and place the result into the original word". The equality sign in a FORTRAN statement should therefore be read as "replace by", e.g.

```
A=B
```

signifies: "replace the contents of the word with the name A by the contents of the word with the name B". It should be noted that the contents of B remain unchanged.

Often it is useful to group several variables in an *array*. An element of an array is referred to by an array name which is of the same type as a variable name followed by a *subscript* in brackets, e.g.

```
A(5),   AX1(J),   M3(J+1).
```

The subscripts are integer constants or variables or simple integer expressions. The size of an array, i.e. the storage space to be reserved has to be indicated at the beginning of the program by means of a DIMENSION statement. Such a statement has the form

```
DIMENSION A(50).
```

This indicates that the array A is of length 50. The dimension of several arrays can be given in one dimension statement, e.g.

```
DIMENSION A(50),   AX1(700),   M3(15).
```

Arrays can have several subscripts, e.g. A(5,3). We will however mostly use single subscripts, i.e. one-dimensional arrays.

The DIMENSION statement is only an indication of the size of arrays to organize storage allocation. It is not executed by the machine in the same sense as an arithmetic statement. The *flow* of the program proceeds from the first *executable statement* onwards in the order in which the statements occur, unless it is changed by a *control statement*. The simplest of these is the GO TO statement which appears for example in the form

```
GO TO 25.
```

This signifies that the next statement to be executed is not the one immediately following the GO TO statement but the one with the statement number 25. A very important control statement is the arithmetic IF statement. An example is

```
IF(A—5.0*B)70,55,60.
```

Here the flow of the program is changed depending on the value of the expression in brackets which can be any arithmetic expression. If this value is equal to zero the next statement executed will be the one with the statement number 55. If it is smaller or larger than zero the flow is transferred to statement numbers 70 or 60, respectively.

Before turning to the first example we have to introduce two more control statements. The logically last statement, i.e. the last statement to be executed is

```
STOP.
```

It tells the computer to halt operation or – if the machine is run under a supervisory system – to turn to the next problem. The physically last statement, i.e. the last card of each program is

```
END.
```

Example A-1: *Program for the computation of the factorials* 1!, 2!, ..., 10!

```
C
C      PROGRAM CALCULATES FIRST 10 FACTORIALS
C
       DIMENSION FACT(10)
       N=1
       A=1.
       B=1.
     1 A=A*B
       FACT(N)=A
       N=N+1
       B=B+1.
       IF(N—10)1,1,2
     2 STOP
       END
```

First the dimension of the array FACT is chosen appropriate to the problem. Then the initial values of the subscript and of two factors, of which the factorial will be composed, are set. In the statement numbered "1" both factors are multiplied (it seems somewhat silly to construct $1! = 1 \cdot 1$ but it is mathematically true). The result is stored away in the array FACT with the appropriate index. Now the subscript and the second factor are increased by one. If the subscript is less than or equal to 10 the sequence is repeated starting with statement "1". In this way the calculations

$$
\begin{aligned}
1! &= 1 \cdot 1, \\
2! &= 1! \cdot 2, \\
3! &= 2! \cdot 3, \\
&\vdots \\
10! &= 9! \cdot 10
\end{aligned}
$$

are performed. After the last calculation the subscript N is greater than 10. Therefore the program proceeds to statement "2" which finishes the operation.

The repetition of part of a program (a very common procedure) can be achieved more conveniently by the use of the so called DO statement, e.g.

```
DO 10, J=1,10,
DO 115, IX2=M,L
```

or generally

DO $st, \quad v = i, k.$

It has the following effect. The part of the program contained between the DO statement and a statement appearing later and carrying the statement number *st* is repeated a number of times. The first time the integer variable v takes the value i. After each step v is increased by one. The last step is performed with $v = k$. Our example now takes the simpler form

```
C
C     PROGRAM CALCULATES FIRST 10 FACTORIALS
C
      DIMENSION FACT(10)
      A=1.
      B=1.
      DO 5 N=1,10
```

```
      A=A*B
      FACT(N)=A
    5 B=B+1.
      STOP
      END
```

It is sometimes convenient to use as the end of a "DO-loop", or as some other numbered statement, a statement that "does nothing", i.e. simply advances the program to the next statement. This is done by the statement

```
CONTINUE.
```

Of course it is important to introduce data into a program and communicate the results back to the user. This is done by *input-output statements* (I/O statements). They have the form

```
READ(5,100)   list
WRITE(6,200)  list
```

for input (reading) and output (writing), respectively. The first number in brackets indicates the I/O device to be used, e.g. the card punch, card reader, printer, a certain magnetic tape unit etc. The actual value of these numbers depends on the computer used. In our examples we shall always use "5" for input and "6" for output. The second number is the statement number of a FORMAT statement accompanying each I/O statement (except binary input or output which we do not discuss). The FORMAT statement controls the layout of the data on punched cards or printed paper, i.e. the blank spaces between numbers, the number of digits etc. Although important for actual programming, the details of FORMATs are not necessary for our purposes. We only sketch the most important features. The FORMAT statement takes the general form

FORMAT $(c_1, c_2/c_3, c_4, \ldots, c_n)$.

The symbols in brackets are format codes that specify the appearance of information. We use only three types of code.

(a) F-code, e.g.

```
F 7.3,
3F 10.5.
```

The first code would cause a real number to be written (or read) using 7 places on the printer (or card reader) of which three are behind the decimal point (e.g. 3.142 or 273.100). The second code would cause 3 numbers to be written on the same line each using 10 places with 5 places behind the decimal point.

(b) I-code, e.g.

```
    I3,
   10I5.
```

These codes cause integer numbers to be written using as many places as defined behind the symbol I.

(c) H-code, e.g.

```
      7H NUMBER.
```

Such a code will write as many characters of explanatory text as specified before the symbol H. The text immediately follows this symbol.

The different codes are separated by commas. If a new line (or card) is started, the separation is done by a slash (/). The I/O list consists of a string of variable names, e.g.

```
N,A,M1(3).
```

The contents of the words with the corresponding names are then read or written. If a subscripted variable appears in the list without a subscript then the whole array is written. E.g. the statements

```
    WRITE (6,500) FACT
500 FORMAT (11H FACTORIALS/5F10.1/5F10.1)
```

would cause the 10 resulting factorials of example A-1 to be written in the following form:

```
FACTORIALS

       1.0       2.0       6.0      24.0     120.0
     720.0    5040.0   40320.0  362880.0 3628800.0
```

Finally if only part of an array is read or written, we can include an expression of the form

```
(FACT(J), J=2,7)
```

into the I/O list. The similarity with the DO statement already indicates that now the words FACT(2), FACT(3), ... FACT(7) are written or read.

Often a larger number of numerical data, e.g. tabulated values or important constants have to be used that are fixed throughout the program and need not be read in during its execution. It is rather lengthy to define them one by one in arithmetic statements like

```
DIMENSION PRIME (5)

PRIME(1) = 1.0
PRIME(2) = 2.0
PRIME(3) = 3.0
PRIME(4) = 5.0
PRIME(5) = 7.0
```

More convenient is the use of a DATA statement. It is defined only in FORTRAN IV. In the IBM version of this language it takes the form

DATA list / $c_1, c_2, \ldots, c_n$ /.

Here list stands for an I/O list referring to the (exactly n) variable names and $c_1, c_2, \ldots c_n$ are the n constants to be defined. To introduce, for example, the number π and the first 5 prime numbers into a program we can use the statements

```
DIMENSION PRIME (5)
DATA PI,PRIME / 3.14159,1.0,2.0,3.0,5.0,7.0/
```

Finally we still have to introduce the important concept of *subprograms*. Whenever a particular set of calculations is repeated in the course of a problem, it is more convenient to place these calculations into a subprogram which can be referred to repeatedly. The FORTRAN language distinguishes between two types of subprograms: *subroutines* and FUNCTION *subprograms*. A subroutine begins with a statement of the type

```
SUBROUTINE XZPROG(A,N,X,B).
```

The name after the word SUBROUTINE is the name of the subprogram. The list in the brackets comprises the arguments, which are variable names. By means of this list, data are transferred to the subprogram and the results back to the main program. The logically last statement in a subprogram is not STOP but

```
RETURN.
```

It directs the program flow back to the main program. The main program gives control to the subroutine by the CALL statement, e.g.

```
CALL XZPROG(X(5),7,12.3,B).
```

The arguments correspond in type (real or integer) and order to those of the SUBROUTINE statement. They need not, however, have the same names and can be constants. If arguments are subscripted variables only the dimensions in the main program are important. DIMENSION statements must however also be given in the subroutine.

Example A-2: SUBROUTINE *subprogram for the computation of a vector product*

In the course of a program the vector product V of two 3-component vectors $\boldsymbol{A}$ and $\boldsymbol{B}$ is to be computed frequently. Explicitly

$$V_1 = A_2B_3 - B_2A_3,$$
$$V_2 = A_3B_1 - B_3A_1,$$
$$V_3 = A_1B_2 - B_1A_2.$$

The result is to be stored in the array V.

Main program:

```
      DIMENSION A(3),B(3),V(3)
C       .
C       .
C       .
      CALL VECT (A,B,V)
C       .
C       .
C       .
      STOP
      END
```

Subprogram:

```
SUBROUTINE VECT (X,Y,RESULT)
DIMENSION X(1),Y(1),RESULT(1)
RESULT(1)=X(2)*Y(3)—X(3)*Y(2)
RESULT(2)=X(3)*Y(1)—X(1)*Y(3)
RESULT(3)=X(1)*Y(2)—X(2)*Y(1)
RETURN
END
```

If the result of a subprogram is a single number, as in the case of the scalar product

$$S = A_1B_1 + A_2B_2 + A_3B_3,$$

we can use the FUNCTION subprogram. The reference to a FUNCTION subprogram can be incorporated directly into an arithmetic statement, e.g.

```
A=FNPRG(AX,BY,12.3)
```

places the result of a FUNCTION subprogram into word A. More complicated expressions like

```
A=(2.*FNPRG(AZ,B,FM)+3.14)**2
```

are also admissable. The subprogram itself begins with a FUNCTION statement, e.g.

```
FUNCTION FNPRG(Q,R,S).
```

containing the name of the subprogram and an argument list. Whereas a SUBROUTINE need not have an argument a FUNCTION must have at least one argument so that the compiler can recognize it. Sometimes the argument is never used in the FUNCTION subprogram ("dummy" argument cf. program 5-1). The result of the subprogram is defined by an arithmetic statement with the name of the function appearing on the left-hand side, e.g.

```
FNPRG=Q**2+R**2.
```

As in the case of a subroutine, the logically last statement of a FUNCTION has to be the RETURN statement.

Example A-3: FUNCTION *subprogram for the computation of a scalar product*

The scalar product of the 3-component vectors A and B is to be calculated and to be placed in word S of the main program.
Main program:

```
      DIMENSION A(3),B(3)
C     .
C     .
C     .
```

```
      S=SCAL (A,B)
C     .
C     .
C     .
      STOP
      END
```

Subprogram:

```
FUNCTION SCAL (X,Y)
DIMENSION X(1),Y(1)
SCAL=X(1)*Y(1)+X(2)*Y(2)+X(3)*Y(3)
RETURN
END
```

Some functions are directly at the disposal of the FORTRAN programmer from the FORTRAN library. The most important ones are SIN(X), COS(X), TAN(X), EXP(X), ALOG(X), SQRT(X) and ABS(X). They give the trigonometrical functions sin, cos, tan, the exponential, the natural logarithm, the square root and the absolute value of the variable X, respectively. The mathematical expression

$$a = e^{\sqrt{(\sin^2 x - \cos^2 x)}}$$

can therefore be calculated using the statement

```
A=EXP(SQRT(SIN(X)**2—COS(X)**2))
```

Further examples of rather simple programs are contained in appendix B, §4.

APPENDIX B

SHORT REVIEW OF MATRIX CALCULUS

In this book extensive use is made of the vector and matrix notation. Although in physics this notation is by now fairly common knowledge, this is unfortunately not true of other branches of science. In this appendix therefore the most important definitions and formulae are presented. We should like to stress the point that matrix calculus does not contain any "new" mathematics at all. It is merely a convenient method of expressing a number of (linear) equations simultaneously. It provides a much more compact notation and, at the same time, helps to express problems more clearly. The same holds for the use of matrices in computer programs. Rather than handling all equations of a set one by one, a simple reference to a subroutine is sufficient. In this way programs are kept concise provided that a small library of matrix routines is at hand. In the following sections we give a short review of matrix calculus and write down the required subroutines which are used in the programming examples of the main text.

B-1. Definitions of matrices and vectors

A matrix is a rectangular arrangement of a number of elements. The elements are real or complex numbers. We restrict ourselves to real elements. An example of a matrix would be

$$X = \begin{pmatrix} 1 & -3 & 5.2 \\ 6 & 2 & 7 \end{pmatrix}. \tag{1.1}$$

This matrix has 2 *rows* and 3 *columns*. A general matrix with m rows and n columns (an $(m \times n)$ matrix) is written

$$A = \begin{pmatrix} a_{11} & a_{12} & \cdots & a_{1n} \\ a_{21} & a_{22} & \cdots & a_{2n} \\ \vdots & & & \\ a_{m1} & a_{m2} & \cdots & a_{mn} \end{pmatrix}. \tag{1.2}$$

Obviously the number of elements is nm. The *transpose* of a matrix is obtained by interchanging rows and columns, e.g. the transpose of the matrix (1.1) is

$$X^{\mathrm{T}} = \begin{pmatrix} 1 & 6 \\ -3 & 2 \\ 5.2 & 7 \end{pmatrix}. \tag{1.3}$$

The transpose of the general matrix (1.2) is

$$A^{\mathrm{T}} = \begin{pmatrix} a_{11} & a_{21} & \dots & a_{m1} \\ a_{12} & a_{22} & \dots & a_{m2} \\ \vdots & & & \\ a_{1n} & a_{2n} & \dots & a_{mn} \end{pmatrix}. \tag{1.4}$$

An element of the transposed matrix is simply the element of the original matrix with inverted indices

$$(A^{\mathrm{T}})_{ij} = a_{ij}^{\mathrm{T}} = a_{ji}. \tag{1.5}$$

Matrices with equal number of rows and columns are particularly important. They are called *square matrices*

$$A = \begin{pmatrix} a_{11} & a_{12} & \dots & a_{1n} \\ a_{21} & a_{22} & \dots & a_{2n} \\ \vdots & & & \\ a_{n1} & a_{n2} & \dots & a_{nn} \end{pmatrix}. \tag{1.6}$$

The elements of square matrices with repeated indices, e.g. a_{11}, a_{33}, are called *diagonal elements*. A matrix where only the diagonal elements are different from zero is called a *diagonal matrix*, e.g.

$$D = \begin{pmatrix} 3 & 0 & 0 \\ 0 & 2.5 & 0 \\ 0 & 0 & 7 \end{pmatrix}.$$

A diagonal matrix with n rows and columns with all diagonal elements equal to one, i.e.

$$I = \begin{pmatrix} 1 & 0 & 0 & \dots & 0 \\ 0 & 1 & 0 & \dots & 0 \\ 0 & 0 & 1 & \dots & 0 \\ \vdots & & & & \\ 0 & 0 & 0 & \dots & 1 \end{pmatrix} \tag{1.7}$$

is called the *unit matrix of order n*. Its elements take the values of the

Kronecker-symbol

$$\delta_{ij} = \begin{cases} 1 & \text{for } i = j, \\ 0 & \text{for } i \neq j. \end{cases} \tag{1.8}$$

The transpose of a square matrix can be simply visualized as the reflection of the elements on the diagonal. If the original and the transposed matrix are identical in all elements, i.e. if

$$a_{ij} = a_{ji} \quad \text{for any } i, j, \tag{1.9}$$

the matrix is called *symmetric*. This definition is meaningful only for square matrices. An example is

$$\begin{pmatrix} 5 & -2 & 2.1 \\ -2 & 3.7 & 6 \\ 2.1 & 6 & 9 \end{pmatrix}.$$

A matrix with only one column, e.g.

$$\boldsymbol{a} = \begin{pmatrix} a_{11} \\ a_{21} \\ \vdots \\ a_{m1} \end{pmatrix} = \begin{pmatrix} a_1 \\ a_2 \\ \\ a_m \end{pmatrix} \tag{1.10}$$

is called a *column vector*. For brevity the elements are usually written with only one index. Similarly a row vector is a matrix with only one row

$$\boldsymbol{b} = (b_{11}, b_{12}, \ldots, b_{1n}) = (b_1, b_2, \ldots, b_n). \tag{1.11}$$

To distinguish vectors from general matrices we denote them by symbols like $\boldsymbol{a}$, $\boldsymbol{b}$, ..., $\boldsymbol{x}$. Of course the transpose of a row vector is a column vector and vice versa, i.e.

$$\boldsymbol{a}^{\mathrm{T}} = (a_1, a_2, \ldots, a_m), \qquad \boldsymbol{b}^{\mathrm{T}} = \begin{pmatrix} b_1 \\ b_2 \\ \vdots \\ b_n \end{pmatrix}. \tag{1.12}$$

Finally we can of course define a matrix with only one row and one column which we write

$$a = (a).$$

It contains only one element and is called a *scalar*. In matrix algebra all simple numbers (like constant factors in equations) are treated as scalars.

B-2. Equality, addition, subtraction and multiplication of matrices

One of the peculiarities of matrix algebra is that relations between two matrices are defined only if certain relations exist between the sizes, i.e. the number of rows and columns of the matrices.

Two matrices A and B are equal only if they have the same number of rows and the same number of columns and if all respective elements are equal, i.e.

$$a_{ij} = b_{ij} \quad \text{for all } i, j. \tag{2.1}$$

Correspondingly the addition of two matrices can only be performed if both have the same dimension, e.g. both are $(m \times n)$ matrices. Then the expression

$$C = A + B \tag{2.2}$$

signifies that each element of the matrix C is obtained by adding the corresponding elements of matrices A and B

$$c_{ij} = a_{ij} + b_{ij}. \tag{2.3}$$

Of course the subtraction of two matrices is defined accordingly, i.e.

$$C = A - B \tag{2.4}$$

is equivalent to

$$c_{ij} = a_{ij} - b_{ij} \quad \text{for all } i, j. \tag{2.5}$$

There are two types of multiplications involving matrices. First a matrix can be multiplied with a constant (a scalar) k by multiplying each element of the matrix by this constant

$$kA = k\begin{pmatrix} a_{11} & a_{12} & \dots & a_{1n} \\ a_{21} & a_{22} & \dots & a_{2n} \\ \vdots & & & \\ a_{m1} & a_{m2} & \dots & a_{mn} \end{pmatrix} = \begin{pmatrix} ka_{11} & ka_{12} & \dots & ka_{1n} \\ ka_{21} & ka_{22} & \dots & ka_{1n} \\ \vdots & & & \\ ka_{m1} & ka_{m2} & \dots & ka_{mn} \end{pmatrix}. \tag{2.6}$$

This multiplication is commutative, i.e.

$$kA = Ak.$$

The multiplication of two matrices A and B is only defined if A is an $(m \times l)$ matrix and B is an $(l \times n)$ matrix, i.e. if the number of columns of A is equal to the number of rows of B. The element c_{ij} of the product matrix C is then defined as

TABLE B-1
Matrix multiplication

$$
m\left\{\overbrace{\begin{pmatrix} a_{11} & a_{12} & \dots & a_{1l} \\ a_{21} & a_{22} & \dots & a_{2l} \\ \vdots & & & \\ \boxed{a_{i1} \quad a_{i2} \quad \dots \quad a_{il}} & & & \\ \vdots & & & \\ a_{m1} & a_{m2} & \dots & a_{ml} \end{pmatrix}}^{l}\right. \quad \left.\overbrace{\begin{pmatrix} b_{11} & b_{12} & \dots & b_{1j} & \dots & b_{1n} \\ b_{21} & b_{22} & \dots & b_{2j} & \dots & b_{2n} \\ \vdots & & & \vdots & & \\ b_{l1} & b_{l2} & \dots & b_{lj} & \dots & b_{ln} \end{pmatrix}}^{n}\right\} l
$$

$$c_{ij} = a_{i1}\, b_{1j} + a_{i2}\, b_{2j} + \dots + a_{il}\, b_{lj} = \sum_{k=1}^{l} a_{ik}\, b_{kj}$$

$$c_{ij} = \sum_{k=1}^{l} a_{ik}\, b_{kj}. \tag{2.7}$$

Obviously the product is an $(m \times n)$ matrix. The process of matrix multiplication can be visualized as follows (table B-1). One places the ith row of matrix A on top of the jth column of matrix B, forms the products of the coinciding pairs of elements and finally sums up all these products.

Example B-1: *Product of a* (3×2) *matrix with a* (2×2) *matrix*

$$\begin{pmatrix} 1 & 2 \\ 3 & 0 \\ 4 & 1 \end{pmatrix} \begin{pmatrix} 3 & 0 \\ 1 & 2 \end{pmatrix} = \begin{pmatrix} 5 & 4 \\ 9 & 0 \\ 13 & 2 \end{pmatrix}.$$

From the definition (2.7) we see that this multiplication is not commutative. In fact if AB is defined it does not follow that BA is defined also; it is clear that both products are defined only in the special case of $m = n$. Clearly we can always form the product of an $(m \times n)$ matrix with its transpose which is of dimension $(n \times m)$. The result will be an $(m \times m)$ matrix

$$D = A\,A^{\mathrm{T}}. \tag{2.8}$$

Because of (2.7) and (1.5) the element d_{ij} of the product matrix becomes

$$d_{ij} = \sum_{k=1}^{n} a_{ik} a_{kj}^{\mathrm{T}} = \sum_{k=1}^{n} a_{ik} a_{jk}. \tag{2.9}$$

Similarly the product

$$F = A^{\mathrm{T}} A \tag{2.10}$$

is always defined. The resulting matrix F with

$$f_{ij} = \sum_{k=1}^{m} a_{ik}^{\mathrm{T}} a_{kj} = \sum_{k=1}^{m} a_{ki} a_{kj} \tag{2.11}$$

is a square $(n \times n)$ matrix. Of course we have in general

$$A A^{\mathrm{T}} \neq A^{\mathrm{T}} A. \tag{2.12}$$

even if A is a square matrix, i.e. $m = n$. Products of the type (2.8) or (2.10) have the important property that they yield symmetric matrices since

$$d_{ij} = \sum_{k=1}^{n} a_{ik} a_{jk} = \sum_{k=1}^{n} a_{jk} a_{ik} = d_{ji}.$$

Multiplication of a matrix by the unit matrix leaves the original matrix unchanged. If e.g. A is an $(m \times n)$ matrix and I the unit matrix of order n we have

$$A I = A, \tag{2.13}$$

because

$$\sum_{k=1}^{n} a_{ik} \delta_{kj} = a_{ij}.$$

Therefore the product of a square matrix with the unit matrix of equal order is commutative.

The transpose of a product of matrices can be written as the product of the transposed matrices performed in the inverse order, i.e.

$$(A B C \cdots Z)^{\mathrm{T}} = Z^{\mathrm{T}} \cdots C^{\mathrm{T}} B^{\mathrm{T}} A^{\mathrm{T}}. \tag{2.14}$$

This important rule can easily be proved using the definitions (2.7) and (1.5).

We now consider the multiplication of matrices with vectors. An $(m \times n)$ matrix A can be multiplied with a column vector of n elements $\boldsymbol{x}$ from the right. The result is a column vector with m elements. According to eq. (2.7) we have

$$Ax = \begin{pmatrix} a_{11} & a_{12} & \dots & a_{1n} \\ a_{21} & a_{22} & \dots & a_{2n} \\ \vdots & & & \\ a_{m1} & a_{m2} & \dots & a_{mn} \end{pmatrix} \begin{pmatrix} x_1 \\ x_2 \\ \vdots \\ x_n \end{pmatrix} =$$

$$\begin{pmatrix} a_{11}x_1 + a_{12}x_2 + \dots + a_{1n}x_n \\ a_{21}x_1 + a_{22}x_2 + \dots + a_{2n}x_n \\ \vdots \\ a_{m1}x_1 + a_{m2}x_2 + \dots + a_{mn}x_n \end{pmatrix} = \begin{pmatrix} b_1 \\ b_2 \\ \vdots \\ b_m \end{pmatrix} = \boldsymbol{b}. \tag{2.15}$$

Since we have defined two matrices as being equal if all corresponding elements are equal, we see that the matrix equation

$$A\boldsymbol{x} = \boldsymbol{b} \tag{2.16}$$

is equivalent to the system of m simultaneous linear equations

$$\begin{aligned} a_{11}x_1 + a_{12}x_2 + \dots + a_{1n}x_n &= b_1, \\ a_{21}x_1 + a_{22}x_2 + \dots + a_{2n}x_n &= b_2, \\ \vdots \\ a_{m1}x_1 + a_{m2}x_2 + \dots + a_{mn}x_n &= b_m, \end{aligned} \tag{2.17}$$

which can be solved to yield the n unknowns $x_1, x_2, \dots, x_n$.

Of course an $(m \times n)$ matrix can also be multiplied from the left by a row vector of m elements. The result will be a row vector of n elements

$$\boldsymbol{y}A = \boldsymbol{c}. \tag{2.18}$$

This equation is equivalent to a system of n linear equations for the m unknowns $\boldsymbol{y} = (y_1, y_2, \dots, y_m)$.

Finally we consider the multiplication of two vectors. The product is only defined if both have the same number of elements and one factor is a row vector and the other a column vector. The multiplication of a row vector with a column vector

$$\boldsymbol{ab} = (a_1, a_2, \dots, a_n) \begin{pmatrix} b_1 \\ b_2 \\ \vdots \\ b_n \end{pmatrix} = \sum_{k=1}^{n} a_k b_k = c \tag{2.19}$$

yields a scalar. It is called the *scalar product* of the two vectors. The scalar product of a vector with its transpose is called the *square* of the vector

$$\boldsymbol{a}\,\boldsymbol{a}^{\mathrm{T}} = (a_1, a_2, \ldots, a_n)\begin{pmatrix} a_1 \\ a_2 \\ \vdots \\ a_n \end{pmatrix} = \sum_{k=1}^{n} a_k^2 = a^2. \tag{2.20}$$

We can also form the product of an n-component column vector with an n-component row vector. This product which yields a square matrix of order n is called the *dyadic product*

$$\begin{pmatrix} a_1 \\ a_2 \\ \vdots \\ a_n \end{pmatrix}(b_1, b_2, \ldots, b_n) = \begin{pmatrix} a_1 b_1 & a_1 b_2 & \ldots & a_1 b_n \\ a_2 b_1 & a_2 b_2 & \ldots & a_2 b_n \\ \vdots & & & \\ a_n b_1 & a_n b_2 & \ldots & a_n b_n \end{pmatrix}. \tag{2.21}$$

B-3. Determinant and inverse of a square matrix; solution of matrix equations

A *determinant* is a single number which can be thought of as being the result of a simple calculation involving the elements of the determinant, which are numbers written in an arrangement with equal number of rows and columns. Determinants of order 2 and 3 are defined as

$$\det A = \begin{vmatrix} a_{11} & a_{12} \\ a_{21} & a_{22} \end{vmatrix} = a_{11}a_{22} - a_{12}a_{21} \tag{3.1}$$

and

$$\begin{aligned} \det A = \begin{vmatrix} a_{11} & a_{12} & a_{13} \\ a_{21} & a_{22} & a_{23} \\ a_{31} & a_{32} & a_{33} \end{vmatrix} &= a_{11}\,a_{22}\,a_{33} - a_{11}\,a_{23}\,a_{32} \\ &+ a_{12}\,a_{23}\,a_{31} - a_{12}\,a_{21}\,a_{33} \\ &+ a_{13}\,a_{21}\,a_{32} - a_{13}\,a_{22}\,a_{31}, \end{aligned} \tag{3.2}$$

which can be written as

$$\begin{aligned} \det A &= a_{11}(a_{22}\,a_{33} - a_{23}\,a_{32}) \\ &- a_{12}(a_{21}\,a_{33} - a_{23}\,a_{31}) \\ &+ a_{13}(a_{21}\,a_{32} - a_{22}\,a_{31}). \end{aligned} \tag{3.3}$$

A general determinant of order n is written as

$$\det A = \begin{vmatrix} a_{11} & a_{12} & \ldots & a_{1n} \\ a_{21} & a_{22} & \ldots & a_{2n} \\ \vdots & & & \\ a_{n1} & a_{n2} & \ldots & a_{nn} \end{vmatrix}. \tag{3.4}$$

It should always be borne in mind that in contrast to a matrix, a determinant is simply a number written in a particular form to exhibit the way in which it is derived from some other numbers.

The *cofactor* A_{ij} of the element a_{ij} of a determinant is the determinant of order $(n - 1)$, which one obtains by deleting the ith row and the jth column of the original determinant, multiplied by $(-1)^{i+j}$

$$A_{ij} = (-1)^{i+j} \begin{vmatrix} a_{11} & a_{12} & \cdots & a_{1,j-1} & a_{1,j+1} & \cdots & a_{1n} \\ a_{21} & a_{22} & \cdots & a_{2,j-1} & a_{2,j+1} & \cdots & a_{2n} \\ \vdots & & & & & & \\ a_{i-1,1} & a_{i-1,2} & \cdots & a_{i-1,j-1} & a_{i-1,j+1} & \cdots & a_{i-1,n} \\ a_{i+1,1} & a_{i+1,2} & \cdots & a_{i+1,j-1} & a_{i+1,j+1} & \cdots & a_{i+1,n} \\ \vdots & & & & & & \\ a_{n1} & a_{n2} & \cdots & a_{n,j-1} & a_{n,j+1} & \cdots & a_{nn} \end{vmatrix}. \tag{3.5}$$

Determinants of higher order can be expressed as the sum of all elements of a particular row or column each multiplied by its cofactor

$$\det A = \sum_{k=1}^{n} a_{ik} A_{ik} = \sum_{k=1}^{n} a_{kj} A_{kj}. \tag{3.6}$$

It is easy to show that the result is independent of the choice of a particular row i or column j. Eq. (3.3) already illustrates the definition (3.6) for $n = 3$. Determinants of any order can be calculated by decomposing the cofactors further until they are of order 2.

Although determinants are a priori not related to matrices, we can call the expression (3.4) the determinant of the square matrix

$$A = \begin{pmatrix} a_{11} & a_{12} & \cdots & a_{1n} \\ a_{21} & a_{22} & \cdots & a_{2n} \\ \vdots & & & \\ a_{n1} & a_{n2} & \cdots & a_{nn} \end{pmatrix}.$$

A matrix with a vanishing determinant is called *singular*. We can construct from A another matrix by replacing each element ij by the cofactor of the element ji. This is called the *adjoint matrix* of A

$$A^{\dagger} = \begin{pmatrix} A_{11} & A_{21} & \cdots & A_{n1} \\ A_{12} & A_{22} & \cdots & A_{n2} \\ \vdots & & & \\ A_{1n} & A_{2n} & \cdots & A_{nn} \end{pmatrix}. \tag{3.7}$$

We are now able to define the *inverse* A^{-1} of a square matrix. This is the matrix that fulfills the equation

$$AA^{-1} = I. \tag{3.8}$$

An inverse is only defined for *non-singular* matrices.

The computation of the inverse of a matrix, which is very important in data analysis, is a rather lengthy problem. We will illustrate it using a matrix of order 2. Calling the inverse matrix X, we have

$$AX = \begin{pmatrix} a_{11} & a_{12} \\ a_{21} & a_{22} \end{pmatrix} \begin{pmatrix} x_{11} & x_{12} \\ x_{21} & x_{22} \end{pmatrix} = \begin{pmatrix} 1 & 0 \\ 0 & 1 \end{pmatrix} \tag{3.9}$$

or

$$\begin{pmatrix} a_{11}x_{11} + a_{12}x_{21} & a_{11}x_{12} + a_{12}x_{22} \\ a_{21}x_{11} + a_{22}x_{21} & a_{21}x_{12} + a_{22}x_{22} \end{pmatrix} = \begin{pmatrix} 1 & 0 \\ 0 & 1 \end{pmatrix}. \tag{3.10}$$

This expression is equivalent to the 4 equations

$$\begin{aligned} a_{11}x_{11} + a_{12}x_{21} &= 1, \\ a_{11}x_{12} + a_{12}x_{22} &= 0, \\ a_{21}x_{11} + a_{22}x_{21} &= 0, \\ a_{21}x_{12} + a_{22}x_{22} &= 1. \end{aligned} \tag{3.11}$$

By successive substitution we easily find that

$$\begin{aligned} x_{11} &= \frac{a_{22}}{a_{11}a_{22} - a_{12}a_{21}}, \\ x_{12} &= \frac{-a_{12}}{a_{11}a_{22} - a_{12}a_{21}}, \\ x_{21} &= \frac{-a_{21}}{a_{11}a_{22} - a_{12}a_{21}}, \\ x_{22} &= \frac{a_{11}}{a_{11}a_{22} - a_{12}a_{21}}, \end{aligned} \tag{3.12}$$

or, in matrix notation, that

$$X = A^{-1} = \frac{1}{\det A} \begin{pmatrix} a_{22} & -a_{12} \\ -a_{21} & a_{11} \end{pmatrix}. \tag{3.13}$$

The matrix on the right-hand side is the adjoint of the original matrix, i.e.

$$A^{-1} = \frac{A^{\dagger}}{\det A}. \tag{3.14}$$

It can be shown that this equality holds for square matrices of any order. It is obvious from (3.14) that the inverse of a singular matrix, i.e. a matrix with vanishing determinant, is undefined.

In practice the construction of the adjoint matrix, i.e. the determination of all cofactors, is very time consuming. We consider the problem from a different angle and go back to eq. (3.8)

$$AA^{-1} = AX = I.$$

We discuss this equation in a more general sense by allowing the matrix on the right-hand side to be different from the unit matrix, i.e.

$$AX = B.$$

Now X is of course no longer the inverse of A, except for $B = I$. It is, however, advantageous to treat the general case. Eq. (3.15) is equivalent to a set of n^2 simultaneous linear equations just as eq. (3.9) was equivalent to the four equations (3.11). As the first n equations we can choose those corresponding to the first column of B

$$\begin{aligned} a_{11} x_{11} + a_{12} x_{21} + \dots + a_{1n} x_{n1} &= b_{11}, \\ a_{21} x_{11} + a_{22} x_{21} + \dots + a_{2n} x_{n1} &= b_{21}, \\ &\vdots \\ a_{n1} x_{11} + a_{n2} x_{21} + \dots + a_{nn} x_{n1} &= b_{n1}. \end{aligned} \tag{3.15}$$

This set is equivalent to the set of equations (2.17) if we group the $x_{11}, x_{21}, \dots, x_{n1}$ and $b_{11}, b_{21}, \dots, b_{n1}$ into column vectors or, more generally,

$$\boldsymbol{x}_i = \begin{pmatrix} x_{1i} \\ x_{2i} \\ \vdots \\ x_{ni} \end{pmatrix}, \qquad \boldsymbol{b}_i = \begin{pmatrix} b_{1i} \\ b_{2i} \\ \vdots \\ b_{ni} \end{pmatrix}. \tag{3.16}$$

The n^2 equations therefore can be grouped into n sets

$$A\boldsymbol{x}_i = \boldsymbol{b}_i, \qquad i = 1, 2, \dots n, \tag{3.17}$$

each with a different set of unknowns. The problem therefore reduces to the solution of equations of the type

$$A\boldsymbol{x} = \boldsymbol{b} \tag{3.18}$$

for the unknown vector $\boldsymbol{x}$. Writing down all the elements we have

$$\begin{aligned} a_{11} x_1 + a_{12} x_2 + \dots + a_{1n} x_n &= b_1, \\ a_{21} x_1 + a_{22} x_2 + \dots + a_{2n} x_n &= b_2, \\ &\vdots \\ a_{n1} x_1 + a_{n2} x_2 + \dots + a_{nn} x_n &= b_n. \end{aligned} \tag{3.19}$$

We solve this system of simultaneous linear equations using the so-called *Gaussian algorithm.*

We define $n - 1$ multipliers

$$m_i = \frac{a_{i1}}{a_{11}}, \qquad i = 2, 3, \ldots, n.$$

Then the first equation is multiplied in turn by $m_2, m_3, \ldots, m_n$ and the resulting equation is subtracted from the 2nd, 3rd, nth equation of the set (3.19). In the resulting set

$$\begin{aligned} a_{11}x_1 + a_{12}x_2 + a_{13}x_3 + \ldots + a_{1n}x_n &= b_1, \\ a'_{22}x_2 + a'_{23}x_3 + \ldots + a'_{2n}x_n &= b'_2, \\ &\vdots \\ a'_{n2}x_2 + a'_{n3}x_3 + \ldots + a'_{nn}x_n &= b'_n, \end{aligned} \tag{3.20}$$

the unknown x_1 has been eliminated from all but the first equation. The coefficients a'_{ij}, b'_i are given by

$$\begin{aligned} a'_{ij} &= a_{ij} - m_i a_{1j}, \\ b'_i &= b_i - m_i b_1. \end{aligned} \tag{3.21}$$

The procedure is now repeated with the last $n - 1$ equations by defining

$$m'_i = \frac{a'_{i2}}{a'_{22}}, \qquad i = 3, 4, \ldots, n,$$

multiplying the second equation of (3.20) by the appropriate m'_i and subtracting it from the 3rd, 4th, ..., nth equation.

In general the kth step of the process is performed using multipliers

$$m_i^{(k-1)} = \frac{a_{ik}^{(k-1)}}{a_{kk}^{(k-1)}}, \qquad i = k + 1, k + 2, \ldots, n, \tag{3.22}$$

and calculating the new coefficients

$$\begin{aligned} a_{ij}^{(k)} &= a_{ij}^{(k-1)} - m_i^{(k-1)} a_{kj}^{(k-1)}, \\ b_i^{(k)} &= b_i^{(k-1)} - m_i^{(k-1)} b_k^{(k-1)}. \end{aligned} \tag{3.23}$$

After $n - 1$ steps the following *triangular system* is obtained

$$\begin{aligned} a_{11}x_1 + a_{12}x_2 + \ldots + a_{1,n-1}x_{n-1} + a_{1n}x_n &= b_1, \\ a'_{22}x_2 + \ldots + a'_{2,n-1}x_{n-1} + a'_{2n}x_n &= b'_2, \\ &\vdots \\ a_{n-1,n-1}^{(n-2)}x_{n-1} + a_{n-1,n}^{(n-2)}x_n &= b_{n-1}^{(n-2)}, \\ a_{nn}^{(n-1)}x_n &= b_n^{(n-1)}. \end{aligned} \tag{3.24}$$

The last equation contains only x_n which can therefore be immediately read off. By substitution into the next higher equation we get x_{n-1} and so on. In general

$$x_i = \frac{1}{a_{ii}^{(i-1)}} \left\{ b_i^{(i-1)} - \sum_{l=i+1}^{n} a_{il}^{(i-1)} x_l \right\}. \tag{3.25}$$

It is interesting to note that the reduction of the coefficients of the matrix does not depend on the right-hand side $\boldsymbol{b}$. Several systems with different right-hand sides can therefore be processed at the same time, i.e. instead of (3.15) we can consider the general system (3.17).

TABLE B-2

Gaussian algorithm applied to example B-2

Reduction

	Matrix A			Matrix B			Multiplier
	1	2	3	1	0	1	–
a_{ij}, b_{ij}	2	1	−2	0	1	0	2
	1	1	2	0	0	1	1
a'_{ij}, b'_{ij}		−3	−8	−2	1	0	–
		−1	−1	−1	0	1	$\frac{1}{3}$
a''_{ij}, b''_{ij}			$\frac{5}{8}$	$-\frac{1}{3}$	$-\frac{1}{3}$	1	–

Back substitution

	$j = 1$	$j = 2$	$j = 3$
x_{3j}	$-\frac{1}{5}$	$-\frac{1}{5}$	$\frac{3}{5}$
x_{2j}	$-\frac{1}{3}(-2-8\times\frac{1}{5}) = \frac{6}{5}$	$-\frac{1}{3}(1-\frac{8}{5}) = \frac{1}{5}$	$-\frac{1}{3}(0+8\times\frac{3}{5}) = -\frac{8}{5}$
x_{1j}	$1-2\times\frac{6}{5}+3\times\frac{1}{5} = -\frac{4}{5}$	$-2\times\frac{1}{5}+3\times\frac{1}{5} = \frac{1}{5}$	$2\times\frac{8}{5}-3\times\frac{3}{5} = \frac{7}{5}$

Example B-2: *Inversion of a* (3 × 3) *matrix*

In this case we have $B = I$. The actual calculations of the example

$$\begin{pmatrix} 1 & 2 & 3 \\ 2 & 1 & -2 \\ 1 & 1 & 2 \end{pmatrix} X = I$$

are performed in the scheme of table B-2. The result is

$$X = \begin{pmatrix} -\frac{4}{5} & \frac{1}{5} & \frac{7}{5} \\ \frac{6}{5} & \frac{1}{5} & -\frac{8}{5} \\ -\frac{1}{5} & -\frac{1}{5} & \frac{3}{5} \end{pmatrix}.$$

Following the single steps of the calculation given by eqs. (3.22)–(3.25) we see that divisions are performed at two stages (eqs. (3.22) and (3.25)). Each time the divisor is a coefficient

$$a_{ii}^{(i-1)}, \qquad i = 1, 2, \ldots, n-1,$$

i.e. the coefficient at the upper left of the system, called the "pivot", at the stage $i - 1$ in the process of reduction. If this coefficient equals zero our method will break down. In such a case we simply exchange the ith row of the system with any lower row whose first coefficient is different from zero. Of course the system itself is insensitive to a change in the sequence of the individual equations.

Only in the case when all the coefficients in one column vanish, does this method also fail. In that case, however, the matrix A is singular and no solution exists. In practice (at least in computer use where the additional work is negligible) it is always advantageous to perform such an exchange if the pivot is not the largest coefficient (in absolute value) in the first column of the reduced system. In this way we are always dealing with the numerically largest possible divisor. It can be shown that this procedure keeps rounding errors small. The subroutine MTXEQU reproduced in the next section uses this procedure.

We conclude this section by some general remarks on systems of linear equations. In (3.18) the matrix A was a square ($n \times n$) matrix. Let us consider the more general system

$$f_{j1}x_1 + f_{j2}x_2 + \ldots + f_{jn}x_n = g_j, \qquad j = 1, 2, \ldots, m, \tag{3.26}$$

i.e.

$$F\boldsymbol{x} = \boldsymbol{g}, \tag{3.27}$$

with F being an $(m \times n)$ matrix and g being an m-vector.

We assume that the m equations of the system are linearly independent. It is well known that for $m < n$ the system has no unique solution since there are fewer equations than unknowns; the system is then said to be *underdetermined.* For $m > n$, on the other hand, the system is *overdetermined.* Then the m equations cannot be fulfilled simultaneously. By introducing additional conditions, however, "best" values $\tilde{\boldsymbol{x}}$ of the unknowns $\boldsymbol{x}$ can be obtained. It may seem reasonable for example to keep the sum of squares of the differences between the right-hand sides and the left-hand sides of (3.26) as small as possible:

$$\sum_{j=1}^{m} (f_{j1}x_1 + f_{j2}x_2 + \ldots + f_{jn}x_n - g_j)^2 = \text{min.},$$

or, in matrix notation

$$(F\boldsymbol{x} - \boldsymbol{g})^{\mathrm{T}}(F\boldsymbol{x} - \boldsymbol{g}) = \text{min.} \tag{3.28}$$

This is the simplest least squares condition. To obtain this solution it is necessary for the total differential of (3.28) with respect to the $\boldsymbol{x}$ to vanish, i.e.

$$2F^{\mathrm{T}}(F\boldsymbol{x} - \boldsymbol{g}) = 0. \tag{3.29}$$

Now the system (3.29) has the same form as (3.18) since $F^{\mathrm{T}}F$ is an $(n \times n)$ matrix and $F^{\mathrm{T}}\boldsymbol{g}$ an n-vector. The solution is

$$\tilde{\boldsymbol{x}} = (F^{\mathrm{T}}F)^{-1}F^{\mathrm{T}}\boldsymbol{g} \tag{3.30}$$

In ch. 9 the solution of overdetermined systems is discussed under conditions more general than (3.28).

B-4. FORTRAN programs for matrix handling

In this section a small subroutine library is assembled for the most important matrix operations. By a careful study of the programs the reader will at the same time become more familiar with the FORTRAN language and with matrix calculus. The routines presented are used in the examples of ch. 9.

Subroutine libraries for matrix manipulations exist at most computer installations. They are written directly in "machine language", i.e. composed of the specific instructions that are particular to a given computer. In this way subscript handling can be done efficiently and execution time can be very significantly reduced. Matrix routines in machine language should be used whenever available.

We must first define a convention for the storage of the elements of a matrix in a singly subscripted array (each element being specified by two indices). The use of two-dimensional arrays, although possible, is not suitable for certain technical reasons. We shall store matrices row by row, i.e. an $(m \times n)$ matrix in an array A in the order indicated in table B-3.

TABLE B-3
Storage of an $(m \times n)$ matrix in a FORTRAN array

Word of array A	A(1)	A(2)	... A(N)	A(N + 1)	A(N + 2)	... A(2 * N)	... A(M * N)
Matrix element stored in this word	a_{11}	a_{12}	... a_{1n}	a_{21}	a_{22}	... a_{2n}	... a_{mn}

First row (A(1) ... A(N)); Second row (A(N + 1) ... A(2 * N))

Therefore, if the indices of an element a_{ij} are represented by the FORTRAN variables I and J, the subscript K of the corresponding word of the array A is given by the statement

```
K=(I—1)*N+J
```

We further agree to call A, B, C the matrices on which the operation is performed and R the resulting matrix. The symbols I, J, K denote running variables, and L, N, M represent the sizes (numbers of rows and columns) of the matrices.

Most of the programs are self explanatory. In some cases comments have been placed between the statements to facilitate understanding. Mnemonic names have been given to the different subroutines to indicate the operation performed by them. Some of the subroutines do not correspond to actual matrix calculations but are convenient for the handling of matrices on computers. If not specified otherwise the matrix resulting from the operation of each subroutine is stored in the array R.

Program B-1: *Subroutine* MTXTRA *transferring the* $(m \times n)$ *matrix* A *into the array* R

```
      SUBROUTINE MTXTRA (A,R,M,N)
      DIMENSION A(1),R(1)
      NSTEP=M*N
      DO 1 J=1,NSTEP
    1 R(J)=A(J)
      RETURN
      END
```

A very similar program can be used to multiply a matrix by a scalar:

Program B-2: *Subroutine* MTXMSC *multiplying the* ($m \times n$) *matrix* A *by the scalar* S

```
      SUBROUTINE MTXMSC(A,R,S,M,N)
      DIMENSION A(1),R(1)
      NSTEP=M*N
      DO 10 J=1,NSTEP
   10 R(J)=S*A(J)
      RETURN
      END
```

Program B-3: *Subroutine* MTXUNT *placing a unit matrix of order n into the array* R

```
      SUBROUTINE MTXUNT(R,N)
      DIMENSION R(1)
      DO 3 I=1,N
      DO 3 J=1,N
      K=(I—1)*N+J
      IF(I—J) 1,2,1
    1 R(K)=0.
      GO TO 3
    2 R(K)=1.
    3 CONTINUE
      RETURN
      END
```

This program provides an example of nested DO-loops typical for matrix operations. While the index of the outer loop (the number of a particular row) is kept at a fixed value, the index of the inner loop runs through all possible column numbers. In this way all elements of the matrix R are obtained. If row and column index are equal the element is set equal to one, otherwise it is set equal to zero.

The next subroutine can be conveniently used to print out a matrix.

Program B-4: *Subroutine* MTXWRT *printing out the* ($m \times n$) *matrix* A

One row is printed to a line if $n \leqslant 12$. If A has more than 12 columns each row is written in consecutive lines, each containing at most 12 elements

```
      SUBROUTINE MTXWRT (A,M,N)
      DIMENSION A(1)
      DO 1 I=1,M
      KBEG=(I—1)*N+1
      KEND=KBEG+N—1
    1 WRITE (6,1000)(A(K),K=KBEG,KEND)
      RETURN
 1000 FORMAT (12 F10.5)
      END
```

Program B-5: *Subroutine* MTXTRP *placing the transpose of the* $(m \times n)$ *matrix* A *in* R

```
      SUBROUTINE MTXTRP (A,R,M,N)
      DIMENSION A(1),R(1)
      DO 1 I=1,M
      DO 1 J=1,N
      KA=(I—1)*N+J
      KR=(J—1)*M+I
    1 R(KR)=A(KA)
      RETURN
      END
```

Program B-6: *Subroutine* MTXADD *adding the two* $(m \times n)$ *matrices* A *and* B

```
      SUBROUTINE MTXADD(A,B,R,M,N)
      DIMENSION A(1),B(1),R(1)
      NSTEP = M*N
      DO 1 J=1,NSTEP
    1 R(J)=A(J)+B(J)
      RETURN
      END
```

Note that a single DO-loop is sufficient in this case. Of course matrix subtraction can be performed in a completely analogous manner.

Program B-7: *Subroutine* MTXSUB *subtracting the* $(m \times n)$ *matrix* B *from the* $(m \times n)$ *matrix* A

```
      SUBROUTINE MTXSUB(A,B,R,M,N)
      DIMENSION A(1),B(1),R(1)
      NSTEP=M*N
      DO 1 J=1,NSTEP
    1 R(J)=A(J)—B(J)
      RETURN
      END
```

Program B-8: *Subroutine* MTXMLT *multiplying the* $(m \times l)$ *matrix* A *by the* $(l \times n)$ *matrix* B

```
      SUBROUTINE MTXMLT (A,B,R,M,L,N)
      DIMENSION A(1),B(1),R(1)
      DO 2 I=1,M
      DO 2 J=1,N
      KR=(I—1)*N+J
      R(KR)=0.
      DO 1 I1=1,L
  100 KA=(I—1)*L+I1
  200 KB=(I1—1)*N+J
    1 R(KR)=R(KR)+A(KA)*B(KB)
    2 CONTINUE
      RETURN
      END
```

Frequently we have to evaluate products of the form AB^{T} where A is an $(m \times l)$ matrix and B an $(n \times l)$ matrix. It is easy to see that this problem could be solved by the subroutine MTXMLT if only the statement 200 is replaced. We therefore write another subroutine.

Program B-9: *Subroutine* MTXMBT *multiplying the* $(m \times l)$ *matrix* A *by the transpose of the* $(n \times l)$ *matrix* B

```
      SUBROUTINE MTXMBT (A,B,R,M,L,N)
C     .
C     .
C     .
  100 KA=(I—1)*L+I1
```

```
  200 KB=(J—1)*L+I1
C      .
C      .
C      .
       END
```

Correspondingly we can write yet another program for matrix multiplication which is identical to MTXMLT except for the statement 100.

Program B-10: *Subroutine* MTXMAT *multiplying the transpose of the* $(l \times m)$ *matrix* A *by the* $(l \times n)$ *matrix* B

```
       SUBROUTINE MTXMAT (A,B,R,M,L,N)
C      .
C      .
C      .
  100 KA=(I1—1)*M+I
  200 KB=(I1—1)*N+J
C      .
C      .
C      .
       END
```

The longest subroutine solves the matrix equation (3.17)

$$AX = B.$$

Here A is a symmetric $(n \times n)$ matrix; B and X, the latter being the matrix of unknowns, are $(n \times m)$ matrices. The Gaussian algorithm with row exchange described in §3 is used.

Program B-11: *Subroutine* MTXEQU *solving a matrix equation*

The equation is solved for the $(n \times m)$ matrix X of unknowns. The matrix A is overwritten. To save storage space the solution X is written into the original array B

```
      SUBROUTINE MTXEQU(A,B,N,M)
      DIMENSION A(1),B(1)
C
C      WRITE ORIGINAL MATRICES A AND B                       ........
C                                                            ........
      WRITE(6,1000)                                          ........
      CALL MTXWRT (A,N,N)                                    ........
      WRITE(6,1100)                                          ........
      CALL MTXWRT (B,N,M)                                    ........
```

```
C      REDUCTION OF MATRIX A
C      STEPS OF REDUCTION ARE COUNTED BY INDEX K
C
       KMAX =N-1
       DO 90 K=1,KMAX
C
C      SEARCH FOR LARGEST COEFFICIENT OF A (DENOTED BY AMAX) IN
C      FIRST COLUMN OF REDUCED SYSTEM
C
       AMAX=0.
       J2=K
       DO 20 J1=K,N
       IK=(J1-1)*N+K
       IF(ABS(AMAX)-ABS(A(IK)))10,20,20
    10 AMAX=A(IK)
       J2=J1
    20 CONTINUE
C
C      EXCHANGE ROW NUMBER K WITH ROW NUMBER J2, IF NECESSARY
C
       IF (J2-K) 30,60,30
    30 DO 40 J=K,N
       J3=(K-1)*N+J
       J4=(J2-1)*N+J
       SAVE=A(J3)
       A(J3)=A(J4)
    40 A(J4)=SAVE
       DO 50 J=1,M
       J3=(K-1)*M+J
       J4=(J2-1)*M+J
       SAVE=B(J3)
       B(J3)=B(J4)
    50 B(J4)=SAVE
C
C       WRITE A AND B (ROWS EXCHANGED) BEFORE STEP K
C
       WRITE(6,1200)
       WRITE(6,1000)
       CALL MTXWRT (A,N,N)
       WRITE(6,1100)
       CALL MTXWRT (B,N,M)
C
C      ACTUAL REDUCTION
C
    60 K1=K+1
       KK=(K-1)*N+K
       DO 80 I=K1,N
       IK=(I-1)*N+K
       DO 70 J=K1,N
       IJ=(I-1)*N+J
       KJ=(K-1)*N+J
    70 A(IJ)=A(IJ)-A(KJ)*A(IK)/A(KK)
       DO 80 J=1,M
       IJ=(I-1)*M+J
       KJ=(K-1)*M+J
    80 B(IJ)=B(IJ)-B(KJ)*A(IK)/A(KK)
C
C      WRITE A AND B AFTER STEP K OF REDUCTION
C
       WRITE(6,1300) K
       WRITE(6,1000)
       CALL MTXWRT (A,N,N)
       WRITE(6,1100)
       CALL MTXWRT (B,N,M)
    90 CONTINUE
C
C      BACKSUBSTITUTION
```

```
      NN=N**2
      DO 110 J=1,M
      NJ=(N-1)*M+J
      B(NJ)=B(NJ)/A(NN)
      I1MAX=N-1
      IF(I1MAX) 110,110,95
   95 DO 110 I1=1,I1MAX
      I=N-I1
      IJ=(I-1)*M+J
      II=(I-1)*N+I
      I2=I+1
      DO 100 L=I2,N
      IL=(I-1)*N+L
      LJ=(L-1)*M+J
  100 B(IJ)=B(IJ)-A(IL)*B(LJ)
      B(IJ)=B(IJ)/A(II)
  110 CONTINUE
C                                                          ........
C     WRITE RESULT                                         ........
C                                                          ........
      WRITE(6,1400)                                        ........
      CALL MTXWRT (B,N,M)                                  ........
      RETURN
 1000 FORMAT (//11H    MATRIX A/)                          ........
 1100 FORMAT (//11H    MATRIX B/)                          ........
 1200 FORMAT (//17H    ROWS EXCHANGED)                     ........
 1300 FORMAT (//5H STEP,I2,13H OF REDUCTION)               ........
 1400 FORMAT (//11H    MATRIX X/)                          ........
      WRITE(6,1000)                                        ........
      END
```

The comments in the program together with the following example should suffice to explain its operation.

Example B-3: *Solution of a matrix equation*

The equation

$AX = B$

is solved for $n = 3$, $m = 4$. The initial matrices A and B, the results of intermediate steps and the resulting matrix X are reproduced in table B-4. They were printed by the statements marked by .... on the right margin of the program. If no print-out is wanted these statements could be removed without changing the performance of the program. It is obvious from (3.24) that the coefficients in the lower left part of the triangular system, i.e. a'_{12}, a''_{31}, a''_{32}, ... need not be calculated, since by the construction of the algorithm we know that they will vanish. In the array A which originally contains the matrix C and is gradually filled with the coefficients, the lower left part contains numbers left over from previous steps and should be disregarded.

To invert an $(n \times n)$ matrix A we can use the subroutine MTXEQU if B initially contains the unit matrix of order n. Since the inversion of a matrix is a particularly frequent operation we perform it in a special subroutine.

TABLE B-4
Print-out from program MTXEQU for example B-3

```
MATRIX A

4.00000   2.00000   3.00000   1.00000
1.00000   2.00000   3.00000   4.00000
3.00000   4.00000   1.00000   2.00000
2.00000   1.00000   3.00000   4.00000

MATRIX B

2.00000   1.00000   0.0
1.00000   4.00000   0.50000
2.00000   3.00000   0.0
0.0       0.0       1.00000

STEP 1 OF REDUCTION

MATRIX A

4.00000   2.00000   3.00000   1.00000
1.00000   1.50000   2.25000   3.75000
3.00000   2.50000  -1.25000   1.25000
2.00000  -0.0       1.50000   3.50000

MATRIX B

2.00000   1.00000   0.0
0.50000   3.75000   0.50000
0.50000   2.25000  -0.0
-1.00000 -0.50000   1.00000

ROWS EXCHANGED

MATRIX A

4.00000   2.00000   3.00000   1.00000
1.00000   2.50000  -1.25000   1.25000
3.00000   1.50000   2.25000   3.75000
2.00000  -0.0       1.50000   3.50000

MATRIX B

2.00000   1.00000   0.0
0.50000   2.25000  -0.0
0.50000   3.75000   0.50000
-1.00000 -0.50000   1.00000

STEP 2 OF REDUCTION

MATRIX A

4.00000   2.00000   3.00000   1.00000
1.00000   2.50000  -1.25000   1.25000
3.00000   1.50000   3.00000   3.00000
2.00000  -0.0       1.50000   3.50000

MATRIX B

2.00000   1.00000   0.0
0.50000   2.25000  -0.0
0.20000   2.40000   0.50000
-1.00000 -0.50000   1.00000

STEP 3 OF REDUCTION

MATRIX A

4.00000   2.00000   3.00000   1.00000
1.00000   2.50000  -1.25000   1.25000
3.00000   1.50000   3.00000   3.00000
2.00000  -0.0       1.50000   2.00000

MATRIX B

2.00000   1.00000   0.0
0.50000   2.25000  -0.0
0.20000   2.40000   0.50000
-1.10000 -1.70000   0.75000

MATRIX X

-0.21667  -1.85000   0.20833
0.78333   2.15000  -0.29167
0.61667   1.65000  -0.20833
-0.55000  -0.85000   0.37500
```

Program B-12: *Subroutine* MTXINV *placing the inverse of the* $(n \times n)$ *matrix* A *in* R

The array A is overwritten

```
SUBROUTINE MTXINV(A,R,N)
DIMENSION A(1),R(1)
CALL MTXUNT(R,N)
CALL MTXEQU(A,R,N,N)
RETURN
END
```

APPENDIX C

ELEMENTS OF COMBINATORIAL ANALYSIS

We consider n distinguishable objects $a_1, a_2, \dots, a_n$ and we want to know the number of possibilities P_n^k in which any k of these can be arranged in a string. Such arrangements are called *permutations*. In the case $n = 4$, $k = 2$ these permutations are

$$\begin{array}{lll} a_1a_2, & a_1a_3, & a_1a_4, \\ a_2a_1, & a_2a_3, & a_2a_4, \\ a_3a_1, & a_3a_2, & a_3a_4, \\ a_4a_1, & a_4a_2, & a_4a_3, \end{array}$$

i.e. $P_k^n = 12$. The answer to the general problem can be deduced from this scheme. There are n different possibilities of occupying the first position of the string. Once one of these has been chosen there are $n-1$ objects left, i.e. there are $n-1$ possibilities for the second position etc. Therefore

$$P_k^n = n(n-1)(n-2)\cdots(n-k+1). \tag{1}$$

This result can be written as

$$P_k^n = \frac{n!}{(n-k)!} \tag{2}$$

with

$$n! = 1 \cdot 2 \cdots n; \quad 0! = 1, \quad 1! = 1. \tag{3}$$

Often the order of the k objects within a permutation is unimportant (the same k objects can be arranged in $k!$ ways within a string), but only the number of different possible selections of k objects out of n is of interest. Such a selection is called a *combination*. The number of combinations of k objects out of n is then

$$C_k^n = \frac{P_k^n}{k!} = \frac{n!}{k!(n-k)!} = \binom{n}{k}. \tag{4}$$

For the *binomial coefficients* $\binom{n}{k}$ we have the recursion formula

$$\binom{n-1}{k} + \binom{n-1}{k-1} = \binom{n}{k}. \tag{5}$$

It is verified by the simple calculation

$$\frac{(n-1)!}{k!(n-k-1)!} + \frac{(n-1)!}{(k-1)!(n-k)!}$$

$$= \frac{(n-k)(n-1)! + k(n-1)!}{k!(n-k)!}$$

$$= \frac{n!}{k!(n-k)!}.$$

This recursion formula is the basis for the construction of Pascal's triangle which is presented in fig. C-1 because of its beauty.

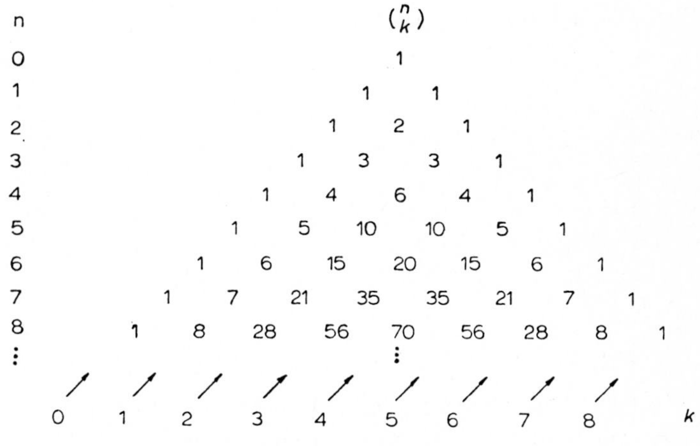

Fig. C-1. Pascal's triangle.

The name binomial coefficients stems from the famous *binomial theorem*

$$(a+b)^n = \sum_{k=0}^{n} \binom{n}{k} a^k b^{n-k}. \tag{6}$$

Its proof (by complete induction) is left to the reader. To derive an important property of the coefficients we write down (6) for $b = 1$, i.e.

$$(a + 1)^n = \sum_{k=0}^{n} \binom{n}{k} a^k.$$

We now apply the theorem a second time

$$(a + 1)^{n+m} = (a + 1)^n (a + 1)^m,$$

$$\sum_{l=0}^{n+m} \binom{n + m}{l} a^l = \sum_{j=0}^{n} \binom{n}{j} a^j \sum_{k=0}^{m} \binom{m}{k} a^k.$$

Considering only the term with a^l we find by comparing coefficients that

$$\binom{n + m}{l} = \sum_{j=0}^{l} \binom{n}{j} \binom{m}{l - j}. \tag{7}$$

Appendix D

EULER'S GAMMA-FUNCTION

If x is a real number with $x + 1 > 0$, we define Euler's *gamma-function* by

$$\Gamma(x+1) = \int_0^\infty t^x \mathrm{e}^{-t}\,\mathrm{d}t. \tag{1}$$

By partial integration eq. (1) becomes

$$\int_0^\infty t^x \mathrm{e}^{-t}\,\mathrm{d}t = \left[-t^x \mathrm{e}^{-t}\right]_0^\infty + x\int_0^\infty t^{x-1}\mathrm{e}^{-t}\,\mathrm{d}t = x\int_0^\infty t^{x-1}\mathrm{e}^{-t}\,\mathrm{d}t.$$

Therefore

$$\Gamma(x+1) = x\,\Gamma(x). \tag{2}$$

This is the so-called *functional equation* of the gamma-function. From (1) we get immediately

$$\Gamma(1) = 1.$$

Using (2) we have the general result

$$\Gamma(n+1) = n!, \qquad n = 0, 1, 2, \ldots. \tag{3}$$

We now substitute t in eq. (1) by $\frac{1}{2}u^2$ (i.e. also $\mathrm{d}t$ by $u\,\mathrm{d}u$). Then

$$\Gamma(x+1) = (\tfrac{1}{2})^x \int_0^\infty u^{2x+1} \mathrm{e}^{-\frac{1}{2}u^2}\,\mathrm{d}u.$$

We consider the special case $x = -\frac{1}{2}$

$$\Gamma(\tfrac{1}{2}) = \sqrt{2}\int_0^\infty \mathrm{e}^{-\frac{1}{2}u^2}\,\mathrm{d}u = \frac{1}{\sqrt{2}}\int_{-\infty}^\infty \mathrm{e}^{-\frac{1}{2}u^2}\,\mathrm{d}u. \tag{4}$$

The integration can be carried out as follows. We first consider

$$A = \int_{-\infty}^{\infty}\int_{-\infty}^{\infty} e^{-\frac{1}{2}(x^2+y^2)}\,dx\,dy = \int_{-\infty}^{\infty} e^{-\frac{1}{2}x^2}\,dx \int_{-\infty}^{\infty} e^{-\frac{1}{2}y^2}\,dy = 2\{\Gamma(\tfrac{1}{2})\}^2.$$

On the other hand the integral A can also be transformed to polar coordinates

$$A = \int_{0}^{2\pi}\int_{0}^{\infty} e^{-\frac{1}{2}r^2}\,r\,dr\,d\varphi = \int_{0}^{2\pi} d\varphi \int_{0}^{\infty} e^{-\frac{1}{2}r^2}\,r\,dr = 2\pi\,\Gamma(1) = 2\pi.$$

We therefore have

$$\Gamma(\tfrac{1}{2}) = \sqrt{\pi}. \tag{5}$$

Using (2) we can then obtain other values of the function for half-integer arguments.

The exact calculation of the gamma-function for arguments other than integer or half-integer is more complicated. Since we do not need it in this book we only present a curve representing the function for low values of x (fig. D-1).

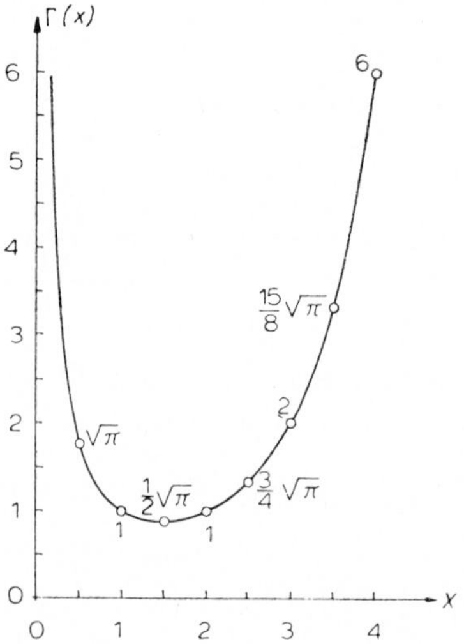

Fig. D-1. Euler's gamma-function.

APPENDIX E

COLLECTION OF IMPORTANT FORMULAE

Probabilities (ch. 2)

$A, B, \ldots$ are *events*, $\bar{A}$ is the event "*not A*"

$(A + B)$ and (AB) signify connection of events by *logical* (*exclusive*) *or* and *logical and*, respectively

$P(A)$ is the *probability* of the event A

$P(B|A) = P(AB)/P(A)$ is the probability of B under the *condition* A (conditional probability)

$P(\bar{A}) = 1 - P(A)$ for every event A

$P(A + B + \ldots + Z) = P(A) + P(B) + \ldots + P(Z)$ for events $A, B, \ldots, Z$ *mutually exclusive*

$P(AB \ldots Z) = P(A)P(B)\ldots P(Z)$ for events A, B, Z *independent*

One random variable x (ch. 3), see also tables E-1, E-2, E-3

Distribution function:

$$F(x) = P(\mathsf{x} < x)$$

Moments of order l:

(a) about the point c:

$$\alpha_l = E\{(\mathsf{x} - c)^l\}$$

(b) about the origin (central moments):

$$\lambda_l = E\{\mathsf{x}^l\}$$

(c) about the mean:

$$\mu_l = E\{(\mathsf{x} - \hat{\mathsf{x}})^l\}$$

Variance:

$$\sigma^2(\mathsf{x}) = \mathrm{var}\,(\mathsf{x}) = \mu_2 = E\{(\mathsf{x} - \hat{\mathsf{x}})^2\}$$

Standard deviation or *error* of x:

$$\Delta\mathsf{x} = \sigma(\mathsf{x}) = +\sqrt{\sigma^2(\mathsf{x})}$$

TABLE E-1

Expectation value of discrete and continuous variables

	x discrete	x continuous; $F(x)$ continuous, differentiable
Probability density	–	$f(x) = F'(x) = \dfrac{\mathrm{d}F(x)}{\mathrm{d}x}, \int_{-\infty}^{\infty} f(x)\,\mathrm{d}x = 1$
Mean of x (expectation value)	$\hat{x} = E(x) = \sum_i x_i P(x = x_i)$	$\hat{x} = E(x) = \int_{-\infty}^{\infty} x f(x)\,\mathrm{d}x$
Mean of function $H(x)$	$E\{H(x)\} = \sum_i H(x_i) P(x = x_i)$	$E\{H(x)\} = \int_{-\infty}^{\infty} H(x) f(x)\,\mathrm{d}x$

Skewness:

$$\gamma = \mu_3/\sigma^3$$

Reduced variable:

$$u = (x - \hat{x})/\sigma(x); \quad E(u) = 0, \sigma^2(u) = 1$$

Mode x_m defined by:

$$P(x = x_m) = \max$$

Median $x_{0.5}$ defined by:

$$F(x_{0.5}) = P(x < x_{0.5}) = 0.5$$

Fractile x_q defined by:

$$F(x_q) = P(x < x_q) = q; \quad 0 \leqslant q \leqslant 1$$

Chebychev's inequality (ch. 3, §4)

$$P(|x - \hat{x}| > k\sigma(x)) < k^{-2}; \quad k \text{ positive real}$$

Several random variables $x = (x_1, x_2, \ldots, x_n)$ (ch. 4)

Distribution function:

$$F(x_1, x_2, \ldots, x_n) = P(x_1 < x_1, x_2 < x_2, \ldots, x_n < x_n)$$

Joint probability density (only for $F(\boldsymbol{x})$ continuous and differentiable in all variables):

$$f(\boldsymbol{x}) = f(x_1, x_2, \ldots, x_n) = \partial^n F(x_1, x_2, \ldots, x_n)/\partial x_1\, \partial x_2 \ldots \partial x_n = \partial F(\boldsymbol{x})/\partial \boldsymbol{x}$$

Marginal distribution of variable x_i:

$$g_i(x_i) = \int_{-\infty}^{\infty} \int_{-\infty}^{\infty} \cdots \int_{-\infty}^{\infty} f(x_1, x_2, \ldots, x_n)\, \mathrm{d}x_1\, \mathrm{d}x_2 \ldots \mathrm{d}x_{i-1}\, \mathrm{d}x_{i+1} \ldots \mathrm{d}x_n$$

Expectation value of function $H(x)$:

$$E\{H(x)\} = \int H(\boldsymbol{x}) f(\boldsymbol{x})\, \mathrm{d}\boldsymbol{x}$$

Expectation value of variable x_i:

$$\hat{x}_i = E(x_i) = \int x_i f(\boldsymbol{x})\,\mathrm{d}\boldsymbol{x} = \int_{-\infty}^{\infty} x_i g_i(x_i)\,\mathrm{d}x_i$$

Variables $x_1, x_2, \ldots, x_n$ are *independent*, if:

$$f(x_1, x_2, \ldots, x_n) = g_1(x_1) g_2(x_2) \ldots g_n(x_n)$$

Moments of order $l_1, l_2, \ldots, l_n$:

(a) about $\boldsymbol{c} = (c_1, c_2, \ldots, c_n)$:

$$\alpha_{l_1 l_2 \ldots l_n} = E\{(x_1 - c_1)^{l_1} (x_2 - c_2)^{l_2} \ldots (x_n - c_n)^{l_n}\}$$

(b) about the origin:

$$\lambda_{l_1 l_2 \ldots l_n} = E\{x_1^{l_1} x_2^{l_2} \ldots x_n^{l_n}\}$$

(c) about $\hat{\boldsymbol{x}}$

$$\mu_{l_1 l_2 \ldots l_n} = E\{(x_1 - \hat{x}_1)^{l_1} (x_2 - \hat{x}_2)^{l_2} \ldots (x_n - \hat{x}_n)^{l_n}\}$$

Variance of x_i:

$$\sigma^2(x_i) = E\{(x_i - \hat{x}_i)^2\} = c_{ii}$$

Covariance between x_i and x_j:

$$\operatorname{cov}(x_i, x_j) = E\{(x_i - \hat{x}_i)(x_j - \hat{x}_j)\} = c_{ij}$$

For x_i, x_j *independent:*

$$\operatorname{cov}(x_i, x_j) = 0$$

Covariance matrix:

$$C = E\{(\boldsymbol{x} - \hat{\boldsymbol{x}})(\boldsymbol{x} - \hat{\boldsymbol{x}})^{\mathrm{T}}\}$$

Correlation coefficient:

$$\rho(x_1, x_2) = \operatorname{cov}(x_1, x_2)/\sigma(x_1)\sigma(x_2); \quad -1 \leqslant \rho \leqslant 1$$

$$\sigma^2(cx_i) = c^2\sigma^2(x_i);$$

$$\sigma^2(ax_i + bx_j) = a^2\sigma^2(x_i) + b^2\sigma^2(x_j) + 2\,ab \operatorname{cov}(x_i, x_j);$$

a, b, c constants

Transformation of variables (ch. 4, §4)

Original variables:

$$\boldsymbol{x} = (x_1, x_2, \ldots, x_n), \quad \text{probability density } f(\boldsymbol{x})$$

TABLE E-2
Distributions of discrete variables

Distribution	Probability to observe $x = k$ $(x_1 = k_1, x_2 = k_2, \ldots, x_l = k_l)$	Mean	Variance (elements of covariance matrix)	Comments
Binomial	$W_k^n = \binom{n}{k} p^k q^{n-k}$	$\hat{x} = np$	$\sigma^2(x) = npq$	$q = 1 - p$
Multinomial	$W_{k_1,k_2,k_l}^n = \dfrac{n!}{\prod_{j=1}^{l} k_j!} \prod_{j=1}^{l} p_j^{kj}$	$\hat{x}_j = np_j$	$c_{ij} = np_i(\delta_{ij} - p_j)$	$\sum_{j=1}^{l} p_j = 1$
Hypergeometric	$W_k = \binom{K}{k}\binom{L}{l} : \binom{N}{n}$	$\hat{x} = n \dfrac{K}{N}$	$\sigma^2(x) = \dfrac{nKL(N-n)}{N^2(N-1)}$	$L = N - K, \quad l = n - k$
Poisson	$f(k) = \dfrac{\lambda^k}{k!} e^{-\lambda}$	$\hat{x} = \lambda$	$\sigma^2(x) = \lambda$	table F-1

TABLE E-3

Distribution	Probability density
Uniform	$f(x) = \begin{cases} 0;\ x < a,\ x \geqslant b \\ \dfrac{1}{b-a};\ a \leqslant x < b \end{cases}$
Normal or Gaussian	$\phi(x) = \dfrac{1}{\sqrt{(2\pi)}b} \exp\left(-\dfrac{(x-a)^2}{2b^2}\right)$
Normalized Gaussian	$\phi_0(x) = \dfrac{1}{\sqrt{(2\pi)}} \exp(-\tfrac{1}{2}x^2)$
Multivariate Gaussian	$\phi(\boldsymbol{x}) = k \exp\{-\tfrac{1}{2}(\boldsymbol{x}-\boldsymbol{a})^{\mathrm{T}} B(\boldsymbol{x}-\boldsymbol{a})\}$
χ^2	$f(\chi^2) = \dfrac{1}{\Gamma(\frac{1}{2}f)2^{\frac{1}{2}f}} (\chi^2)^{\frac{1}{2}f-1} \exp(-\tfrac{1}{2}\chi^2)$
Fisher's F	$f(F) = \left(\dfrac{f_1}{f_2}\right)^{\frac{1}{2}f_1} \dfrac{\Gamma(\frac{1}{2}(f_1+f_2))}{\Gamma(\frac{1}{2}f_1)\,\Gamma(\frac{1}{2}f_2)} F^{\frac{1}{2}f_1-1} \left(1+\dfrac{f_1}{f_2}F\right)^{-\frac{1}{2}(f_1+f_2)}$
Student's t	$f(t) = \dfrac{\Gamma(\frac{1}{2}(f+1))}{\Gamma(\frac{1}{2}f)\sqrt{\pi}\sqrt{f}}\left(1+\dfrac{t^2}{f}\right)^{-\frac{1}{2}(f+1)}$

Distributions of continuous variables

Mean	Variance (covariance matrix)	Comments
$\hat{x} = \frac{1}{2}(b + a)$	$\sigma^2(x) = (b - a)^2/12$	
$\hat{x} = a$	$\sigma^2(x) = b^2$	
$\hat{x} = 0$	$\sigma^2(x) = 1$	tables F-2, F-3
$\hat{\mathbf{x}} = \mathbf{a}$	$C = B^{-1}$	$k=(2\pi)^{-\frac{1}{2}n}(\det B)^{-\frac{1}{2}}$, n being number of variables
$E(\chi^2) = f$	$\sigma^2(\chi^2) = 2f$	f degrees of freedom; tables F-4, F-5
$E(F) = \dfrac{f_2}{f_2 - 2}$, $f_2 > 2$	$\sigma^2(F) = \dfrac{2f_2(f_1^2 + f_2 - 2)}{f_1(f_2 - 2)^2(f_2 - 4)}$, $f_2 > 4$	f_1, f_2 degrees of freedom; table F-6
$E(t) = 0$	$\sigma^2(t) = \dfrac{f}{f - 2}, f > 2$	f degrees of freedom; table F-7

Transformed variables:

$$y = (y_1, y_2, \ldots, y_n)$$

Connection:

$$y_1 = y_1(x), \quad y_2 = y_2(x), \quad \ldots, \quad y_n = y_n(x)$$

Probability density:

$$g(y) = |J| f(x)$$

with *Jacobian*

$$J = J\left(\frac{x_1, x_2, \ldots, x_n}{y_1, y_2, \ldots, y_n}\right) = \begin{vmatrix} \frac{\partial x_1}{\partial y_1} & \frac{\partial x_2}{\partial y_1} & \cdots & \frac{\partial x_n}{\partial y_1} \\ \vdots & & & \\ \frac{\partial x_1}{\partial y_n} & \frac{\partial x_2}{\partial y_n} & & \frac{\partial x_n}{\partial y_n} \end{vmatrix}$$

Propagation of errors (ch. 4, §5)

Original variables x have covariance matrix C_x
Covariance matrix of transformed variables y:

$$C_y = T C_x T^{\mathrm{T}} \text{ with } T = \begin{pmatrix} \frac{\partial y_1}{\partial x_1} & \frac{\partial y_2}{\partial x_1} & \cdots & \frac{\partial y_n}{\partial x_1} \\ \frac{\partial y_1}{\partial x_2} & \frac{\partial y_2}{\partial x_2} & \cdots & \frac{\partial y_n}{\partial x_2} \\ \vdots & & & \\ \frac{\partial y_1}{\partial x_m} & \frac{\partial y_2}{\partial x_m} & \cdots & \frac{\partial y_n}{\partial x_m} \end{pmatrix}$$

Formula is exact only if connection between y and x is linear, good approximation if only small deviation from linearity in neighbourhood of order of standard deviations around $\hat{x}$
Only for vanishing covariances in C_x:

$$\sigma(y_i) = \Delta y_i = \sqrt{\left[\sum_{j=1}^{m} \left(\frac{\partial y_i}{\partial x_j}\right)^2 (\Delta x_i)^2\right]}$$

Law of large numbers (ch. 5, §2)

A total of n observations is performed, described by random variable x_i (= 1 if ith observation yields event A, otherwise = 0)
Frequency of A is:

$$h = \frac{1}{n} \sum_{i=1}^{n} x_i$$

For $n \to \infty$ this frequency becomes equal to probability p of A:

$$E(h) = \hat{h} = p, \quad \sigma^2(h) = \frac{1}{n}\, p\,(1 - p)$$

Central limit theorem (ch. 5, §11)

If x_i are independent variables with mean a and variance σ^2 then, for $n \to \infty$, $(1/n)\, \Sigma_{i=1}^{n}\, x_i$ follows the normal distribution with mean a and variance σ^2/n

Convolution (ch. 5, §13), i.e. distribution of the sum $u = x + y$ of two independently distributed random variables x and y:

$$f_u(u) = \int_{-\infty}^{\infty} f_x(x)\, f_y(u - x)\, \mathrm{d}x = \int_{-\infty}^{\infty} f_y(y)\, f_x(u - y)\, \mathrm{d}y$$

– of Poisson distributions with parameters λ_1 and λ_2 is again a Poisson distribution with parameter $\lambda = \lambda_1 + \lambda_2$ (example 5-5)
– of χ^2-distributions with f_1 and f_2 degrees of freedom is again a χ^2-distribution with $f = f_1 + f_2$ degrees of freedom (ch. 6, §5)

Sampling (ch. 6), see also table E-4

Population: infinite (in some cases finite) set of elements described by random variable x which may follow discrete or continuous distribution
(Random) sample of size N: selection of N elements $(x^{(1)}, x^{(2)}, \ldots, x^{(N)})$ from the population (for condition that sampling is random, see ch. 6, §1)

Sample distribution function:
$W_n(x) = n_x/N$ with n_x being the number of elements in sample with $x < x$

TABLE E-4
Sampling from various populations

	Sampling from population described by continuous variable x, sample size n	Sampling from partitioned population; variables $x_1, x_2, \ldots, x_t$, subsample sizes $n_1, n_2, \ldots, n_t$, $\sum_{j=1}^{t} n_j = n$	Sampling without replacement from finite population of size N; population described by discrete variable y, sample by x; sample size n
Population mean	$E(x) = \hat{x}$	(of subpopulations) $\hat{x}_1, \hat{x}_2, \ldots, \hat{x}_t$	$\bar{y} = \frac{1}{N} \sum_{j=1}^{N} y_j$
Population variance	$\sigma^2(x)$	(of subpopulations) $\sigma^2(x_1), \sigma^2(x_2), \ldots, \sigma^2(x_t)$	$\sigma^2(y) = \frac{1}{N-1} \sum_{j=1}^{N} (y_j - \bar{y})^2$
Sample mean	$\bar{x} = \frac{1}{n} \sum_{i=1}^{n} x_i$	$\bar{x} = \frac{1}{n} \sum_{j=1}^{t} n_j \bar{x}_j; \quad \bar{x}_j = \frac{1}{n_j} \sum_{i=1}^{n_j} x_{ij}$	$\bar{x} = \frac{1}{n} \sum_{i=1}^{n} x_i$
Variance of sample mean	$\sigma^2(\bar{x}) = \frac{1}{n} \sigma^2(x)$	$\sigma^2(\bar{x}) = \frac{1}{n} \sum_{j=1}^{t} \frac{n_j}{n} \sigma^2(x_j)$	$\sigma^2(\bar{x}) = \frac{\sigma^2(y)}{n} \left(1 - \frac{n}{N}\right)$
Sample variance	$s^2 = \frac{1}{n-1} \sum_{i=1}^{n} (x_i - \bar{x})^2$		$s^2 = \frac{1}{n-1} \sum_{i=1}^{n} (x_i - \bar{x})^2$

Statistic: any function of elements of the sample

$$S = S(x^{(1)}, x^{(2)}, \ldots, x^{(N)})$$

Estimator: statistic used to estimate a *parameter* λ of the population; an estimator is *unbiased* if $E(S) = \lambda$ and *consistent* if

$$\lim_{N \to \infty} \sigma(S) = 0$$

Maximum likelihood method (ch. 7), see also table E-5

A population may be described by the probability density $f(\mathbf{x}; \boldsymbol{\lambda})$ with $\boldsymbol{\lambda} = (\lambda_1, \lambda_2, \ldots, \lambda_p)$ being a set of parameters. If a sample $\mathbf{x}^{(1)}, \mathbf{x}^{(2)}, \ldots, \mathbf{x}^{(N)}$ is drawn it has the *likelihood function:*

$$L = \prod_{j=1}^{N} f(\mathbf{x}^{(j)}, \boldsymbol{\lambda})$$

and the (*logarithmic*) *likelihood function*

$$l = \ln L$$

To determine the unknown parameters $\boldsymbol{\lambda}$ from the sample the maximum likelihood method consists of choosing that value of $\boldsymbol{\lambda}$ for which L (or l) becomes maximum, i.e. to solve the *likelihood equations:*

$$\partial l / \partial \lambda_i = 0; \qquad i = 1, 2, \ldots, p$$

or, for only one parameter:

$$\mathrm{d}l/\mathrm{d}\lambda = l' = 0$$

Information of a sample:

$$I(\lambda) = E(l'^2) = -E(l'')$$

Information inequality:

$$\sigma^2(S) \geqslant \{1 - B'(\lambda)\}^2 / I(\lambda)$$

S is an estimator for λ, $B(\lambda) = E(S) - \lambda$ its *bias*

Estimator has minimum variance, if:

$l' = A(\lambda)(S - E(S))$ holds, where $A(\lambda)$ does not depend on the sample

The maximum likelihood estimator $\tilde{\lambda}$, i.e. the solution of the likelihood equation is unique and is asymptotically (i.e. for $N \to \infty$) unbiased and of minimum variance

TABLE E-5

Asymptotic properties (i.e. for sample size $N \to \infty$) of likelihood function and maximum likelihood estimators

	One parameter λ	Several parameters $\boldsymbol{\lambda} = (\lambda_1, \lambda_2, \ldots, \lambda_p)$
Likelihood function	$L = \text{const.} \exp\left(-\dfrac{(\lambda - \tilde{\lambda})^2}{2b^2}\right)$	$L = \text{const.} \exp\{\tfrac{1}{2}(\boldsymbol{\lambda} - \tilde{\boldsymbol{\lambda}})^{\mathrm{T}} B (\boldsymbol{\lambda} - \tilde{\boldsymbol{\lambda}})\}$
Variance (covariance matrix) of maximum likelihood estimate	$\sigma^2(\tilde{\lambda}) = b^2 = \dfrac{1}{E(l'^2(\lambda))} = -\dfrac{1}{E(l''(\lambda))}$	$C = B^{-1}; \quad B = \begin{pmatrix} E\left(\dfrac{\partial^2 l}{\partial \lambda_1^2}\right) & E\left(\dfrac{\partial^2 l}{\partial \lambda_1 \partial \lambda_2}\right) & \cdots & E\left(\dfrac{\partial^2 l}{\partial \lambda_1 \partial \lambda_p}\right) \\ E\left(\dfrac{\partial^2 l}{\partial \lambda_2 \partial \lambda_1}\right) & E\left(\dfrac{\partial^2 l}{\partial \lambda_2^2}\right) & \cdots & E\left(\dfrac{\partial^2 l}{\partial \lambda_2 \partial \lambda_p}\right) \\ \vdots & & & \\ E\left(\dfrac{\partial^2 l}{\partial \lambda_p \partial \lambda_1}\right) & E\left(\dfrac{\partial^2 l}{\partial \lambda_p \partial \lambda_2}\right) & \cdots & E\left(\dfrac{\partial^2 l}{\partial \lambda_p^2}\right) \end{pmatrix}_{\boldsymbol{\lambda} = \tilde{\boldsymbol{\lambda}}}$ i.e.: $\sigma^2(\tilde{\lambda}_i) = c_{ii}; \quad \operatorname{cov}(\tilde{\lambda}_j, \tilde{\lambda}_k) = c_{jk}$

Testing of hypotheses (ch. 8), see also table E-6

Null hypothesis $H_0(\boldsymbol{\lambda} = \boldsymbol{\lambda}_0)$: assumption on the value(s) of parameter(s) $\boldsymbol{\lambda}$ describing a population with distribution $f(\boldsymbol{x}; \boldsymbol{\lambda})$

Alternative hypotheses $H_1(\boldsymbol{\lambda} = \boldsymbol{\lambda}_1)$, $H_2(\boldsymbol{\lambda} = \boldsymbol{\lambda}_2)$, ...: other possibilities for $\boldsymbol{\lambda}$ against which null hypothesis is to be tested by considering a sample $X = (x^{(1)}, x^{(2)}, \ldots, x^{(N)})$ taken from the population

Hypothesis is *simple* if parameters are completely defined, e.g. $H_0(\lambda_1 = 1, \lambda_2 = 5)$, otherwise it is *composite*, e.g. $H_0(\lambda_1 = 2, \lambda_2 < 7)$

Test of hypothesis H_0 with *level of significance* α: H_0 is rejected, if $X \in S_c$, where S_c is a *critical region* in sample space and

$$P(X \in S_c \mid H_0) = \alpha$$

Error of the first kind: rejection of H_0, although in fact H_0 true; probability for this error is α

Error of the second kind: H_0 is not rejected although in fact H_1 true; probability for this error $P(X \bar{\in} S_c \mid H_1) = \beta$

Power function:

$$M(S_c, \boldsymbol{\lambda}) = P(X \in S_c \mid H) = P(X \in S_c \mid \boldsymbol{\lambda})$$

Operation characteristic:

$$L(S_c, \boldsymbol{\lambda}) = 1 - M(S_c, \boldsymbol{\lambda})$$

Most powerful test of H_0 *with respect to* H_1 has:

$$M(S_c, \boldsymbol{\lambda}_1) = 1 - \beta = \text{maximum}$$

Uniformly most powerful test is most powerful with respect to all possible H_1

Unbiased test has:

$$M(S_c, \boldsymbol{\lambda}_1) \geqslant \alpha \text{ for all possible } H_1$$

Neyman–Pearson theorem: a test of H_0 relative to H_1 (both hypotheses simple) using critical region S_c is most powerful if:

$$f(X \mid H_0)/f(x \mid H_1) \leqslant c \text{ for } X \in S_c \text{ and } \geqslant c \text{ for } X \bar{\in} S_c,$$

c being a constant that only depends on α

Test statistic $T(X)$: scalar function of the sample X; after a mapping $X \to T(X)$, $S_c(X) \to U$ the question $X \in S_c$ can be reformulated to $T \in U$

Likelihood ratio test: if ω denotes the region of parameter space corresponding to the null hypothesis H_0 and Ω denotes the region corresponding to

TABLE E-6. Frequently used statistical tests on samples drawn from populations following

Null hypothesis		Test statistic
$\lambda = \lambda_0$ $\lambda \leqslant \lambda_0$ $\lambda \geqslant \lambda_0$	(σ known)	$T = \dfrac{\bar{x} - \lambda_0}{\sigma/\sqrt{N}}$, $\bar{x} = \dfrac{1}{N}\sum_{j=1}^{N} x^{(j)}$
$\lambda = \lambda_0$ $\lambda \leqslant \lambda_0$ $\lambda \geqslant \lambda_0$	Student's test (σ unknown)	$T = \dfrac{\bar{x} - \lambda_0}{s/\sqrt{N}}$, $s^2 = \dfrac{1}{N-1}\sum_{j=1}^{N} (x^{(j)} - \bar{x})^2$
$\sigma^2 = \sigma_0^2$ $\sigma^2 \leqslant \sigma_0^2$ $\sigma^2 \geqslant \sigma_0^2$	χ^2-test on variance	$T = (N-1)\dfrac{s^2}{\sigma_0^2}$
$\lambda_1 = \lambda_2$	Student's difference test (2 samples of sizes N_1 and N_2)	$T = \dfrac{\bar{x}_1 - \bar{x}_2}{s_\Delta^2}$, $s_\Delta^2 = \dfrac{1}{N_1 + N_2 - 2}\left(\dfrac{1}{N_1}\sum_{j=1}^{N_1} (x_1^{(j)} - \bar{x}_1)^2 + \dfrac{1}{N_2}\sum_{j=1}^{N_2} (x_2^{(j)} - \bar{x}_2)\right.$
$\sigma_1^2 = \sigma_2^2$ $\sigma_1^2 \leqslant \sigma_2^2$ $\sigma_1^2 \geqslant \sigma_2^2$	F-test (2 samples of sizes N_1 and N_2)	$T = s_1^2/s_2^2$, $s_i^2 = \dfrac{1}{N_i - 1}\sum_{j=1}^{N_i} (x_i^{(j)} - \bar{x}_i)^2,\ i = 1, 2$

the normal distribution $f(x) = \frac{1}{\sqrt{(2\pi)}\,\sigma} \exp\left(-\frac{(x-\lambda)^2}{2\sigma^2}\right)$

Critical region U of test statistic	Degrees of freedom	See table
$\lvert T\rvert > \Omega'(1-\alpha) = \Omega(1-\frac{1}{2}\alpha)$		F-3b
$T > \Omega(1-\alpha)$	–	F-3a
$T < \Omega(\alpha)$		F-3a
$\lvert T\rvert > t_{1-\frac{1}{2}\alpha}$		
$T > t_{1-\alpha}$	$f = N-1$	F-7
$T < -t_{1-\alpha} = t_\alpha$		
$T < \chi^2_{\frac{1}{2}\alpha}$, $T > \chi^2_{1-\frac{1}{2}\alpha}$		
$T > \chi^2_{1-\alpha}$	$f = N-1$	F-5
$T < \chi^2_\alpha$		
$\lvert T\rvert > t_{1-\frac{1}{2}\alpha}$	$f = N_1 + N_2 - 2$	F-7
$T > F_{1-\frac{1}{2}\alpha}$, $T < F_{\frac{1}{2}\alpha}$		
$T > F_{1-\alpha}$	$f_1 = N_1 - 1$	F-6
$T < F_\alpha$	$f_2 = N_2 - 1$	

TABLE E-7

	General case	Constrained measurements
Equations	$f_k(\boldsymbol{x}, \boldsymbol{\eta}) = 0, k = 1, 2, \ldots, m$	$f_k(\boldsymbol{\eta}) = 0$
First approximations	$\boldsymbol{x}_0, \boldsymbol{\eta}_0 = \boldsymbol{y}$	$\boldsymbol{\eta}_0 = \boldsymbol{y}$
Equations in matrix notation, expanded	$\boldsymbol{f} = A\xi + B\delta + \boldsymbol{c} + \ldots$	$\boldsymbol{f} = B\delta + \boldsymbol{c} + \ldots$
A	$\{A\}_{kl} = \left(\dfrac{\partial f_k}{\partial x_l}\right)_{x_0, \eta_0}$	–
B	$\{B\}_{kl} = \left(\dfrac{\partial f_k}{\partial \eta_l}\right)_{x_0, \eta_0}$	$\{B\}_{kl} = \left(\dfrac{\partial f_k}{\partial \eta_l}\right)_{\eta_0}$
c	$\boldsymbol{c} = \boldsymbol{f}(\boldsymbol{x}_0, \boldsymbol{\eta}_0)$	$\boldsymbol{c} = \boldsymbol{f}(\boldsymbol{\eta}_0)$
Covariance matrix of measurements	$C_y = G_y^{-1}$	$C_y = G_y^{-1}$
Corrections	$\tilde{\xi} = -(A^{\mathrm{T}}G_BA)^{-1}A^{\mathrm{T}}G_B\boldsymbol{c},\ G_B = (BG_y^{-1}B^{\mathrm{T}})^{-1}$ $\tilde{\delta} = G_y^{-1}B^{\mathrm{T}}G_B(A\tilde{\xi} + \boldsymbol{c})$	– $\tilde{\delta} = G_y^{-1}B^{\mathrm{T}}G_B\boldsymbol{c},\ G_B = (BG_y^{-1}B^{\mathrm{T}}$
Next step	$\boldsymbol{x}_1 = \boldsymbol{x}_0 + \tilde{\xi},\ \boldsymbol{\eta}_1 = \boldsymbol{\eta}_0 + \tilde{\delta}$, new values for $A, B, \boldsymbol{c}, \tilde{\xi}, \tilde{\delta}$	$\boldsymbol{\eta}_1 = \boldsymbol{\eta}_0 + \tilde{\delta}$, new values for $B, \boldsymbol{c}, \tilde{\delta}$
Solution (after s steps)	$\tilde{\boldsymbol{x}} = \boldsymbol{x}_{s-1} + \tilde{\xi},\ \tilde{\boldsymbol{\eta}} = \boldsymbol{\eta}_{s-1} + \tilde{\delta},\ \tilde{\varepsilon} = \boldsymbol{y} - \tilde{\boldsymbol{\eta}}$	$\tilde{\boldsymbol{\eta}} = \boldsymbol{\eta}_{s-1} + \tilde{\delta},\ \tilde{\varepsilon} = \boldsymbol{y} - \tilde{\boldsymbol{\eta}}$
Minimum function	$M = (B\tilde{\varepsilon})^{\mathrm{T}}G_B(B\tilde{\varepsilon})$	$M = (B\tilde{\varepsilon})^{\mathrm{T}}G_B(B\tilde{\varepsilon})$
Covariance matrices	$G_{\tilde{x}}^{-1} = (A^{\mathrm{T}}G_BA)^{-1}$ $G_{\tilde{\eta}}^{-1} = G_y^{-1} - G_y^{-1}B^{\mathrm{T}}G_BBG_y^{-1}$ $+ G_y^{-1}B^{\mathrm{T}}G_BA(A^{\mathrm{T}}G_BA)^{-1}A^{\mathrm{T}}G_BBG_y^{-1}$	 $G_{\tilde{\eta}}^{-1} = G_y^{-1} - G_y^{-1}B^{\mathrm{T}}G_BBG_y^{-1}$

Different cases of the least squares problem

Indirect measurements	Direct measurements of different accuracy	Direct measurements of equal accuracy
$f_k = \eta_k - g_k(\boldsymbol{x}) = 0$	$f_k = \eta_k - x$	$f_k = \eta_k - x$
$x_0,\ \boldsymbol{\eta}_0 = \boldsymbol{y}$	$x_0 = 0,\ \boldsymbol{\eta}_0 = \boldsymbol{y}$	$x_0 = 0,\ \boldsymbol{\eta}_0 = \boldsymbol{y}$
$\boldsymbol{f} = A\boldsymbol{\xi} + \boldsymbol{\varepsilon} + \boldsymbol{c} + \dots$	$\boldsymbol{f} = A\boldsymbol{\xi} + \boldsymbol{\varepsilon} + \boldsymbol{c}$	$\boldsymbol{f} = A\boldsymbol{\xi} + \boldsymbol{\varepsilon} + \boldsymbol{c}$
$\{A\}_{kl} = \left(\frac{\partial f_k}{\partial x_l}\right)_{x_0}$	$A = -\begin{pmatrix}1\\1\\\vdots\\1\end{pmatrix}$	$A = -\begin{pmatrix}1\\1\\\vdots\\1\end{pmatrix}$
$B = I$	$B = I$	$B = I$
$\boldsymbol{c} = \boldsymbol{y} - \boldsymbol{g}(\boldsymbol{x}_0)$	$\boldsymbol{c} = \boldsymbol{y}$	$\boldsymbol{c} = \boldsymbol{y}$
$C_y = G_y^{-1}$	$C_y = G_y^{-1} = \begin{pmatrix}\sigma_1^2 & & & 0\\ & \sigma_2^2 & & \\ & & \ddots & \\ 0 & & & \sigma_n^2\end{pmatrix}$	$C_y = G_y^{-1} = \sigma^2 I$
$\tilde{\boldsymbol{\xi}} = -(A^{\mathrm{T}}G_yA)^{-1}A^{\mathrm{T}}G_y^{-1}\boldsymbol{c}$	—	—
—	—	—
$\boldsymbol{x}_1 = \boldsymbol{x}_0 + \tilde{\boldsymbol{\xi}}$, new values for A, $\boldsymbol{c}$, $\tilde{\boldsymbol{\xi}}$	—	—
$\tilde{\boldsymbol{x}} = \boldsymbol{x}_{s-1} + \tilde{\boldsymbol{\xi}},\ \tilde{\boldsymbol{\varepsilon}} = A\tilde{\boldsymbol{\xi}} + \boldsymbol{c}$	$\tilde{\xi} = \tilde{x} = \dfrac{\sum_k \frac{y_k}{\sigma_k^2}}{\sum_k \frac{1}{\sigma_k^2}},\ \tilde{\varepsilon}_k = y_k - \tilde{x}$	$\tilde{\xi} = \tilde{x} = \frac{1}{n}\sum_k y_k,\ \tilde{\varepsilon}_k = y_k - \tilde{x}$
$M = \tilde{\boldsymbol{\varepsilon}}^{\mathrm{T}}G_y\tilde{\boldsymbol{\varepsilon}}$	$M = \tilde{\boldsymbol{\varepsilon}}^{\mathrm{T}}G_y\tilde{\boldsymbol{\varepsilon}}$	$M = \frac{1}{\sigma^2}\sum \tilde{\varepsilon}_k^2$
$G_{\tilde{x}}^{-1} = (A^{\mathrm{T}}G_yA)^{-1}$	$G_{\tilde{x}}^{-1} = \sigma^2(\tilde{x}) = \left(\sum_k \frac{1}{\sigma_k^2}\right)^{-1}$	$G_{\tilde{x}}^{-1} = \sigma^2(\tilde{x}) = \frac{\sigma^2}{n}$
$G_{\tilde{\eta}}^{-1} = A(A^{\mathrm{T}}G_yA)A^{\mathrm{T}}$	—	—

all possible parameter values, the test statistic:

$$T = f(x; \tilde{\boldsymbol{\lambda}}^{(\Omega)})/f(x; \tilde{\boldsymbol{\lambda}}^{(\omega)})$$

is used, where $\tilde{\boldsymbol{\lambda}}^{(\Omega)}$ and $\tilde{\boldsymbol{\lambda}}^{(\omega)}$ are the maximum likelihood estimates in the regions Ω and ω, respectively; H_0 is rejected if $T > T_{1-\alpha}$, with

$$P(T > T_{1-\alpha} | H_0) = \int_{T_{1-\alpha}}^{\infty} g(T)\,\mathrm{d}T = \alpha,$$

$g(T)$ being the conditional probability density of T given H_0

Wilks' theorem: (valid with weak requirements on the probability density of the population): if $H_0(\lambda_1 = \lambda_{10}, \ldots, \lambda_r = \lambda_{r0})$, $r < p$, specifies r out of p parameters, then $-2\ln T$ (T being the likelihood ratio statistic) follows a χ^2-distribution with $f = p - r$ degrees of freedom in the limit of infinite sample size; for $r = p$ one has $f = 1$

χ^2-test on goodness of fit: a sample $(x^{(1)}, x^{(2)}, \ldots, x^{(n)})$ yields $n_1, n_2, \ldots, n_r$ elements in the intervals $\xi_1, \xi_2, \ldots, \xi_r$ of variable x; the null hypothesis assumes distribution of population to be given by $f(x)$; test statistic is:

$$T = \sum_{k=1}^{r} \frac{(n_k - np_k)^2}{np_k}; \quad p_k = \int_{\xi_k} f(x)\,\mathrm{d}x$$

Hypothesis is rejected if $T > \chi^2_{1-\alpha}$ with $f = r - 1$ degrees of freedom; if in addition p parameters $\boldsymbol{\lambda}$ estimated from the n_k using maximum likelihood we have

$$T = \sum_{k=1}^{r} \frac{(n_k - np_k(\tilde{\boldsymbol{\lambda}}))^2}{np_k(\tilde{\boldsymbol{\lambda}})}; \quad p_k(\tilde{\boldsymbol{\lambda}}) = \int_{\xi_k} f(x, \tilde{\boldsymbol{\lambda}})\,\mathrm{d}x$$

Critical region is again $T > \chi^2_{1-\alpha}$ with $f = r - p - 1$

The method of least squares (ch. 9), see also table E-7

Given is a set of *m equations*

$$f_k(\mathbf{x}, \boldsymbol{\eta}) = 0; \quad k = 1, \ldots, m$$

connecting the r-vector of *unknowns* $\mathbf{x} = (x_1, x_2, \ldots, x_r)$ with the n-vector of *measurable quantities* $\boldsymbol{\eta} = (\eta_1, \eta_2, \ldots, \eta_n)$. Instead of the $\boldsymbol{\eta}$ other quantities $\mathbf{y}$ are measured that differ from them by the measurement *errors* $\boldsymbol{\varepsilon}$, i.e. $\mathbf{y} = \boldsymbol{\eta} + \boldsymbol{\varepsilon}$. The $\boldsymbol{\varepsilon}$ are assumed to be distributed normally around zero. This is described by the *covariance matrix* $C_y =$

G_y^{-1}. To obtain *solutions* $\tilde{\boldsymbol{x}}$, $\tilde{\boldsymbol{\eta}}$ the f_k are expanded at the *first approximations* $\boldsymbol{x}_0$, $\boldsymbol{\eta}_0 = \boldsymbol{y}$. Only the linear terms of the expansion are kept and second approximations $\boldsymbol{x}_1 = \boldsymbol{x}_0 + \boldsymbol{\xi}$, $\boldsymbol{\eta}_1 = \boldsymbol{\eta}_0 + \boldsymbol{\delta}$ are obtained. The process is then repeated in iterations until a certain convergence criterion is met, e.g. until a scalar function M (given in table E-7) does not decrease any more. If the f_k are linear in $\boldsymbol{x}$ and $\boldsymbol{\eta}$ only one step is necessary. The method can be understood as a prescription to minimize M. The function M belonging to the solution depends on the measurements errors. It is a random variable and follows a χ^2-distribution with $f = n - r$ degrees of freedom. A χ^2-*test* can therefore be performed on it to test the quality of the solution or the assumptions in particular on G_y^{-1}. If the $\boldsymbol{\varepsilon}$ are not distributed normally, but symmetrically around zero, the least squares solution $\tilde{\boldsymbol{x}}$ still has the smallest variance and $E(M) = n - r$ (Gauss–Markov theorem)

Analysis of variance (ch. 11)

Purpose: a random variable (measurement) x is studied under the influence of *external variables*. Different *models* are constructed on the influence of these variables on x. By appropriately constructed F-tests it is then decided whether such an influence exists.

Example E-1. *Crossed two-way classification with several observations.* Two external variables give rise to classification of observations into classes A_i, B_j ($i = 1, \dots, I, j = 1, \dots, J$). Each class A_i, B_j contains K observations x_{ijk} ($k = 1, \dots, K$). *The model*

$$x_{ijk} = \mu + a_i + b_j + (ab)_{ij} + \varepsilon_{ijk}$$

is assumed, where ε_{ijk} is the error of observations, assumed to be normally distributed about zero, and a_i, b_j and $(ab)_{ij}$ are the influences of the *classifications* in A_i, B_j and *interaction* of both. Three null hypotheses

$$H_0^{(A)}\ (a_i = 0, \quad i = 1, \dots, I), \qquad H_0^{(B)}\ (b_j = 0, \quad j = 1, \dots, J),$$

$$H_0^{(AB)}\ ((ab)_{ij} = 0, \quad i = 1, \dots, I, \quad j = 1, \dots, J)$$

can be tested using the quotients $F^{(A)}$, $F^{(B)}$, $F^{(AB)}$ given in the *analysis of variance table* (table 11-6). For other models, see ch. 11.

Linear regression (ch. 12)

Linear regression is the special case of the type "indirect measurements" of a least squares problem (table E-7), where the $g_k(\boldsymbol{x})$ are *linear* in $\boldsymbol{x}$. Only the case $\boldsymbol{x} = (x_1, x_2) = (\alpha, \beta)$ is considered here. The *measurements* are $\boldsymbol{y} = (y_1, y_2, ..., y_n)$, the equations $f_k = y_k - \alpha - \beta t_k - \varepsilon_k$. The t_k are known numbers, the ε_k are distributed about zero with variance σ_y^2. According to table E-7 the *least squares solution* is

$$\tilde{\alpha} = \frac{(\Sigma t_i^2)(\Sigma y_i) - (\Sigma t_i)(\Sigma t_i y_i)}{n\Sigma t_i^2 - (\Sigma t_i)^2} = \bar{y} - \bar{t}\tilde{\beta}$$

$$\tilde{\beta} = \frac{n\Sigma t_i y_i - (\Sigma t_i)(\Sigma y_i)}{n\Sigma t_i^2 - (\Sigma t_i)^2} = \frac{\Sigma t_i' \, y_i'}{\Sigma t_i'^2}$$

$$\bar{t} = \frac{1}{n}\Sigma t_i, \quad \bar{y} = \frac{1}{n}\Sigma y_i, \quad t_i' = t_i - \bar{t}, \quad y_i' = y_i - \bar{y}$$

Σ stands for $\sum_{i=1}^{n}$

The covariance matrix of $\tilde{x} = (\tilde{\alpha}, \tilde{\beta})$ is:

$$C_{\tilde{x}} = \sigma_y^2 \begin{pmatrix} 1 - \dfrac{\bar{t}^2}{\Sigma t_i'^2} & -\dfrac{\bar{t}}{\Sigma t_i'^2} \\ & \\ -\dfrac{\bar{t}}{\Sigma t_i'^2} & \dfrac{1}{\Sigma t_i'^2} \end{pmatrix}$$

Time series analysis (ch. 13)

A series of measurements $y_i(t_i)$, $i = 1, ..., n$ is given which depends (in some unknown way) on the controlled variable t (usually the time). One assumes that the y_i are composed of a *trend* η_i and an error ε_i, i.e. $y_i = \eta_i + \varepsilon_i$. The measurements are performed at equidistant time intervals, i.e. $t_i - t_{i-1} = \text{const}$. In order to reduce the error ε_i for each t_i $(i > k, i \leqslant n - k)$ a *moving average* is obtained by using a total of $2k + 1$ measurements centered at i and fitting a polynomial of order l to them. The result of the fit at the point t_i is the moving average

$$\tilde{\eta}_0(i) = a_{-k}y_{i-k} + a_{-k+1}y_{i-k+1} + ... + a_k y_{i+k}.$$

The coefficients $a_{-k}, ..., a_k$ are given in table 13-1 for small values of k and l. For higher values they can be computed using the equations of ch. 13, § 2. For end points t_i $(i < k, i > n - k)$ the results of the fit are used also outside the center of interval of the $2k + 1$ measurements (cf. ch. 13, § 3).

Appendix F

STATISTICAL TABLES

Table F-1
Poisson distribution

$$f(k) = e^{-\lambda} \lambda^k / k!$$

λ	k = 0	1	2	3	4	5	6	7	8	9
0.1	0.90484	0.09048	0.00452	0.00015	0.00000	0.00000	0.00000	0.00000	0.00000	0.00000
0.2	0.81873	0.16375	0.01637	0.00109	0.00005	0.00000	0.00000	0.00000	0.00000	0.00000
0.3	0.74082	0.22225	0.03334	0.00333	0.00025	0.00002	0.00000	0.00000	0.00000	0.00000
0.4	0.67032	0.26813	0.05363	0.00715	0.00072	0.00006	0.00000	0.00000	0.00000	0.00000
0.5	0.60653	0.30327	0.07582	0.01264	0.00158	0.00016	0.00001	0.00000	0.00000	0.00000
0.6	0.54881	0.32929	0.09879	0.01976	0.00296	0.00036	0.00004	0.00000	0.00000	0.00000
0.7	0.49659	0.34761	0.12166	0.02839	0.00497	0.00070	0.00008	0.00001	0.00000	0.00000
0.8	0.44933	0.35946	0.14379	0.03834	0.00767	0.00123	0.00016	0.00002	0.00000	0.00000
0.9	0.40657	0.36591	0.16466	0.04940	0.01111	0.00200	0.00030	0.00004	0.00000	0.00000
1.0	0.36788	0.36788	0.18394	0.06131	0.01533	0.00307	0.00051	0.00007	0.00001	0.00000
2.0	0.13534	0.27067	0.27067	0.18045	0.09022	0.03609	0.01203	0.00344	0.00086	0.00019
3.0	0.04979	0.14936	0.22404	0.22404	0.16803	0.10082	0.05041	0.02160	0.00810	0.00270
4.0	0.01832	0.07326	0.14652	0.19537	0.19537	0.15629	0.10420	0.05954	0.02977	0.01323
5.0	0.00674	0.03369	0.08422	0.14037	0.17547	0.17547	0.14622	0.10444	0.06528	0.03627
6.0	0.00248	0.01487	0.04462	0.08923	0.13385	0.16062	0.16062	0.13768	0.10326	0.06884
7.0	0.00091	0.00638	0.02234	0.05213	0.09123	0.12772	0.14900	0.14900	0.13038	0.10140
8.0	0.00034	0.00268	0.01073	0.02863	0.05725	0.09160	0.12214	0.13959	0.13959	0.12408
9.0	0.00012	0.00111	0.00500	0.01499	0.03374	0.06073	0.09109	0.11712	0.13175	0.13175
10.0	0.00005	0.00045	0.00227	0.00757	0.01892	0.03783	0.06306	0.09008	0.11260	0.12511
11.0	0.00002	0.00018	0.00101	0.00370	0.01019	0.02242	0.04109	0.06458	0.08879	0.10852
12.0	0.00001	0.00007	0.00044	0.00177	0.00531	0.01274	0.02548	0.04368	0.06552	0.08736
13.0	0.00000	0.00003	0.00019	0.00083	0.00269	0.00699	0.01515	0.02814	0.04573	0.06605
14.0	0.00000	0.00001	0.00008	0.00038	0.00133	0.00373	0.00870	0.01739	0.03044	0.04734
15.0	0.00000	0.00000	0.00003	0.00017	0.00065	0.00194	0.00484	0.01037	0.01944	0.03241
16.0	0.00000	0.00000	0.00001	0.00008	0.00031	0.00098	0.00262	0.00599	0.01199	0.02131
17.0	0.00000	0.00000	0.00001	0.00003	0.00014	0.00049	0.00139	0.00337	0.00716	0.01353
18.0	0.00000	0.00000	0.00000	0.00001	0.00007	0.00024	0.00072	0.00185	0.00416	0.00833
19.0	0.00000	0.00000	0.00000	0.00001	0.00003	0.00012	0.00037	0.00099	0.00236	0.00498
20.0	0.00000	0.00000	0.00000	0.00000	0.00001	0.00005	0.00018	0.00052	0.00131	0.00291
21.0	0.00000	0.00000	0.00000	0.00000	0.00001	0.00003	0.00009	0.00027	0.00071	0.00166

TABLE F-1 (continued)
Poisson distribution

$$f(k) = e^{-\lambda} \lambda^k / k!$$

λ	k = 10	11	12	13	14	15	16	17	18	19
1.0	0.00000	0.00000	0.00000	0.00000	0.00000	0.00000	0.00000	0.00000	0.00000	0.00000
2.0	0.00004	0.00001	0.00000	0.00000	0.00000	0.00000	0.00000	0.00000	0.00000	0.00000
3.0	0.00081	0.00022	0.00006	0.00001	0.00000	0.00000	0.00000	0.00000	0.00000	0.00000
4.0	0.00529	0.00192	0.00064	0.00020	0.00006	0.00002	0.00000	0.00000	0.00000	0.00000
5.0	0.01813	0.00824	0.00343	0.00132	0.00047	0.00016	0.00005	0.00001	0.00000	0.00000
6.0	0.04130	0.02253	0.01126	0.00520	0.00223	0.00089	0.00033	0.00012	0.00004	0.00001
7.0	0.07098	0.04517	0.02635	0.01419	0.00709	0.00331	0.00145	0.00060	0.00023	0.00009
8.0	0.09926	0.07219	0.04813	0.02962	0.01692	0.00903	0.00451	0.00212	0.00094	0.00040
9.0	0.11858	0.09702	0.07276	0.05038	0.03238	0.01943	0.01093	0.00579	0.00289	0.00137
10.0	0.12511	0.11374	0.09478	0.07291	0.05208	0.03472	0.02170	0.01276	0.00709	0.00373
11.0	0.11938	0.11938	0.10943	0.09259	0.07275	0.05335	0.03668	0.02373	0.01450	0.00840
12.0	0.10484	0.11437	0.11437	0.10557	0.09049	0.07239	0.05429	0.03832	0.02555	0.01614
13.0	0.08587	0.10148	0.10994	0.10994	0.10209	0.08847	0.07189	0.05497	0.03970	0.02716
14.0	0.06628	0.08436	0.09842	0.10599	0.10599	0.09892	0.08656	0.07128	0.05544	0.04085
15.0	0.04861	0.06629	0.08286	0.09561	0.10243	0.10243	0.09603	0.08473	0.07061	0.05575
16.0	0.03410	0.04960	0.06613	0.08139	0.09302	0.09922	0.09922	0.09338	0.08300	0.06990
17.0	0.02300	0.03554	0.05035	0.06585	0.07996	0.09062	0.09628	0.09628	0.09093	0.08136
18.0	0.01498	0.02452	0.03678	0.05093	0.06548	0.07857	0.08840	0.09360	0.09360	0.08867
19.0	0.00947	0.01635	0.02589	0.03784	0.05135	0.06504	0.07724	0.08633	0.09112	0.09112
20.0	0.00582	0.01058	0.01762	0.02712	0.03874	0.05165	0.06456	0.07595	0.08439	0.08883
21.0	0.00349	0.00665	0.01164	0.01881	0.02821	0.03950	0.05184	0.06404	0.07472	0.08258
22.0	0.00204	0.00408	0.00749	0.01267	0.01991	0.02920	0.04015	0.05196	0.06350	0.07353
23.0	0.00117	0.00245	0.00469	0.00831	0.01365	0.02092	0.03008	0.04069	0.05200	0.06294
24.0	0.00066	0.00144	0.00288	0.00531	0.00911	0.01457	0.02186	0.03086	0.04115	0.05198
25.0	0.00036	0.00083	0.00173	0.00332	0.00593	0.00989	0.01545	0.02273	0.03157	0.04153
26.0	0.00020	0.00047	0.00102	0.00204	0.00378	0.00655	0.01065	0.01629	0.02352	0.03219
27.0	0.00011	0.00026	0.00059	0.00122	0.00236	0.00425	0.00717	0.01138	0.01707	0.02426
28.0	0.00006	0.00014	0.00034	0.00072	0.00144	0.00270	0.00472	0.00777	0.01209	0.01781
29.0	0.00003	0.00008	0.00019	0.00042	0.00087	0.00168	0.00304	0.00519	0.00836	0.01276
30.0	0.00002	0.00004	0.00010	0.00024	0.00051	0.00103	0.00193	0.00340	0.00566	0.00894

TABLE F-1 (continued)
Poisson distribution

$$f(k) = e^{-\lambda} \lambda^k / k!$$

λ	k = 20	21	22	23	24	25	26	27	28	29
11.0	0.00462	0.00242	0.00121	0.00058	0.00027	0.00012	0.00005	0.00002	0.00001	0.00000
12.0	0.00968	0.00553	0.00302	0.00157	0.00079	0.00038	0.00017	0.00008	0.00003	0.00001
13.0	0.01766	0.01093	0.00646	0.00365	0.00198	0.00103	0.00051	0.00025	0.00011	0.00005
14.0	0.02860	0.01906	0.01213	0.00738	0.00431	0.00241	0.00130	0.00067	0.00034	0.00016
15.0	0.04181	0.02986	0.02036	0.01328	0.00830	0.00498	0.00287	0.00160	0.00086	0.00044
16.0	0.05592	0.04260	0.03099	0.02156	0.01437	0.00920	0.00566	0.00335	0.00192	0.00106
17.0	0.06916	0.05599	0.04326	0.03198	0.02265	0.01540	0.01007	0.00634	0.00385	0.00226
18.0	0.07980	0.06840	0.05597	0.04380	0.03285	0.02365	0.01637	0.01092	0.00702	0.00436
19.0	0.08657	0.07832	0.06764	0.05588	0.04424	0.03362	0.02457	0.01729	0.01173	0.00769
20.0	0.08884	0.08460	0.07691	0.06688	0.05573	0.04459	0.03430	0.02541	0.01815	0.01252
21.0	0.08671	0.08671	0.08277	0.07557	0.06613	0.05555	0.04486	0.03489	0.02617	0.01895
22.0	0.08088	0.08473	0.08473	0.08105	0.07429	0.06538	0.05532	0.04508	0.03542	0.02687
23.0	0.07239	0.07928	0.08288	0.08288	0.07943	0.07308	0.06464	0.05507	0.04523	0.03588
24.0	0.06238	0.07129	0.07777	0.08115	0.08115	0.07791	0.07191	0.06392	0.05479	0.04534
25.0	0.05192	0.06181	0.07023	0.07634	0.07952	0.07952	0.07646	0.07080	0.06321	0.05450
26.0	0.04185	0.05181	0.06123	0.06922	0.07499	0.07799	0.07799	0.07510	0.06974	0.06252
27.0	0.03275	0.04211	0.05167	0.06066	0.06824	0.07370	0.07654	0.07654	0.07381	0.06872
28.0	0.02493	0.03324	0.04231	0.05151	0.06009	0.06731	0.07248	0.07517	0.07517	0.07258
29.0	0.01851	0.02555	0.03369	0.04247	0.05132	0.05953	0.06640	0.07132	0.07387	0.07387
30.0	0.01341	0.01916	0.02613	0.03408	0.04260	0.05111	0.05898	0.06553	0.07021	0.07263
31.0	0.00951	0.01403	0.01977	0.02665	0.03442	0.04268	0.05089	0.05843	0.06469	0.06915
32.0	0.00660	0.01005	0.01463	0.02035	0.02713	0.03473	0.04274	0.05066	0.05789	0.06388
33.0	0.00449	0.00706	0.01059	0.01519	0.02089	0.02757	0.03500	0.04277	0.05041	0.05736
34.0	0.00300	0.00486	0.00751	0.01110	0.01573	0.02140	0.02798	0.03523	0.04278	0.05016
35.0	0.00197	0.00329	0.00523	0.00796	0.01160	0.01625	0.02187	0.02835	0.03544	0.04277
36.0	0.00127	0.00218	0.00358	0.00560	0.00839	0.01209	0.01674	0.02231	0.02869	0.03561
37.0	0.00081	0.00143	0.00240	0.00387	0.00596	0.00882	0.01255	0.01720	0.02273	0.02900
38.0	0.00051	0.00092	0.00159	0.00263	0.00416	0.00632	0.00924	0.01300	0.01764	0.02312
39.0	0.00031	0.00058	0.00104	0.00176	0.00285	0.00445	0.00668	0.00964	0.01343	0.01807
40.0	0.00019	0.00037	0.00066	0.00116	0.00193	0.00308	0.00474	0.00703	0.01004	0.01385

TABLE F-2a

Normal distribution function

$$\psi_0(x) = \frac{1}{\sqrt{(2\pi)}} \int_{-\infty}^{x} e^{-\frac{1}{2}u^2}\, du$$

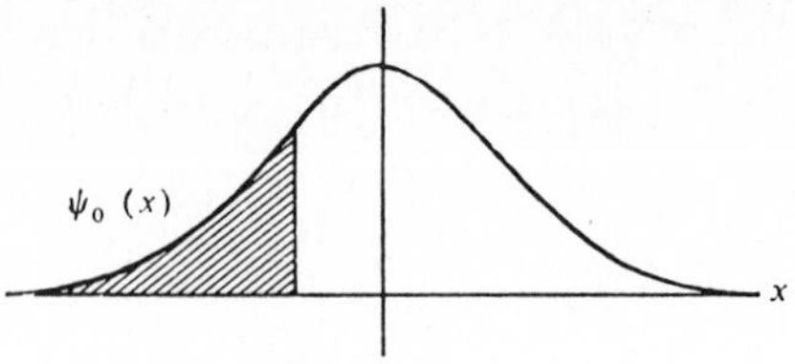

X	0	1	2	3	4	5	6	7	8	9
-3.0	0.001	0.001	0.001	0.001	0.001	0.001	0.001	0.001	0.001	0.001
-2.9	0.002	0.002	0.002	0.002	0.002	0.002	0.002	0.001	0.001	0.001
-2.8	0.003	0.002	0.002	0.002	0.002	0.002	0.002	0.002	0.002	0.002
-2.7	0.003	0.003	0.003	0.003	0.003	0.003	0.003	0.003	0.003	0.003
-2.6	0.005	0.005	0.004	0.004	0.004	0.004	0.004	0.004	0.004	0.004
-2.5	0.006	0.006	0.006	0.006	0.006	0.005	0.005	0.005	0.005	0.005
-2.4	0.008	0.008	0.008	0.008	0.007	0.007	0.007	0.007	0.007	0.006
-2.3	0.011	0.010	0.010	0.010	0.010	0.009	0.009	0.009	0.009	0.008
-2.2	0.014	0.014	0.013	0.013	0.013	0.012	0.012	0.012	0.011	0.011
-2.1	0.018	0.017	0.017	0.017	0.016	0.016	0.015	0.015	0.015	0.014
-2.0	0.023	0.022	0.022	0.021	0.021	0.020	0.020	0.019	0.019	0.018
-1.9	0.029	0.028	0.027	0.027	0.026	0.026	0.025	0.024	0.024	0.023
-1.8	0.036	0.035	0.034	0.034	0.033	0.032	0.031	0.031	0.030	0.029
-1.7	0.045	0.044	0.043	0.042	0.041	0.040	0.039	0.038	0.038	0.037
-1.6	0.055	0.054	0.053	0.052	0.051	0.049	0.048	0.047	0.046	0.046
-1.5	0.067	0.066	0.064	0.063	0.062	0.061	0.059	0.058	0.057	0.056
-1.4	0.081	0.079	0.078	0.076	0.075	0.074	0.072	0.071	0.069	0.068
-1.3	0.097	0.095	0.093	0.092	0.090	0.089	0.087	0.085	0.084	0.082
-1.2	0.115	0.113	0.111	0.109	0.107	0.106	0.104	0.102	0.100	0.099
-1.1	0.136	0.133	0.131	0.129	0.127	0.125	0.123	0.121	0.119	0.117
-1.0	0.159	0.156	0.154	0.152	0.149	0.147	0.145	0.142	0.140	0.138
-0.9	0.184	0.181	0.179	0.176	0.174	0.171	0.169	0.166	0.164	0.161
-0.8	0.212	0.209	0.206	0.203	0.200	0.198	0.195	0.192	0.189	0.187
-0.7	0.242	0.239	0.236	0.233	0.230	0.227	0.224	0.221	0.218	0.215
-0.6	0.274	0.271	0.268	0.264	0.261	0.258	0.255	0.251	0.248	0.245
-0.5	0.309	0.305	0.302	0.298	0.295	0.291	0.288	0.284	0.281	0.278
-0.4	0.345	0.341	0.337	0.334	0.330	0.326	0.323	0.319	0.316	0.312
-0.3	0.382	0.378	0.374	0.371	0.367	0.363	0.359	0.356	0.352	0.348
-0.2	0.421	0.417	0.413	0.409	0.405	0.401	0.397	0.394	0.390	0.386
-0.1	0.460	0.456	0.452	0.448	0.444	0.440	0.436	0.433	0.429	0.425
-0.0	0.500	0.496	0.492	0.488	0.484	0.480	0.476	0.472	0.468	0.464

TABLE F-2a (continued)
Normal distribution function

$$\psi_0(x) = \frac{1}{\sqrt{(2\pi)}} \int_{-\infty}^{x} e^{-\frac{1}{2}u^2} du$$

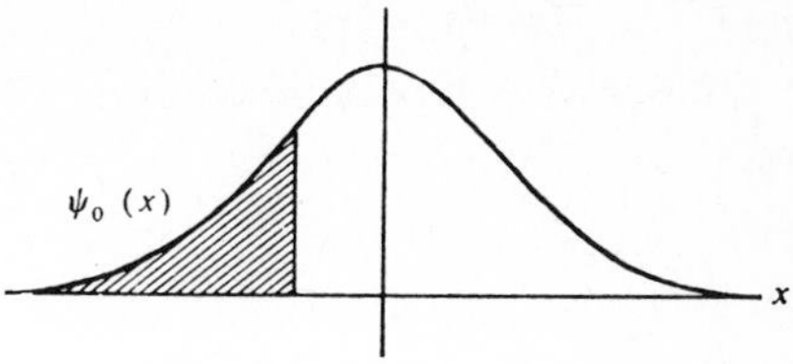

X	0	1	2	3	4	5	6	7	8	9
0.0	0.500	0.504	0.508	0.512	0.516	0.520	0.524	0.528	0.532	0.536
0.1	0.540	0.544	0.548	0.552	0.556	0.560	0.564	0.567	0.571	0.575
0.2	0.579	0.583	0.587	0.591	0.595	0.599	0.603	0.606	0.610	0.614
0.3	0.618	0.622	0.626	0.629	0.633	0.637	0.641	0.644	0.648	0.652
0.4	0.655	0.659	0.663	0.666	0.670	0.674	0.677	0.681	0.684	0.688
0.5	0.691	0.695	0.698	0.702	0.705	0.709	0.712	0.716	0.719	0.722
0.6	0.726	0.729	0.732	0.736	0.739	0.742	0.745	0.749	0.752	0.755
0.7	0.758	0.761	0.764	0.767	0.770	0.773	0.776	0.779	0.782	0.785
0.8	0.788	0.791	0.794	0.797	0.800	0.802	0.805	0.808	0.811	0.813
0.9	0.816	0.819	0.821	0.824	0.826	0.829	0.831	0.834	0.836	0.839
1.0	0.841	0.844	0.846	0.848	0.851	0.853	0.855	0.858	0.860	0.862
1.1	0.864	0.866	0.869	0.871	0.873	0.875	0.877	0.879	0.881	0.883
1.2	0.885	0.887	0.889	0.891	0.893	0.894	0.896	0.898	0.900	0.901
1.3	0.903	0.905	0.907	0.908	0.910	0.911	0.913	0.915	0.916	0.918
1.4	0.919	0.921	0.922	0.924	0.925	0.926	0.928	0.929	0.931	0.932
1.5	0.933	0.934	0.936	0.937	0.938	0.939	0.941	0.942	0.943	0.944
1.6	0.945	0.946	0.947	0.948	0.949	0.951	0.952	0.953	0.954	0.954
1.7	0.955	0.956	0.957	0.958	0.959	0.960	0.961	0.962	0.962	0.963
1.8	0.964	0.965	0.966	0.966	0.967	0.968	0.969	0.969	0.970	0.971
1.9	0.971	0.972	0.973	0.973	0.974	0.974	0.975	0.976	0.976	0.977
2.0	0.977	0.978	0.978	0.979	0.979	0.980	0.980	0.981	0.981	0.982
2.1	0.982	0.983	0.983	0.983	0.984	0.984	0.985	0.985	0.985	0.986
2.2	0.986	0.986	0.987	0.987	0.987	0.988	0.988	0.988	0.989	0.989
2.3	0.989	0.990	0.990	0.990	0.990	0.991	0.991	0.991	0.991	0.992
2.4	0.992	0.992	0.992	0.992	0.993	0.993	0.993	0.993	0.993	0.994
2.5	0.994	0.994	0.994	0.994	0.994	0.995	0.995	0.995	0.995	0.995
2.6	0.995	0.995	0.996	0.996	0.996	0.996	0.996	0.996	0.996	0.996
2.7	0.997	0.997	0.997	0.997	0.997	0.997	0.997	0.997	0.997	0.997
2.8	0.997	0.998	0.998	0.998	0.998	0.998	0.998	0.998	0.998	0.998
2.9	0.998	0.998	0.998	0.998	0.998	0.998	0.998	0.999	0.999	0.999
3.0	0.999	0.999	0.999	0.999	0.999	0.999	0.999	0.999	0.999	0.999

TABLE F-2b
Normal distribution function

$2\psi_0(x) - 1$

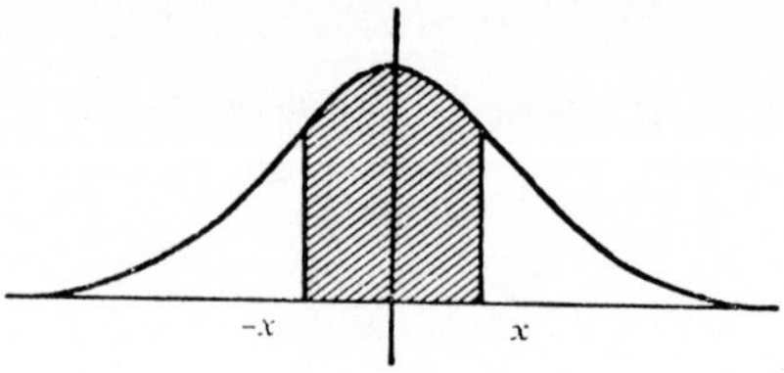

X	0	1	2	3	4	5	6	7	8	9
0.0	0.0	0.008	0.016	0.024	0.032	0.040	0.048	0.056	0.064	0.072
0.1	0.080	0.088	0.096	0.103	0.111	0.119	0.127	0.135	0.143	0.151
0.2	0.159	0.166	0.174	0.182	0.190	0.197	0.205	0.213	0.221	0.228
0.3	0.236	0.243	0.251	0.259	0.266	0.274	0.281	0.289	0.296	0.303
0.4	0.311	0.318	0.326	0.333	0.340	0.347	0.354	0.362	0.369	0.376
0.5	0.383	0.390	0.397	0.404	0.411	0.418	0.425	0.431	0.438	0.445
0.6	0.451	0.458	0.465	0.471	0.478	0.484	0.491	0.497	0.503	0.510
0.7	0.516	0.522	0.528	0.535	0.541	0.547	0.553	0.559	0.565	0.570
0.8	0.576	0.582	0.588	0.593	0.599	0.605	0.610	0.616	0.621	0.627
0.9	0.632	0.637	0.642	0.648	0.653	0.658	0.663	0.668	0.673	0.678
1.0	0.683	0.688	0.692	0.697	0.702	0.706	0.711	0.715	0.720	0.724
1.1	0.729	0.733	0.737	0.742	0.746	0.750	0.754	0.758	0.762	0.766
1.2	0.770	0.774	0.778	0.781	0.785	0.789	0.792	0.796	0.799	0.803
1.3	0.806	0.810	0.813	0.816	0.820	0.823	0.826	0.829	0.832	0.835
1.4	0.838	0.841	0.844	0.847	0.850	0.853	0.856	0.858	0.861	0.864
1.5	0.866	0.869	0.871	0.874	0.876	0.879	0.881	0.884	0.886	0.888
1.6	0.890	0.893	0.895	0.897	0.899	0.901	0.903	0.905	0.907	0.909
1.7	0.911	0.913	0.915	0.916	0.918	0.920	0.922	0.923	0.925	0.927
1.8	0.928	0.930	0.931	0.933	0.934	0.936	0.937	0.939	0.940	0.941
1.9	0.943	0.944	0.945	0.946	0.948	0.949	0.950	0.951	0.952	0.953
2.0	0.954	0.956	0.957	0.958	0.959	0.960	0.961	0.962	0.962	0.963
2.1	0.964	0.965	0.966	0.967	0.968	0.968	0.969	0.970	0.971	0.971
2.2	0.972	0.973	0.974	0.974	0.975	0.976	0.976	0.977	0.977	0.978
2.3	0.979	0.979	0.980	0.980	0.981	0.981	0.982	0.982	0.983	0.983
2.4	0.984	0.984	0.984	0.985	0.985	0.986	0.986	0.986	0.987	0.987
2.5	0.988	0.988	0.988	0.989	0.989	0.989	0.990	0.990	0.990	0.990
2.6	0.991	0.991	0.991	0.991	0.992	0.992	0.992	0.992	0.993	0.993
2.7	0.993	0.993	0.993	0.994	0.994	0.994	0.994	0.994	0.995	0.995
2.8	0.995	0.995	0.995	0.995	0.995	0.996	0.996	0.996	0.996	0.996
2.9	0.996	0.996	0.996	0.997	0.997	0.997	0.997	0.997	0.997	0.997
3.0	0.997	0.997	0.997	0.998	0.998	0.998	0.998	0.998	0.998	0.998

TABLE F-3a
Fractiles of the normal distribution
Values of $\Omega(P) = x_P$, given by

$$P = \frac{1}{\sqrt{(2\pi)}} \int_{-\infty}^{x_P} e^{-\frac{1}{2}u^2}\, du$$

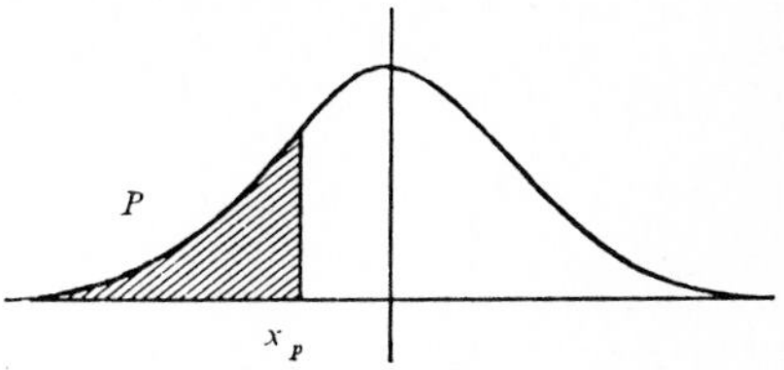

P	0	1	2	3	4	5	6	7	8	9
0.0	-∞	-3.091	-2.879	-2.748	-2.652	-2.576	-2.513	-2.458	-2.409	-2.366
0.01	-2.327	-2.291	-2.258	-2.227	-2.198	-2.171	-2.145	-2.121	-2.097	-2.075
0.02	-2.054	-2.034	-2.015	-1.996	-1.978	-1.960	-1.944	-1.927	-1.911	-1.896
0.03	-1.881	-1.867	-1.853	-1.839	-1.825	-1.812	-1.800	-1.787	-1.775	-1.763
0.04	-1.751	-1.740	-1.728	-1.717	-1.706	-1.696	-1.685	-1.675	-1.665	-1.655
0.05	-1.645	-1.636	-1.626	-1.617	-1.608	-1.599	-1.590	-1.581	-1.572	-1.564
0.06	-1.555	-1.547	-1.539	-1.530	-1.522	-1.514	-1.507	-1.499	-1.491	-1.484
0.07	-1.476	-1.469	-1.461	-1.454	-1.447	-1.440	-1.433	-1.426	-1.419	-1.412
0.08	-1.405	-1.399	-1.392	-1.385	-1.379	-1.372	-1.366	-1.360	-1.353	-1.347
0.09	-1.341	-1.335	-1.329	-1.323	-1.317	-1.311	-1.305	-1.299	-1.293	-1.287
0.10	-1.282	-1.276	-1.270	-1.265	-1.259	-1.254	-1.248	-1.243	-1.237	-1.232
0.11	-1.227	-1.221	-1.216	-1.211	-1.206	-1.200	-1.195	-1.190	-1.185	-1.180
0.12	-1.175	-1.170	-1.165	-1.160	-1.155	-1.150	-1.146	-1.141	-1.136	-1.131
0.13	-1.126	-1.122	-1.117	-1.112	-1.108	-1.103	-1.099	-1.094	-1.089	-1.085
0.14	-1.080	-1.076	-1.071	-1.067	-1.063	-1.058	-1.054	-1.049	-1.045	-1.041
0.15	-1.036	-1.032	-1.028	-1.024	-1.019	-1.015	-1.011	-1.007	-1.003	-0.999
0.16	-0.994	-0.990	-0.986	-0.982	-0.978	-0.974	-0.970	-0.966	-0.962	-0.958
0.17	-0.954	-0.950	-0.946	-0.942	-0.938	-0.935	-0.931	-0.927	-0.923	-0.919
0.18	-0.915	-0.911	-0.908	-0.904	-0.900	-0.896	-0.893	-0.889	-0.885	-0.881
0.19	-0.878	-0.874	-0.870	-0.867	-0.863	-0.859	-0.856	-0.852	-0.849	-0.845
0.20	-0.841	-0.838	-0.834	-0.831	-0.827	-0.824	-0.820	-0.817	-0.813	-0.810
0.21	-0.806	-0.803	-0.799	-0.796	-0.792	-0.789	-0.786	-0.782	-0.779	-0.775
0.22	-0.772	-0.769	-0.765	-0.762	-0.759	-0.755	-0.752	-0.749	-0.745	-0.742
0.23	-0.739	-0.735	-0.732	-0.729	-0.725	-0.722	-0.719	-0.716	-0.712	-0.709
0.24	-0.706	-0.703	-0.700	-0.696	-0.693	-0.690	-0.687	-0.684	-0.681	-0.677
0.25	-0.674	-0.671	-0.668	-0.665	-0.662	-0.659	-0.655	-0.652	-0.649	-0.646
0.26	-0.643	-0.640	-0.637	-0.634	-0.631	-0.628	-0.625	-0.622	-0.619	-0.615
0.27	-0.612	-0.609	-0.606	-0.603	-0.600	-0.597	-0.594	-0.591	-0.588	-0.585
0.28	-0.582	-0.580	-0.577	-0.574	-0.571	-0.568	-0.565	-0.562	-0.559	-0.556
0.29	-0.553	-0.550	-0.547	-0.544	-0.541	-0.538	-0.536	-0.533	-0.530	-0.527
0.30	-0.524	-0.521	-0.518	-0.515	-0.513	-0.510	-0.507	-0.504	-0.501	-0.498
0.31	-0.495	-0.493	-0.490	-0.487	-0.484	-0.481	-0.478	-0.476	-0.473	-0.470
0.32	-0.467	-0.464	-0.462	-0.459	-0.456	-0.453	-0.451	-0.448	-0.445	-0.442
0.33	-0.439	-0.437	-0.434	-0.431	-0.428	-0.426	-0.423	-0.420	-0.417	-0.415
0.34	-0.412	-0.409	-0.407	-0.404	-0.401	-0.398	-0.396	-0.393	-0.390	-0.388
0.35	-0.385	-0.382	-0.379	-0.377	-0.374	-0.371	-0.369	-0.366	-0.363	-0.361
0.36	-0.358	-0.355	-0.353	-0.350	-0.347	-0.345	-0.342	-0.339	-0.337	-0.334
0.37	-0.331	-0.329	-0.326	-0.323	-0.321	-0.318	-0.316	-0.313	-0.310	-0.308
0.38	-0.305	-0.302	-0.300	-0.297	-0.295	-0.292	-0.289	-0.287	-0.284	-0.281
0.39	-0.279	-0.276	-0.274	-0.271	-0.268	-0.266	-0.263	-0.261	-0.258	-0.256
0.40	-0.253	-0.250	-0.248	-0.245	-0.243	-0.240	-0.237	-0.235	-0.232	-0.230
0.41	-0.227	-0.225	-0.222	-0.219	-0.217	-0.214	-0.212	-0.209	-0.207	-0.204
0.42	-0.202	-0.199	-0.196	-0.194	-0.191	-0.189	-0.186	-0.184	-0.181	-0.179
0.43	-0.176	-0.173	-0.171	-0.168	-0.166	-0.163	-0.161	-0.158	-0.156	-0.153
0.44	-0.151	-0.148	-0.146	-0.143	-0.141	-0.138	-0.135	-0.133	-0.130	-0.128
0.45	-0.125	-0.123	-0.120	-0.118	-0.115	-0.113	-0.110	-0.108	-0.105	-0.103
0.46	-0.100	-0.098	-0.095	-0.093	-0.090	-0.088	-0.085	-0.083	-0.080	-0.078
0.47	-0.075	-0.073	-0.070	-0.068	-0.065	-0.063	-0.060	-0.058	-0.055	-0.053
0.48	-0.050	-0.048	-0.045	-0.043	-0.040	-0.038	-0.035	-0.033	-0.030	-0.028
0.49	-0.025	-0.022	-0.020	-0.017	-0.015	-0.012	-0.010	-0.007	-0.005	-0.002

TABLE F-3a (continued)
Fractiles of the normal distribution
Values of $\Omega(P) = x_P$, given by

$$P = \frac{1}{\sqrt{(2\pi)}} \int_{-\infty}^{x_P} e^{-\frac{1}{2}u^2}\, du$$

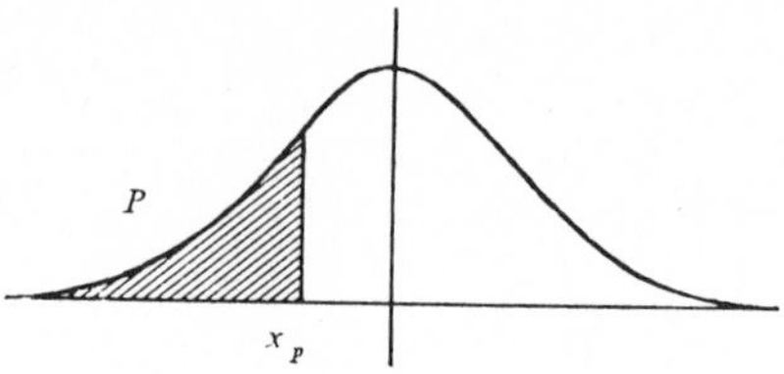

P	0	1	2	3	4	5	6	7	8	9
0.50	0.000	0.002	0.005	0.007	0.010	0.012	0.015	0.017	0.020	0.022
0.51	0.025	0.027	0.030	0.033	0.035	0.038	0.040	0.043	0.045	0.048
0.52	0.050	0.053	0.055	0.058	0.060	0.063	0.065	0.068	0.070	0.073
0.53	0.075	0.078	0.080	0.083	0.085	0.088	0.090	0.093	0.095	0.098
0.54	0.100	0.103	0.105	0.108	0.110	0.113	0.115	0.118	0.120	0.123
0.55	0.125	0.128	0.130	0.133	0.135	0.138	0.141	0.143	0.146	0.148
0.56	0.151	0.153	0.156	0.158	0.161	0.163	0.166	0.168	0.171	0.173
0.57	0.176	0.179	0.181	0.184	0.186	0.189	0.191	0.194	0.196	0.199
0.58	0.202	0.204	0.207	0.209	0.212	0.214	0.217	0.219	0.222	0.225
0.59	0.227	0.230	0.232	0.235	0.237	0.240	0.243	0.245	0.248	0.250
0.60	0.253	0.256	0.258	0.261	0.263	0.266	0.268	0.271	0.274	0.276
0.61	0.279	0.281	0.284	0.287	0.289	0.292	0.295	0.297	0.300	0.302
0.62	0.305	0.308	0.310	0.313	0.316	0.318	0.321	0.323	0.326	0.329
0.63	0.331	0.334	0.337	0.339	0.342	0.345	0.347	0.350	0.353	0.355
0.64	0.358	0.361	0.363	0.366	0.369	0.371	0.374	0.377	0.379	0.382
0.65	0.385	0.388	0.390	0.393	0.396	0.398	0.401	0.404	0.407	0.409
0.66	0.412	0.415	0.417	0.420	0.423	0.426	0.428	0.431	0.434	0.437
0.67	0.439	0.442	0.445	0.448	0.451	0.453	0.456	0.459	0.462	0.464
0.68	0.467	0.470	0.473	0.476	0.478	0.481	0.484	0.487	0.490	0.493
0.69	0.495	0.498	0.501	0.504	0.507	0.510	0.513	0.515	0.518	0.521
0.70	0.524	0.527	0.530	0.533	0.536	0.538	0.541	0.544	0.547	0.550
0.71	0.553	0.556	0.559	0.562	0.565	0.568	0.571	0.574	0.577	0.580
0.72	0.582	0.585	0.588	0.591	0.594	0.597	0.600	0.603	0.606	0.609
0.73	0.612	0.615	0.619	0.622	0.625	0.628	0.631	0.634	0.637	0.640
0.74	0.643	0.646	0.649	0.652	0.655	0.659	0.662	0.665	0.668	0.671
0.75	0.674	0.677	0.680	0.684	0.687	0.690	0.693	0.696	0.700	0.703
0.76	0.706	0.709	0.712	0.716	0.719	0.722	0.725	0.729	0.732	0.735
0.77	0.739	0.742	0.745	0.749	0.752	0.755	0.759	0.762	0.765	0.769
0.78	0.772	0.775	0.779	0.782	0.786	0.789	0.792	0.796	0.799	0.803
0.79	0.806	0.810	0.813	0.817	0.820	0.824	0.827	0.831	0.834	0.838
0.80	0.841	0.845	0.849	0.852	0.856	0.859	0.863	0.867	0.870	0.874
0.81	0.878	0.881	0.885	0.889	0.893	0.896	0.900	0.904	0.908	0.911
0.82	0.915	0.919	0.923	0.927	0.931	0.935	0.938	0.942	0.946	0.950
0.83	0.954	0.958	0.962	0.966	0.970	0.974	0.978	0.982	0.986	0.990
0.84	0.994	0.999	1.003	1.007	1.011	1.015	1.019	1.024	1.028	1.032
0.85	1.036	1.041	1.045	1.049	1.054	1.058	1.063	1.067	1.071	1.076
0.86	1.080	1.085	1.089	1.094	1.099	1.103	1.108	1.112	1.117	1.122
0.87	1.126	1.131	1.136	1.141	1.146	1.150	1.155	1.160	1.165	1.170
0.88	1.175	1.180	1.185	1.190	1.195	1.200	1.206	1.211	1.216	1.221
0.89	1.227	1.232	1.237	1.243	1.248	1.254	1.259	1.265	1.270	1.276
0.90	1.282	1.287	1.293	1.299	1.305	1.311	1.317	1.323	1.329	1.335
0.91	1.341	1.347	1.353	1.360	1.366	1.372	1.379	1.385	1.392	1.399
0.92	1.405	1.412	1.419	1.426	1.433	1.440	1.447	1.454	1.461	1.469
0.93	1.476	1.484	1.491	1.499	1.507	1.514	1.522	1.530	1.539	1.547
0.94	1.555	1.564	1.572	1.581	1.590	1.599	1.608	1.617	1.626	1.636
0.95	1.645	1.655	1.665	1.675	1.685	1.696	1.706	1.717	1.728	1.740
0.96	1.751	1.763	1.775	1.787	1.800	1.812	1.825	1.839	1.853	1.867
0.97	1.881	1.896	1.911	1.927	1.944	1.960	1.978	1.996	2.015	2.034
0.98	2.054	2.075	2.097	2.121	2.145	2.171	2.198	2.227	2.258	2.291
0.99	2.327	2.366	2.409	2.458	2.513	2.576	2.652	2.748	2.878	3.090

TABLE F-3b
Fractiles of the normal distribution
Values of $\Omega'(P) = x'_P$, given by

$$P = \frac{1}{\sqrt{(2\pi)}} \int_{-x'_P}^{x'_P} e^{-\frac{1}{2}u^2}\, du$$

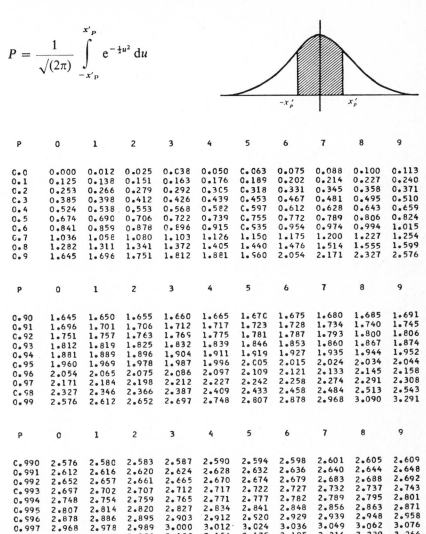

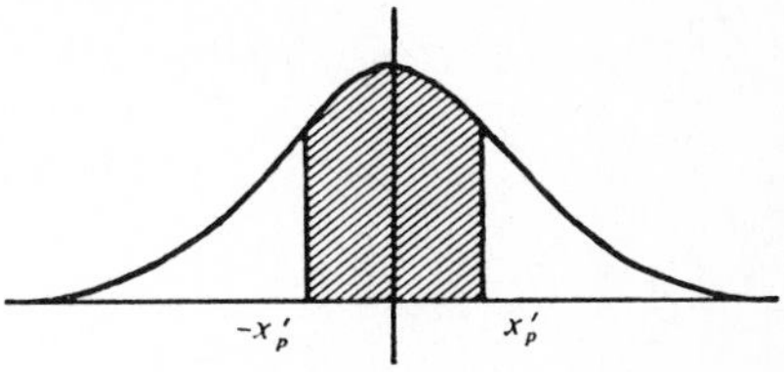

P	0	1	2	3	4	5	6	7	8	9
0.0	0.000	0.012	0.025	0.038	0.050	0.063	0.075	0.088	0.100	0.113
0.1	0.125	0.138	0.151	0.163	0.176	0.189	0.202	0.214	0.227	0.240
0.2	0.253	0.266	0.279	0.292	0.305	0.318	0.331	0.345	0.358	0.371
0.3	0.385	0.398	0.412	0.426	0.439	0.453	0.467	0.481	0.495	0.510
0.4	0.524	0.538	0.553	0.568	0.582	0.597	0.612	0.628	0.643	0.659
0.5	0.674	0.690	0.706	0.722	0.739	0.755	0.772	0.789	0.806	0.824
0.6	0.841	0.859	0.878	0.896	0.915	0.935	0.954	0.974	0.994	1.015
0.7	1.036	1.058	1.080	1.103	1.126	1.150	1.175	1.200	1.227	1.254
0.8	1.282	1.311	1.341	1.372	1.405	1.440	1.476	1.514	1.555	1.599
0.9	1.645	1.696	1.751	1.812	1.881	1.960	2.054	2.171	2.327	2.576

P	0	1	2	3	4	5	6	7	8	9
0.90	1.645	1.650	1.655	1.660	1.665	1.670	1.675	1.680	1.685	1.691
0.91	1.696	1.701	1.706	1.712	1.717	1.723	1.728	1.734	1.740	1.745
0.92	1.751	1.757	1.763	1.769	1.775	1.781	1.787	1.793	1.800	1.806
0.93	1.812	1.819	1.825	1.832	1.839	1.846	1.853	1.860	1.867	1.874
0.94	1.881	1.889	1.896	1.904	1.911	1.919	1.927	1.935	1.944	1.952
0.95	1.960	1.969	1.978	1.987	1.996	2.005	2.015	2.024	2.034	2.044
0.96	2.054	2.065	2.075	2.086	2.097	2.109	2.121	2.133	2.145	2.158
0.97	2.171	2.184	2.198	2.212	2.227	2.242	2.258	2.274	2.291	2.308
0.98	2.327	2.346	2.366	2.387	2.409	2.433	2.458	2.484	2.513	2.543
0.99	2.576	2.612	2.652	2.697	2.748	2.807	2.878	2.968	3.090	3.291

P	0	1	2	3	4	5	6	7	8	9
0.990	2.576	2.580	2.583	2.587	2.590	2.594	2.598	2.601	2.605	2.609
0.991	2.612	2.616	2.620	2.624	2.628	2.632	2.636	2.640	2.644	2.648
0.992	2.652	2.657	2.661	2.665	2.670	2.674	2.679	2.683	2.688	2.692
0.993	2.697	2.702	2.707	2.712	2.717	2.722	2.727	2.732	2.737	2.743
0.994	2.748	2.754	2.759	2.765	2.771	2.777	2.782	2.789	2.795	2.801
0.995	2.807	2.814	2.820	2.827	2.834	2.841	2.848	2.856	2.863	2.871
0.996	2.878	2.886	2.895	2.903	2.912	2.920	2.929	2.939	2.948	2.958
0.997	2.968	2.978	2.989	3.000	3.012	3.024	3.036	3.049	3.062	3.076
0.998	3.090	3.106	3.122	3.138	3.156	3.175	3.195	3.216	3.239	3.264
0.999	3.291	3.320	3.353	3.389	3.431	3.481	3.540	3.615	3.719	3.890

Table F-4

χ^2-distribution function

Values of P, given by

$$P = F(\chi^2, f) = \frac{1}{\Gamma(\frac{1}{2}f)\, 2^{\frac{1}{2}f}} \int_0^{\chi^2} u^{\frac{1}{2}f}\, e^{-\frac{1}{2}u}\, du$$

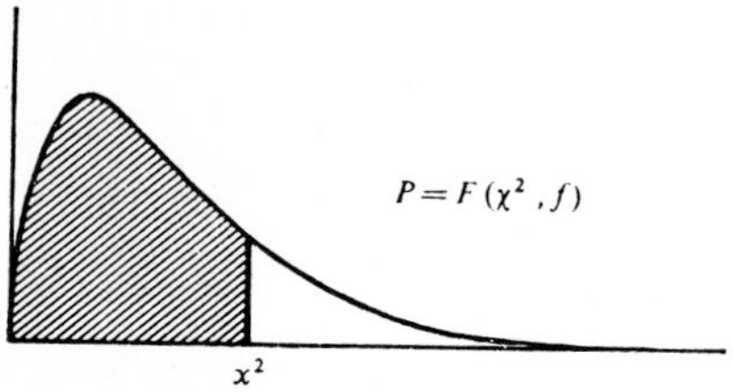

χ^2	f = 1	2	3	4	5	6	7	8	9	10
0.1	0.248	0.049	0.008	0.001	0.000	0.000	0.000	0.000	0.000	0.000
0.2	0.345	0.095	0.022	0.005	0.001	0.000	0.000	0.000	0.000	0.000
0.3	0.416	0.139	0.040	0.010	0.002	0.001	0.000	0.000	0.000	0.000
0.4	0.473	0.181	0.060	0.018	0.005	0.001	0.000	0.000	0.000	0.000
0.5	0.520	0.221	0.081	0.026	0.008	0.002	0.001	0.000	0.000	0.000
0.6	0.561	0.259	0.104	0.037	0.012	0.004	0.001	0.000	0.000	0.000
0.7	0.597	0.295	0.127	0.049	0.017	0.006	0.002	0.000	0.000	0.000
0.8	0.629	0.330	0.151	0.062	0.023	0.008	0.003	0.001	0.000	0.000
0.9	0.657	0.362	0.175	0.075	0.030	0.011	0.004	0.001	0.000	0.000
1.0	0.683	0.393	0.199	0.090	0.037	0.014	0.005	0.002	0.001	0.000
1.1	0.706	0.423	0.223	0.106	0.046	0.018	0.007	0.002	0.001	0.000
1.2	0.727	0.451	0.247	0.122	0.055	0.023	0.009	0.003	0.001	0.000
1.3	0.746	0.478	0.271	0.139	0.065	0.028	0.012	0.004	0.002	0.001
1.4	0.763	0.503	0.294	0.156	0.076	0.034	0.014	0.006	0.002	0.001
1.5	0.779	0.528	0.318	0.173	0.087	0.041	0.018	0.007	0.003	0.001
1.6	0.794	0.551	0.341	0.191	0.099	0.047	0.021	0.009	0.004	0.001
1.7	0.808	0.573	0.363	0.209	0.111	0.055	0.025	0.011	0.005	0.002
1.8	0.820	0.593	0.385	0.228	0.124	0.063	0.030	0.013	0.006	0.002
1.9	0.832	0.613	0.407	0.246	0.137	0.071	0.035	0.016	0.007	0.003
2.0	0.843	0.632	0.428	0.264	0.151	0.080	0.040	0.019	0.009	0.004
3.0	0.917	0.777	0.608	0.442	0.300	0.191	0.115	0.066	0.036	0.019
4.0	0.954	0.865	0.739	0.594	0.451	0.323	0.220	0.143	0.089	0.053
5.0	0.975	0.918	0.828	0.713	0.584	0.456	0.340	0.242	0.166	0.109
6.0	0.986	0.950	0.888	0.801	0.694	0.577	0.460	0.353	0.260	0.185
7.0	0.992	0.970	0.928	0.864	0.779	0.679	0.571	0.463	0.363	0.275
8.0	0.995	0.982	0.954	0.908	0.844	0.762	0.667	0.567	0.466	0.371
9.0	0.997	0.989	0.971	0.939	0.891	0.826	0.747	0.658	0.563	0.468
10.0	0.998	0.993	0.981	0.960	0.925	0.875	0.811	0.735	0.650	0.560
11.0	0.999	0.996	0.988	0.973	0.949	0.912	0.861	0.798	0.724	0.642
12.0	0.999	0.998	0.993	0.983	0.965	0.938	0.899	0.849	0.787	0.715
13.0	1.000	0.998	0.995	0.989	0.977	0.957	0.928	0.888	0.837	0.776
14.0	1.000	0.999	0.997	0.993	0.984	0.970	0.949	0.918	0.878	0.827
15.0	1.000	0.999	0.998	0.995	0.990	0.980	0.964	0.941	0.909	0.868
16.0	1.000	1.000	0.999	0.997	0.993	0.986	0.975	0.958	0.933	0.900
17.0	1.000	1.000	0.999	0.998	0.995	0.991	0.983	0.970	0.951	0.926
18.0	1.000	1.000	1.000	0.999	0.997	0.994	0.988	0.979	0.965	0.945
19.0	1.000	1.000	1.000	0.999	0.998	0.996	0.992	0.985	0.975	0.960
20.0	1.000	1.000	1.000	1.000	0.999	0.997	0.994	0.990	0.982	0.971

TABLE F-4 (continued)
χ^2-distribution function
Values of P, given by

$$P = F(\chi^2, f) = \frac{1}{\Gamma(\frac{1}{2}f)\,2^{\frac{1}{2}f}} \int_0^{\chi^2} u^{\frac{1}{2}f}\, e^{-\frac{1}{2}u}\, du$$

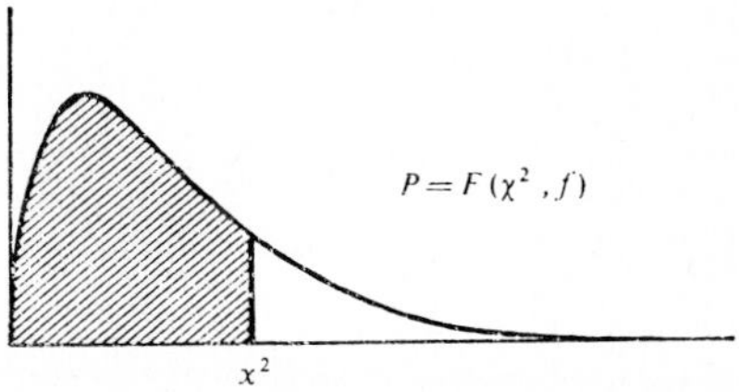

χ^2	f = 11	12	13	14	15	16	17	18	19	20
5.0	0.069	0.042	0.025	0.014	0.008	0.004	0.002	0.001	0.001	0.000
6.0	0.127	0.084	0.054	0.034	0.020	0.012	0.007	0.004	0.002	0.001
7.0	0.201	0.142	0.098	0.065	0.042	0.027	0.016	0.010	0.006	0.003
8.0	0.287	0.215	0.156	0.111	0.076	0.051	0.033	0.021	0.013	0.008
9.0	0.378	0.297	0.227	0.169	0.122	0.087	0.060	0.040	0.027	0.017
10.0	0.470	0.384	0.306	0.238	0.180	0.133	0.096	0.068	0.047	0.032
11.0	0.557	0.471	0.389	0.314	0.247	0.191	0.143	0.106	0.076	0.054
12.0	0.636	0.554	0.472	0.394	0.321	0.256	0.200	0.153	0.114	0.084
13.0	0.707	0.631	0.552	0.473	0.398	0.327	0.264	0.208	0.161	0.123
14.0	0.767	0.699	0.626	0.550	0.474	0.401	0.333	0.271	0.216	0.170
15.0	0.818	0.759	0.693	0.622	0.549	0.475	0.405	0.338	0.277	0.224
16.0	0.859	0.809	0.751	0.687	0.618	0.547	0.476	0.407	0.343	0.283
17.0	0.892	0.850	0.801	0.744	0.681	0.614	0.546	0.477	0.410	0.347
18.0	0.918	0.884	0.842	0.793	0.737	0.676	0.611	0.544	0.478	0.413
19.0	0.939	0.911	0.877	0.835	0.786	0.731	0.671	0.608	0.543	0.478
20.0	0.955	0.933	0.905	0.870	0.828	0.780	0.726	0.667	0.605	0.542
21.0	0.967	0.950	0.927	0.898	0.863	0.821	0.774	0.721	0.663	0.603
22.0	0.976	0.962	0.945	0.921	0.892	0.857	0.815	0.768	0.716	0.659
23.0	0.982	0.972	0.958	0.940	0.916	0.886	0.851	0.809	0.763	0.711
24.0	0.987	0.980	0.969	0.954	0.935	0.910	0.881	0.845	0.804	0.758
25.0	0.991	0.985	0.977	0.965	0.950	0.930	0.905	0.875	0.839	0.799
26.0	0.994	0.989	0.983	0.974	0.962	0.946	0.926	0.900	0.870	0.834
27.0	0.995	0.992	0.988	0.981	0.971	0.959	0.942	0.921	0.895	0.865
28.0	0.997	0.994	0.991	0.986	0.978	0.968	0.955	0.938	0.917	0.891
29.0	0.998	0.996	0.993	0.990	0.984	0.976	0.965	0.952	0.934	0.912
30.0	0.998	0.997	0.995	0.992	0.988	0.982	0.974	0.963	0.948	0.930
31.0	0.999	0.998	0.997	0.994	0.991	0.987	0.980	0.971	0.960	0.945
32.0	0.999	0.999	0.998	0.996	0.994	0.990	0.985	0.978	0.969	0.957
33.0	0.999	0.999	0.998	0.997	0.995	0.993	0.989	0.983	0.976	0.966
34.0	1.000	0.999	0.999	0.998	0.997	0.995	0.992	0.987	0.982	0.974
35.0	1.000	1.000	0.999	0.999	0.998	0.996	0.994	0.991	0.986	0.980
36.0	1.000	1.000	0.999	0.999	0.998	0.997	0.995	0.993	0.989	0.985
37.0	1.000	1.000	1.000	0.999	0.999	0.998	0.997	0.995	0.992	0.988
38.0	1.000	1.000	1.000	0.999	0.999	0.998	0.998	0.996	0.994	0.991
39.0	1.000	1.000	1.000	1.000	0.999	0.999	0.998	0.997	0.996	0.993
40.0	1.000	1.000	1.000	1.000	1.000	0.999	0.999	0.998	0.997	0.995

TABLE F-5
Fractiles of the χ^2-distribution

Values of $\chi_P^2(f)$, given by

$$P = \frac{1}{\Gamma(\frac{1}{2}f)\,2^{\frac{1}{2}f}} \int_0^{\chi_P^2(f)} u^{-\frac{1}{2}f}\, e^{-\frac{1}{2}u}\, du$$

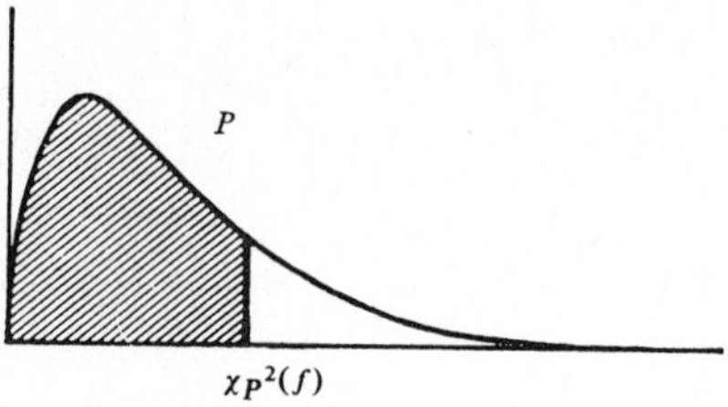

	P								
f	0.100	0.500	0.700	0.800	0.900	0.950	0.990	0.995	0.999
1	0.02	0.45	1.07	1.64	2.71	3.84	6.63	7.88	10.83
2	0.21	1.39	2.41	3.22	4.61	5.99	9.21	10.60	13.82
3	0.58	2.37	3.66	4.64	6.25	7.81	11.34	12.84	16.27
4	1.06	3.36	4.88	5.99	7.78	9.49	13.28	14.86	18.47
5	1.61	4.35	6.06	7.29	9.24	11.07	15.09	16.75	20.51
6	2.20	5.35	7.23	8.56	10.64	12.59	16.81	18.55	22.46
7	2.83	6.35	8.38	9.80	12.02	14.07	18.48	20.28	24.32
8	3.49	7.34	9.52	11.03	13.36	15.51	20.09	21.95	26.12
9	4.17	8.34	10.66	12.24	14.68	16.92	21.67	23.59	27.88
10	4.87	9.34	11.78	13.44	15.99	18.31	23.21	25.19	29.59
11	5.58	10.34	12.90	14.63	17.27	19.68	24.72	26.76	31.26
12	6.30	11.34	14.01	15.81	18.55	21.03	26.22	28.30	32.91
13	7.04	12.34	15.12	16.98	19.81	22.36	27.69	29.82	34.53
14	7.79	13.34	16.22	18.15	21.06	23.68	29.14	31.32	36.12
15	8.55	14.34	17.32	19.31	22.31	25.00	30.58	32.80	37.70
16	9.31	15.34	18.42	20.47	23.54	26.30	32.00	34.27	39.25
17	10.09	16.34	19.51	21.61	24.77	27.59	33.41	35.72	40.79
18	10.86	17.34	20.60	22.76	25.99	28.87	34.81	37.16	42.31
19	11.65	18.34	21.69	23.90	27.20	30.14	36.19	38.58	43.82
20	12.44	19.34	22.77	25.04	28.41	31.41	37.57	40.00	45.31
21	13.24	20.34	23.86	26.17	29.61	32.67	38.93	41.40	46.80
22	14.04	21.34	24.94	27.30	30.81	33.92	40.29	42.80	48.27
23	14.85	22.34	26.02	28.43	32.01	35.17	41.64	44.18	49.73
24	15.66	23.34	27.10	29.55	33.20	36.42	42.98	45.56	51.18
25	16.47	24.34	28.17	30.68	34.38	37.65	44.31	46.93	52.62
26	17.29	25.34	29.25	31.79	35.56	38.89	45.64	48.29	54.05
27	18.11	26.34	30.32	32.91	36.74	40.11	46.96	49.64	55.48
28	18.94	27.34	31.39	34.03	37.92	41.34	48.28	50.99	56.89
29	19.77	28.34	32.46	35.14	39.09	42.56	49.59	52.34	58.30
30	20.60	29.34	33.53	36.25	40.26	43.77	50.89	53.67	59.70
40	29.05	39.33	44.17	47.27	51.80	55.76	63.70	66.76	73.39
50	37.69	49.33	54.72	58.16	63.17	67.51	76.16	79.49	86.66
60	46.46	59.33	65.23	68.97	74.40	79.08	88.38	91.95	99.61
70	55.33	69.33	75.69	79.71	85.53	90.53	100.43	104.21	112.32
80	64.28	79.33	86.12	90.40	96.58	101.88	112.33	116.32	124.84
90	73.29	89.33	96.52	101.05	107.56	113.15	124.12	128.30	137.21
100	82.36	99.33	106.91	111.67	118.50	124.34	135.81	140.17	149.45

TABLE F-6

F-test

Values of fractiles $F_P(f_1, f_2)$, given by

$$P = \int_0^{F_P} f(F)\,\mathrm{d}F = 0.5 \qquad f(F) = \left(\frac{f_1}{f_2}\right)^{\frac{1}{2}f_1} \frac{\Gamma(\frac{1}{2}(f_1+f_2))}{\Gamma(\frac{1}{2}f_1)\,\Gamma(\frac{1}{2}f_2)} F^{\frac{1}{2}f_1-1}\left(1+\frac{f_1}{f_2}F\right)^{-\frac{1}{2}(f_1+f_2)}$$

f_2 \ f_1	1	2	3	4	5	6	8	10	12	14	16	20	30	40	50	∞
1	1.00	1.50	1.71	1.82	1.89	1.94	2.00	2.04	2.07	2.09	2.10	2.12	2.15	2.16	2.17	2.20
2	0.67	1.00	1.13	1.21	1.25	1.28	1.32	1.35	1.36	1.37	1.38	1.39	1.41	1.42	1.42	1.44
3	0.59	0.88	1.00	1.06	1.10	1.13	1.16	1.18	1.20	1.21	1.21	1.23	1.24	1.25	1.25	1.27
4	0.55	0.83	0.94	1.00	1.04	1.06	1.09	1.11	1.13	1.13	1.14	1.15	1.16	1.17	1.18	1.19
5	0.53	0.80	0.91	0.96	1.00	1.02	1.05	1.07	1.09	1.09	1.10	1.11	1.12	1.13	1.13	1.15
6	0.51	0.78	0.89	0.94	0.98	1.00	1.03	1.05	1.06	1.07	1.08	1.08	1.10	1.10	1.11	1.12
7	0.51	0.77	0.87	0.93	0.96	0.98	1.01	1.03	1.04	1.05	1.06	1.07	1.08	1.08	1.09	1.10
8	0.50	0.76	0.86	0.91	0.95	0.97	1.00	1.02	1.03	1.04	1.04	1.05	1.07	1.07	1.07	1.09
9	0.49	0.75	0.85	0.91	0.94	0.96	0.99	1.01	1.02	1.03	1.03	1.04	1.05	1.06	1.06	1.08
10	0.49	0.74	0.85	0.90	0.93	0.95	0.98	1.00	1.01	1.02	1.03	1.03	1.05	1.05	1.06	1.07
11	0.49	0.74	0.84	0.89	0.93	0.95	0.98	0.99	1.01	1.01	1.02	1.03	1.04	1.05	1.05	1.06
12	0.48	0.73	0.84	0.89	0.92	0.94	0.97	0.99	1.00	1.01	1.01	1.02	1.03	1.04	1.04	1.06
13	0.48	0.73	0.83	0.88	0.92	0.94	0.97	0.98	1.00	1.00	1.01	1.02	1.03	1.04	1.04	1.05
14	0.48	0.73	0.83	0.88	0.91	0.94	0.96	0.98	0.99	1.00	1.01	1.01	1.03	1.03	1.04	1.05
15	0.48	0.73	0.83	0.88	0.91	0.93	0.96	0.98	0.99	1.00	1.00	1.01	1.02	1.03	1.03	1.05
16	0.48	0.72	0.82	0.88	0.91	0.93	0.96	0.97	0.99	0.99	1.00	1.01	1.02	1.03	1.03	1.04
17	0.47	0.72	0.82	0.87	0.91	0.93	0.96	0.97	0.98	0.99	1.00	1.01	1.02	1.02	1.03	1.04
18	0.47	0.72	0.82	0.87	0.90	0.93	0.95	0.97	0.98	0.99	1.00	1.00	1.02	1.02	1.02	1.04
19	0.47	0.72	0.82	0.87	0.90	0.92	0.95	0.97	0.98	0.99	0.99	1.00	1.01	1.02	1.02	1.04
20	0.47	0.72	0.82	0.87	0.90	0.92	0.95	0.97	0.98	0.99	0.99	1.00	1.01	1.02	1.02	1.03
25	0.47	0.71	0.81	0.86	0.89	0.92	0.94	0.96	0.97	0.98	0.98	0.99	1.00	1.01	1.01	1.03
30	0.47	0.71	0.81	0.86	0.89	0.91	0.94	0.96	0.97	0.97	0.98	0.99	1.00	1.01	1.01	1.02
35	0.46	0.71	0.80	0.86	0.89	0.91	0.94	0.95	0.96	0.97	0.98	0.99	1.00	1.00	1.01	1.02
40	0.46	0.71	0.80	0.85	0.89	0.91	0.93	0.95	0.96	0.97	0.97	0.98	0.99	1.00	1.00	1.02
45	0.46	0.70	0.80	0.85	0.88	0.90	0.93	0.95	0.96	0.97	0.97	0.98	0.99	1.00	1.00	1.01
50	0.46	0.70	0.80	0.85	0.88	0.90	0.93	0.95	0.96	0.97	0.97	0.98	0.99	1.00	1.00	1.01
∞	0.45	0.69	0.79	0.84	0.87	0.89	0.92	0.93	0.95	0.95	0.96	0.97	0.98	0.98	0.99	1.00

TABLE F-6 (continued)

F-test

Values of fractiles $F_P(f_1, f_2)$, given by

$$P = \int_0^{F_P} f(F)\,\mathrm{d}F = 0.9 \qquad f(F) = \left(\frac{f_1}{f_2}\right)^{\frac{1}{2}f_1} \frac{\Gamma(\frac{1}{2}(f_1+f_2))}{\Gamma(\frac{1}{2}f_1)\,\Gamma(\frac{1}{2}f_2)} F^{\frac{1}{2}f_1-1}\left(1+\frac{f_1}{f_2}F\right)^{-\frac{1}{2}(f_1+f_2)}$$

f_2 \ f_1	1	2	3	4	5	6	8	10	12	14	16	20	30	40	50	∞
1	39.86	49.50	53.59	55.83	57.24	58.20	59.44	60.19	60.70	61.07	61.35	61.74	62.26	62.53	62.70	63.33
2	8.53	9.00	9.16	9.24	9.29	9.33	9.37	9.39	9.41	9.42	9.43	9.44	9.46	9.47	9.47	9.49
3	5.54	5.46	5.39	5.34	5.31	5.29	5.25	5.23	5.22	5.20	5.20	5.18	5.17	5.16	5.15	5.13
4	4.54	4.32	4.19	4.11	4.05	4.01	3.96	3.92	3.90	3.88	3.86	3.84	3.82	3.80	3.80	3.76
5	4.06	3.78	3.62	3.52	3.45	3.40	3.34	3.30	3.27	3.25	3.23	3.21	3.17	3.16	3.15	3.10
6	3.78	3.46	3.29	3.18	3.11	3.05	2.98	2.94	2.90	2.88	2.86	2.84	2.80	2.78	2.77	2.72
7	3.59	3.26	3.07	2.96	2.88	2.83	2.75	2.70	2.67	2.64	2.62	2.59	2.56	2.54	2.52	2.47
8	3.46	3.11	2.92	2.81	2.73	2.67	2.59	2.54	2.50	2.48	2.45	2.42	2.38	2.36	2.35	2.29
9	3.36	3.01	2.81	2.69	2.61	2.55	2.47	2.42	2.38	2.35	2.33	2.30	2.25	2.23	2.22	2.16
10	3.29	2.92	2.73	2.61	2.52	2.46	2.38	2.32	2.28	2.26	2.23	2.20	2.16	2.13	2.12	2.06
11	3.23	2.86	2.66	2.54	2.45	2.39	2.30	2.25	2.21	2.18	2.16	2.12	2.08	2.05	2.04	1.97
12	3.18	2.81	2.61	2.48	2.39	2.33	2.24	2.19	2.15	2.12	2.09	2.06	2.01	1.99	1.97	1.90
13	3.14	2.76	2.56	2.43	2.35	2.28	2.20	2.14	2.10	2.07	2.04	2.01	1.96	1.93	1.92	1.85
14	3.10	2.73	2.52	2.39	2.31	2.24	2.15	2.10	2.05	2.02	2.00	1.96	1.91	1.89	1.87	1.80
15	3.07	2.70	2.49	2.36	2.27	2.21	2.12	2.06	2.02	1.99	1.96	1.92	1.87	1.85	1.83	1.76
16	3.05	2.67	2.46	2.33	2.24	2.18	2.09	2.03	1.99	1.95	1.93	1.89	1.84	1.81	1.79	1.72
17	3.03	2.64	2.44	2.31	2.22	2.15	2.06	2.00	1.96	1.93	1.90	1.86	1.81	1.78	1.76	1.69
18	3.01	2.62	2.42	2.29	2.20	2.13	2.04	1.98	1.93	1.90	1.87	1.84	1.78	1.75	1.74	1.66
19	2.99	2.61	2.40	2.27	2.18	2.11	2.02	1.96	1.91	1.88	1.85	1.81	1.76	1.73	1.71	1.63
20	2.97	2.59	2.38	2.25	2.16	2.09	2.00	1.94	1.89	1.86	1.83	1.79	1.74	1.71	1.69	1.61
25	2.92	2.53	2.32	2.18	2.09	2.02	1.93	1.87	1.82	1.79	1.76	1.72	1.66	1.63	1.61	1.52
30	2.88	2.49	2.28	2.14	2.05	1.98	1.88	1.82	1.77	1.74	1.71	1.67	1.61	1.57	1.55	1.46
35	2.85	2.46	2.25	2.11	2.02	1.95	1.85	1.79	1.74	1.70	1.67	1.63	1.57	1.53	1.51	1.41
40	2.84	2.44	2.23	2.09	2.00	1.93	1.83	1.76	1.71	1.68	1.65	1.61	1.54	1.51	1.48	1.38
45	2.82	2.42	2.21	2.07	1.98	1.91	1.81	1.74	1.70	1.66	1.63	1.58	1.52	1.48	1.46	1.35
50	2.81	2.41	2.20	2.06	1.97	1.90	1.80	1.73	1.68	1.64	1.61	1.57	1.50	1.46	1.44	1.33
∞	2.71	2.30	2.08	1.94	1.85	1.77	1.67	1.60	1.55	1.50	1.47	1.42	1.34	1.30	1.26	1.00

TABLE F-6 (continued)

F-test

Values of fractiles $F_P(f_1, f_2)$, given by

$$P = \int_0^{F_P} f(F)\,\mathrm{d}F = 0.95 \qquad f(F) = \left(\frac{f_1}{f_2}\right)^{\frac{1}{2}f_1} \frac{\Gamma(\frac{1}{2}(f_1+f_2))}{\Gamma(\frac{1}{2}f_1)\,\Gamma(\frac{1}{2}f_2)} F^{\frac{1}{2}f_1-1}\left(1+\frac{f_1}{f_2}F\right)^{-\frac{1}{2}(f_1+f_2)}$$

f_2 \ f_1	1	2	3	4	5	6	8	10	12	14	16	20	30	40	50	∞
1	161.4	199.5	215.6	224.5	230.2	234.0	238.9	241.9	243.9	245.4	246.5	248.0	250.1	251.0	252.0	254.4
2	18.51	19.00	19.16	19.25	19.30	19.33	19.37	19.40	19.41	19.42	19.43	19.44	19.46	19.47	19.47	19.50
3	10.13	9.55	9.28	9.12	9.01	8.94	8.85	8.79	8.75	8.71	8.69	8.66	8.62	8.59	8.58	8.53
4	7.71	6.94	6.59	6.39	6.26	6.16	6.04	5.96	5.91	5.87	5.84	5.80	5.75	5.72	5.70	5.63
5	6.61	5.79	5.41	5.19	5.05	4.95	4.82	4.74	4.68	4.64	4.60	4.56	4.50	4.46	4.44	4.37
6	5.99	5.14	4.76	4.53	4.39	4.28	4.15	4.06	4.00	3.96	3.92	3.87	3.81	3.77	3.75	3.67
7	5.59	4.74	4.35	4.12	3.97	3.87	3.73	3.64	3.57	3.53	3.49	3.44	3.38	3.34	3.32	3.23
8	5.32	4.46	4.07	3.84	3.69	3.58	3.44	3.35	3.28	3.24	3.20	3.15	3.08	3.04	3.02	2.93
9	5.12	4.26	3.86	3.63	3.48	3.37	3.23	3.14	3.07	3.03	2.99	2.94	2.86	2.83	2.80	2.71
10	4.96	4.10	3.71	3.48	3.33	3.22	3.07	2.98	2.91	2.86	2.83	2.77	2.70	2.66	2.64	2.54
11	4.84	3.98	3.59	3.36	3.20	3.09	2.95	2.85	2.79	2.74	2.70	2.65	2.57	2.53	2.51	2.40
12	4.75	3.89	3.49	3.26	3.11	3.00	2.85	2.75	2.69	2.64	2.60	2.54	2.47	2.43	2.40	2.30
13	4.67	3.81	3.41	3.18	3.03	2.92	2.77	2.67	2.60	2.55	2.51	2.46	2.38	2.34	2.31	2.21
14	4.60	3.74	3.34	3.11	2.96	2.85	2.70	2.60	2.53	2.48	2.44	2.39	2.31	2.27	2.24	2.13
15	4.54	3.68	3.29	3.06	2.90	2.79	2.64	2.54	2.48	2.42	2.38	2.33	2.25	2.20	2.18	2.07
16	4.49	3.63	3.24	3.01	2.85	2.74	2.59	2.49	2.42	2.37	2.33	2.28	2.19	2.15	2.12	2.01
17	4.45	3.59	3.20	2.96	2.81	2.70	2.55	2.45	2.38	2.33	2.29	2.23	2.15	2.10	2.08	1.96
18	4.41	3.55	3.16	2.93	2.77	2.66	2.51	2.41	2.34	2.29	2.25	2.19	2.11	2.06	2.04	1.92
19	4.38	3.52	3.13	2.90	2.74	2.63	2.48	2.38	2.31	2.26	2.21	2.16	2.07	2.03	2.00	1.88
20	4.35	3.49	3.10	2.87	2.71	2.60	2.45	2.35	2.28	2.22	2.18	2.12	2.04	1.99	1.97	1.84
25	4.24	3.39	2.99	2.76	2.60	2.49	2.34	2.24	2.16	2.11	2.07	2.01	1.92	1.87	1.84	1.71
30	4.17	3.32	2.92	2.69	2.53	2.42	2.27	2.16	2.09	2.04	1.99	1.93	1.84	1.79	1.76	1.62
35	4.12	3.27	2.87	2.64	2.49	2.37	2.22	2.11	2.04	1.99	1.94	1.88	1.79	1.74	1.70	1.56
40	4.08	3.23	2.84	2.61	2.45	2.34	2.18	2.08	2.00	1.95	1.90	1.84	1.74	1.69	1.66	1.51
45	4.06	3.20	2.81	2.58	2.42	2.31	2.15	2.05	1.97	1.92	1.87	1.81	1.71	1.66	1.63	1.47
50	4.03	3.18	2.79	2.56	2.40	2.29	2.13	2.03	1.95	1.89	1.85	1.78	1.69	1.63	1.60	1.44
∞	3.84	3.00	2.60	2.37	2.21	2.10	1.94	1.83	1.75	1.69	1.64	1.57	1.46	1.39	1.35	1.00

TABLE F-6 (continued)

F-test

Values of fractiles $F_P(f_1, f_2)$, given by

$$P = \int_0^{F_P} f(F)\,\mathrm{d}F = 0.975 \qquad f(F) = \left(\frac{f_1}{f_2}\right)^{\frac{1}{2}f_1} \frac{\Gamma(\frac{1}{2}(f_1+f_2))}{\Gamma(\frac{1}{2}f_1)\,\Gamma(\frac{1}{2}f_2)} F^{\frac{1}{2}f_1-1}\left(1+\frac{f_1}{f_2}F\right)^{-\frac{1}{2}(f_1+f_2)}$$

f_2 \ f_1	1	2	3	4	5	6	8	10	12	14	16	20	30	40	50	∞
1	647.8	799.4	864.0	399.4	921.6	936.8	956.3	968.3	976.3	982.1	986.5	992.7	1001.0	1006.0	1008.0	1017.9
2	38.50	39.00	39.17	39.25	39.30	39.33	39.37	39.39	39.41	39.42	39.43	39.44	39.46	39.47	39.48	39.50
3	17.44	16.04	15.44	15.10	14.88	14.73	14.54	14.42	14.34	14.28	14.23	14.17	14.08	14.04	14.01	13.90
4	12.22	10.65	9.98	9.60	9.36	9.20	8.98	8.84	8.75	8.68	8.63	8.56	8.46	8.41	8.38	8.26
5	10.01	8.43	7.76	7.39	7.15	6.98	6.76	6.62	6.52	6.46	6.40	6.33	6.23	6.18	6.14	6.02
6	8.81	7.26	6.60	6.23	5.99	5.82	5.60	5.46	5.37	5.30	5.24	5.17	5.07	5.01	4.98	4.85
7	8.07	6.54	5.89	5.52	5.29	5.12	4.90	4.76	4.67	4.60	4.54	4.47	4.36	4.31	4.28	4.14
8	7.57	6.06	5.42	5.05	4.82	4.65	4.43	4.30	4.20	4.13	4.08	4.00	3.89	3.84	3.81	3.67
9	7.21	5.71	5.08	4.72	4.48	4.32	4.10	3.96	3.87	3.80	3.74	3.67	3.56	3.51	3.47	3.33
10	6.94	5.46	4.83	4.47	4.24	4.07	3.85	3.72	3.62	3.55	3.50	3.42	3.31	3.26	3.22	3.08
11	6.72	5.26	4.63	4.28	4.04	3.88	3.66	3.53	3.43	3.36	3.30	3.23	3.12	3.06	3.03	2.88
12	6.55	5.10	4.47	4.12	3.89	3.73	3.51	3.37	3.28	3.21	3.15	3.07	2.96	2.91	2.87	2.72
13	6.41	4.97	4.35	4.00	3.77	3.60	3.39	3.25	3.15	3.08	3.03	2.95	2.84	2.78	2.74	2.60
14	6.30	4.86	4.24	3.89	3.66	3.50	3.29	3.15	3.05	2.98	2.92	2.84	2.73	2.67	2.64	2.49
15	6.20	4.77	4.15	3.80	3.58	3.41	3.20	3.06	2.96	2.89	2.84	2.76	2.64	2.59	2.55	2.40
16	6.12	4.69	4.08	3.73	3.50	3.34	3.12	2.99	2.89	2.82	2.76	2.68	2.57	2.51	2.47	2.32
17	6.04	4.62	4.01	3.66	3.44	3.28	3.06	2.92	2.82	2.75	2.70	2.62	2.50	2.44	2.41	2.25
18	5.98	4.56	3.95	3.61	3.38	3.22	3.01	2.87	2.77	2.70	2.64	2.56	2.44	2.38	2.35	2.19
19	5.92	4.51	3.90	3.56	3.33	3.17	2.96	2.82	2.72	2.65	2.59	2.51	2.39	2.33	2.30	2.13
20	5.87	4.46	3.86	3.51	3.29	3.13	2.91	2.77	2.68	2.60	2.55	2.46	2.35	2.29	2.25	2.09
25	5.69	4.29	3.69	3.35	3.13	2.97	2.75	2.61	2.51	2.44	2.38	2.30	2.18	2.12	2.08	1.91
30	5.57	4.18	3.59	3.25	3.03	2.87	2.65	2.51	2.41	2.34	2.28	2.20	2.07	2.01	1.97	1.79
35	5.48	4.11	3.52	3.18	2.96	2.80	2.58	2.44	2.34	2.27	2.21	2.12	2.00	1.93	1.89	1.70
40	5.42	4.05	3.46	3.13	2.90	2.74	2.53	2.39	2.29	2.21	2.15	2.07	1.94	1.88	1.83	1.64
45	5.38	4.01	3.42	3.09	2.86	2.70	2.49	2.35	2.25	2.17	2.11	2.03	1.90	1.83	1.79	1.59
50	5.34	3.97	3.39	3.05	2.83	2.67	2.46	2.32	2.22	2.14	2.08	1.99	1.87	1.80	1.75	1.55
∞	5.02	3.69	3.12	2.79	2.57	2.41	2.19	2.05	1.94	1.87	1.80	1.71	1.57	1.48	1.43	1.00

TABLE F-6 (continued)

F-test

Values of fractiles $F_P(f_1, f_2)$, given by

$$P = \int_0^{F_P} f(F)\,\mathrm{d}F = 0.99$$

$$f(F) = \left(\frac{f_1}{f_2}\right)^{\frac{1}{2}f_1} \frac{\Gamma(\frac{1}{2}(f_1+f_2))}{\Gamma(\frac{1}{2}f_1)\,\Gamma(\frac{1}{2}f_2)} F^{\frac{1}{2}f_1-1}\left(1+\frac{f_1}{f_2}F\right)^{-\frac{1}{2}(f_1+f_2)}$$

f_2 \ f_1	1	2	3	4	5	6	8	10	12	14	16	20	30	40	50	∞
1	4052.	5000.	5403.	5625.	5764.	5859.	5982.	6056.	6106.	6143.	6169.	6209.	6261.	6287.	6303.	6366.
2	98.49	99.01	99.18	99.31	99.30	99.33	99.32	99.49	99.41	99.46	99.41	99.43	99.42	99.47	44.59	99.50
3	34.11	30.83	29.46	28.70	28.23	27.90	27.49	27.23	27.05	26.92	26.83	26.69	26.51	26.41	26.36	26.13
4	21.19	17.99	16.69	15.98	15.53	15.21	14.80	14.55	14.38	14.25	14.15	14.02	13.84	13.75	13.69	13.46
5	16.26	13.27	12.06	11.39	10.97	10.67	10.29	10.05	9.89	9.77	9.68	9.55	9.38	9.29	9.24	9.02
6	13.74	10.92	9.78	9.15	8.75	8.47	8.10	7.87	7.72	7.61	7.52	7.40	7.23	7.14	7.09	6.88
7	12.25	9.55	8.45	7.85	7.46	7.19	6.84	6.62	6.47	6.36	6.28	6.16	5.99	5.91	5.86	5.65
8	11.25	8.65	7.59	7.01	6.63	6.37	6.03	5.82	5.67	5.56	5.48	5.36	5.20	5.12	5.07	4.86
9	10.56	8.02	6.99	6.42	6.06	5.80	5.47	5.26	5.11	5.01	4.92	4.81	4.65	4.57	4.52	4.31
10	10.04	7.56	6.55	6.00	5.64	5.39	5.06	4.85	4.71	4.60	4.52	4.41	4.25	4.17	4.12	3.91
11	9.64	7.21	6.22	5.67	5.32	5.07	4.74	4.54	4.40	4.29	4.21	4.10	3.94	3.86	3.81	3.60
12	9.33	6.93	5.95	5.41	5.06	4.82	4.50	4.30	4.16	4.05	3.97	3.86	3.70	3.62	3.57	3.36
13	9.07	6.70	5.74	5.21	4.86	4.62	4.30	4.10	3.96	3.86	3.78	3.66	3.51	3.43	3.38	3.17
14	8.86	6.51	5.56	5.04	4.70	4.46	4.14	3.94	3.80	3.70	3.62	3.51	3.35	3.27	3.22	3.00
15	8.68	6.36	5.42	4.89	4.56	4.32	4.00	3.80	3.67	3.56	3.49	3.37	3.21	3.13	3.08	2.87
16	8.53	6.23	5.29	4.77	4.44	4.20	3.89	3.69	3.55	3.45	3.37	3.26	3.10	3.02	2.97	2.75
17	8.40	6.11	5.19	4.67	4.34	4.10	3.79	3.59	3.46	3.35	3.27	3.16	3.00	2.92	2.87	2.65
18	8.29	6.01	5.09	4.58	4.25	4.01	3.71	3.51	3.37	3.27	3.19	3.08	2.92	2.84	2.78	2.57
19	8.18	5.93	5.01	4.50	4.17	3.94	3.63	3.43	3.30	3.19	3.12	3.00	2.84	2.76	2.71	2.49
20	8.10	5.85	4.94	4.43	4.10	3.87	3.56	3.37	3.23	3.13	3.05	2.94	2.78	2.69	2.64	2.42
25	7.77	5.57	4.68	4.18	3.86	3.63	3.32	3.13	2.99	2.89	2.81	2.70	2.54	2.45	2.40	2.17
30	7.56	5.39	4.51	4.02	3.70	3.47	3.17	2.98	2.84	2.74	2.66	2.55	2.39	2.30	2.25	2.01
35	7.42	5.27	4.40	3.91	3.59	3.37	3.07	2.88	2.74	2.64	2.56	2.44	2.28	2.19	2.14	1.89
40	7.31	5.18	4.31	3.83	3.51	3.29	2.99	2.80	2.66	2.56	2.48	2.37	2.20	2.11	2.06	1.80
45	7.23	5.11	4.25	3.77	3.45	3.23	2.94	2.74	2.61	2.51	2.43	2.31	2.14	2.05	2.00	1.74
50	7.17	5.06	4.20	3.72	3.41	3.19	2.89	2.70	2.56	2.46	2.38	2.27	2.10	2.01	1.95	1.68
∞	6.63	4.61	3.78	3.32	3.02	2.80	2.51	2.32	2.18	2.08	2.00	1.88	1.70	1.59	1.52	1.00

TABLE F-7

Fractiles for Student's test

Values of $t_P(f)$, given by

$$P = \int_{-\infty}^{t_P(f)} f(t)\,\mathrm{d}t, \qquad f(t) = \frac{\Gamma\left(\frac{1}{2}(f+1)\right)}{\Gamma\left(\frac{1}{2}f\right)\sqrt{\pi}\sqrt{f}}\left(1 + \frac{t^2}{f}\right)^{-\frac{1}{2}(f+1)}$$

For one-tailed test, use $P = 1 - \alpha$; for two-tailed test, use $P = 1 - \frac{1}{2}\alpha$

	P				
f	0.900	0.950	0.975	0.990	0.995
1	3.08	6.31	12.71	31.82	63.66
2	1.89	2.92	4.30	6.97	9.93
3	1.64	2.35	3.18	4.54	5.84
4	1.53	2.13	2.78	3.75	4.60
5	1.48	2.02	2.57	3.36	4.03
6	1.44	1.94	2.45	3.14	3.71
7	1.41	1.89	2.36	3.00	3.50
8	1.40	1.86	2.31	2.90	3.36
9	1.38	1.83	2.26	2.82	3.25
10	1.37	1.81	2.23	2.76	3.17
11	1.36	1.80	2.20	2.72	3.11
12	1.36	1.78	2.18	2.68	3.06
13	1.35	1.77	2.16	2.65	3.01
14	1.35	1.76	2.15	2.63	2.98
15	1.34	1.75	2.13	2.60	2.95
16	1.34	1.75	2.12	2.58	2.92
17	1.33	1.74	2.11	2.57	2.90
18	1.33	1.73	2.10	2.55	2.88
19	1.33	1.73	2.09	2.54	2.86
20	1.33	1.73	2.09	2.53	2.85
30	1.31	1.70	2.04	2.46	2.75
40	1.30	1.68	2.02	2.42	2.70
50	1.30	1.68	2.01	2.40	2.68
60	1.30	1.67	2.00	2.39	2.66
70	1.29	1.67	1.99	2.38	2.65
80	1.29	1.66	1.99	2.37	2.64
90	1.29	1.66	1.99	2.37	2.63
100	1.29	1.66	1.98	2.36	2.63
200	1.29	1.65	1.97	2.35	2.60
500	1.28	1.65	1.97	2.33	2.59
1000	1.28	1.65	1.96	2.33	2.58

TABLE F-8
Random numbers

The numbers in this table are distributed uniformly between 0 and 1. The symbol "0." should be added in front of every sequence of n digits. In this way, the first line of the table can yield the two-digit random numbers

0.65, 0.65, 0.36, 0.25, 0.76, …,

or the six-digit random numbers

0.656536, 0.257684, 0.576442, …

65653	62576	84576	44271	04442	28209	29279	21788	67216	07206
38292	64893	44725	84314	03356	61304	37621	73988	05335	66118
48694	97100	44349	92193	54018	94369	80053	30996	65495	14006
94580	41424	97316	11080	90635	44088	48816	96099	37244	58574
16247	70316	75668	21160	45949	85251	97963	20516	41429	63928
10703	88863	36851	21336	96356	86105	49425	21608	84816	14426
23209	09413	47597	00864	76810	53077	27172	85342	67499	36912
13981	51677	84228	40273	83585	39054	82052	40826	06484	71469
70461	79539	43083	42644	68115	24894	36324	93901	36487	73809
14468	22521	04913	26791	16529	58052	99544	74797	52880	44106
88712	35318	13495	63108	57193	75183	36362	41523	21882	57581
48547	73051	01377	50801	92416	97284	51958	36189	49509	71345
82489	52829	74573	71977	60706	16435	52256	65623	23431	49971
88949	83954	03179	63483	52288	42379	83677	20651	70813	39017
96787	29561	06282	71647	73336	95195	11144	10109	60354	71142
83662	61696	17214	48022	33199	66996	03184	16136	68157	63715
68878	39830	19071	55961	64121	81077	09365	26501	74715	09783
86255	29486	00613	38307	24318	01144	87999	17701	14208	25942
27774	33168	49039	95717	32950	36244	20912	99276	07446	51188
40110	79964	18797	93103	89438	98701	87263	35266	26230	39986
03844	63184	44506	98377	89710	52863	09783	82929	09527	10799
79050	77107	51192	13187	18393	91670	84480	81850	30779	48017
11090	34384	06494	29507	18598	46021	08740	38248	50825	60719
06887	94847	07097	88960	69880	18642	82925	29772	32306	25889
64579	54470	45610	83429	90080	89617	26978	55316	89091	36697
18365	79917	14212	66016	68188	14981	76191	22315	48169	88176
95532	79604	17839	90594	83011	82716	49195	50727	61604	13076
24016	26410	42315	16195	16331	52232	66411	28375	72548	79917
26566	40139	01738	49176	79411	33878	88572	26522	61986	33215
41417	49560	24605	01589	88091	14240	92620	27558	31767	42584
69597	34319	79542	68380	94396	50953	56154	78340	64657	82879
15357	46235	39192	19032	61464	97493	31784	13263	93523	41767
08893	77452	84674	10973	03767	23842	09146	40295	59459	94090
29413	29661	13250	12544	56014	23185	34982	01226	92519	44078
31791	94047	78159	22532	31753	87731	40608	54067	58929	66964
71421	25853	12323	41256	36633	48491	61241	31030	35008	30773
69563	40420	16448	34908	61411	54293	73059	49711	40732	96994
15369	19269	77286	90298	46208	64568	71535	48092	44738	35598
10943	45273	73146	31416	30185	98358	18485	25686	87750	95323
82182	35184	71469	12151	29690	68773	45428	53611	12807	94346
50808	55737	77145	61237	73115	87548	67257	15607	88322	89471
41920	46285	00425	85984	12075	98594	82889	09988	13921	93634
36516	76387	29670	90541	76210	42391	68454	29206	59149	92038
19887	90975	66869	82436	92789	14808	53749	89217	51561	06409
74401	88722	62721	77826	02472	14389	64087	55023	53347	24876
69132	90909	23262	21390	18978	21354	57322	51743	94557	01656
58921	38619	01427	60987	53076	69571	39741	12305	16162	86223
71874	55237	84555	10201	00206	09422	54681	43282	67563	15839
86967	79253	92811	43585	26211	65001	54108	39636	50840	48313
32316	59072	63591	49897	27057	13267	36088	97123	57943	73553

REFERENCES AND BIBLIOGRAPHY

References quoted in the text

BARTLETT, M. S., 1953, Phil. Mag. **44**, 249.

BÖCK, R., 1960, Application of a Generalized Method of Least Squares for Kinematical Analysis of Tracks of Bubble Chamber Photographs, CERN 60-30 (CERN, Geneva, unpublished).

BOOTH, A. D., 1957, Numerical Methods (Butterworths, London).

BORTKIEWICZ, L. VON, 1898, Das Gesetz der kleinen Zahlen (Teubner, Leipzig).

ENGELMANN, R., H. FILTHUTH, G. ALEXANDER, U. KARSHON, A. SHAPIRA and G. YEKUTIELI, 1966, Nuovo Cimento **45**, 1038.

FISZ, M., 1963, Probability Theory and Mathematical Statistics (Wiley, New York).

FLETCHER, R., 1965, Computer J. **8**, 33.

FREEMAN, H., 1963, Introduction to Statistical Inference (Addison-Wesley, Reading, Mass.).

GELFAND, I. M., 1961, Soviet Phys.-Dokl. **6**, 192.

GILBERT, N., 1973, Biometrical Interpretation (Clarendon Press, Oxford).

HAMMERSLEY, J. M. and D. C. HANDSCOMB, 1964, Monte Carlo Methods (Methuen, London and Wiley, New York).

HUMPHREY, W. E., 1962, University of California Radiation Laboratory, Prog. Note P-6 (unpublished).

KENDALL, M. G. and A. STUART, 1963–1968, The Advanced Theory of Statistics, 2nd ed., Vol. 1 (1963), Vol. 2 (1967), Vol. 3 (1968) (Charles Griffin, London).

KÖBEL, J., 1570, Geometrei (Christian Egenolphs Erben, Frankfurt).

KOLMOGOROV, A., 1933 Ergeb. Math. **2**, 3; English edition: 1954, Foundations of the Theory of Probability (Chelsea, New York).

LEON, A., 1966a, In: Lavi, A. and T. P. Vogl, eds., Recent Advances in Optimization Techniques (Wiley, New York), p. 23.

LEON, A., 1966b, ibid. p. 599.

OREAR, J., 1958, Notes on Statistics for Physicists, UCRL-8417 (University of California Radiation Laboratory, Berkeley, unpublished).

POWELL, M. J. D., 1964, Computer J. **7**, 155.

REGENER, V. H., 1951, Phys. Rev. **84**, 161.

ROSENBROCK, H. H., 1960,Computer J. **3**, 175.

ROSENFELD, A. H., A. BARBERO-GALTIERI, W. J. PODOLSKI, L. R. PRICE, P. SÖDING, CH. G. WOHL, M. ROOS and W. J. WILLIS, 1967, Rev. Mod. Phys. **33**, 1.

ROSENFELD, A. H. and W. E. HUMPHREY, 1963, Ann. Rev. Nucl. Sci. **13**, 103.

SHEPPEY, G. C., 1966, A Program to Find Local Minima of General Functions of Many Variables (MINROS) and A Program to Find Local Minima ..., Using Conjugate Directions (MINCON), CERN 6600 Computer Library, sections D 504 and D 505 (CERN, Geneva, unpublished).

SOKAL, R. R. and F. J. ROHLF, 1969, Biometry (Freeman, San Francisco).
SOLLA PRICE, D. J. DE, 1965, Little Science, Big Science (Columbia University Press, New York).
SOLMITZ, F. T., 1964, Ann. Rev. Nucl. Sci. **14**, 375.
WILKS, 1938, Ann. Math. Statist. **9**, 60.

Bibliography

A short list of works is given below which are related to the subject of this book. It is meant to point out a number of books of differing degree of difficulty for additional reading and reference. A very extensive survey of statistical literature is given by M. G. KENDALL and A. G. DOIG, 1962, Bibliography of Statistical Literature (Oliver and Boyd, London).

Probability theory

BREIMAN, L., 1968, Probability (Addison-Wesley, Reading, Mass.).
CRAMÈR, H., 1955, The Elements of Probability Theory (Wiley, New York).
GNEDENKO, B. V., 1967, The Theory of Probability, 4th ed. (Chelsea, New York).
KAI LAI CHUNG, 1968, A Course in Probability Theory (Harcourt, Brace and World, New York).
KRICKEBERG, K., 1963, Wahrscheinlichkeitsrechnung (Teubner, Stuttgart).

Mathematical statistics

COOPER, B. E., 1969, Statistics for Experimentalists (Pergamon, Oxford).
CRAMÈR, H., 1946, Mathematical Methods of Statistics (University Press, Princeton).
DIXON, W. J. and F. J. MASSEY, 1969, Introduction to Statistical Analysis (McGraw-Hill, New York).
DUMAS DE RAULY, D., 1968, L'Estimation Statistique (Gauthier-Villars, Paris).
FELLER, W., 1968, An Introduction to Probability Theory and its Applications, 2 Vols. (Wiley, New York).
FISZ, M., 1963, Probability Theory and Mathematical Statistics (Wiley, New York).
FRASER, D. A. S., 1958, Statistics; An Induction (Wiley, New York).
FREEMAN, H., 1963, Introduction to Statistical Inference (Addison-Wesley, Reading, Mass.).
KENDALL, M. G. and A. STUART, 1963–1968, The Advanced Theory of Statistics, 2nd ed., Vol. 1 (1963), Vol. 2 (1967), Vol. 3 (1968) (Charles Griffin, London).
MANDEL, J., The Statistical Analysis of Experimental Data (Interscience, New York).
MOORE, P. G., A. C. SHIRLEY and D. E. EDWARDS, 1972, Standard Statistical Calculations (Pitman, Bath).
RAHMAN, N. A., 1972, Practical Exercises in Probability and Statistics (Griffin, London).
RAJ, D., 1968, Sampling Theory (McGraw-Hill, New York).
SACHS, L., 1968, Statistische Auswertungsmethoden (Springer, Berlin).

SMILLIE, K. W., An Introduction to Regression and Correlation (Ryerson Press, Toronto, and Academic Press, New York).

SVERDRUP, E., 1967, Laws and Chance Variations, 2 Vols. (North-Holland, Amsterdam).

VAN DER WAERDEN, B. L., 1965, Mathematische Statistik, 2nd ed. (Springer, Berlin).

WILKS, S. S., 1962, Mathematical Statistics (Wiley, New York).

YAMANE, T., 1967, Elementary Sampling Theory (Prentice-Hall, Englewood Cliffs, N.J.).

YAMANE, T., 1967, Statistics, An Introductory Analysis (Harper and Row, New York).

FORTRAN programming and applications to statistics

BURFORD, R. L., 1968, Statistics: A Computer Approach (Charles E. Merrill, Columbus).

HEMMERLE, W. J., 1967, Statistical Computations on a Digital Computer (Blaisdell, Waltham, Mass.).

LOHNES, P. R. and W. W. COOLEY, 1968, Introduction to Statistical Procedures: with Computer Exercises (Wiley, New York).

STUART, F., 1969, FORTRAN Programming (Wiley, New York).

Special applications

BUSLENKO, N. P. and J. A. SCHREIDER, 1974, Die Monte-Carlo-Methode und ihre Verwirklichung mit elektronischen Digitalrechnern (Teubner, Leipzig).

CHRIST, C. F., 1966, Economic Models and Methods (Wiley, New York).

HAMMERSLEY, J. M. and D. C. HANDSCOMB, 1964, Monte Carlo Methods (Methuen, London and Wiley, New York).

HANNAN, E. J., 1970, Multiple Time Series (Wiley, New York).

SOKAL, R. R. and F. J. ROHLF, 1969, Biometry (Freeman, San Francisco).

Statistical tables

ABRAMOWITZ, M. and I. A. STEGUN, eds., 1965, Handbook of Mathematical Functions (Dover, New York).

FISHER, R. A. and F. YATES, 1957, Statistical Tables for Biological, Agricultural and Medical Research (Oliver and Boyd, London).

GRAF, U. and H. J. HENNING, 1953, Formeln und Tabellen zur Mathematischen Statistik (Springer, Berlin).

HALD, A., 1960, Statistical Tables and Formulas (Wiley, New York).

LINDLEY, D. V. and J. C. P. MILLER, 1961, Cambridge Elementary Statistical Tables (University Press, Cambridge).

OWEN, D. B., 1962, Handbook of Statistical Tables (Addison-Wesley, Reading, Mass.).

PEARSON, E. S. and H. O. HARTLEY, 1958, Biometrica Tables for Statisticians (University Press, Cambridge).

WETZEL, W., M. D. JÖHNK and P. NAEVE, 1967, Statistische Tabellen (Walter de Gruyter, Berlin).

In order to find tables of statistical quantities used less frequently in the literature the use of the following excellent compilation is recommended

GREENWOOD, J. A. and H. O. HARTLEY, 1962, Guide to Tables in Mathematical Statistics (University Press, Princeton).

SUBJECT INDEX

The upright numbers refer to pages in the main text, italic numbers to the collection of formulae (appendix E) and bold-face numbers to the statistical tables (appendix F)

INDEX TO FORTRAN STATEMENTS AND FORTRAN PROGRAMS USED IN THIS BOOK